Towards Efficient Designing of Safe Nanomaterials
Innovative Merge of Computational Approaches and Experimental Techniques

RSC Nanoscience & Nanotechnology

Series Editors:
Professor Paul O'Brien, *University of Manchester, UK*
Professor Sir Harry Kroto FRS, *University of Sussex, UK*
Professor Ralph Nuzzo, *University of Illinois at Urbana-Champaign, USA*

Titles in the Series:
1: Nanotubes and Nanowires
2: Fullerenes: Principles and Applications
3: Nanocharacterisation
4: Atom Resolved Surface Reactions: Nanocatalysis
5: Biomimetic Nanoceramics in Clinical Use: From Materials to Applications
6: Nanofluidics: Nanoscience and Nanotechnology
7: Bionanodesign: Following Nature's Touch
8: Nano-Society: Pushing the Boundaries of Technology
9: Polymer-based Nanostructures: Medical Applications
10: Metallic and Molecular Interactions in Nanometer Layers, Pores and Particles: New Findings at the Yoctolitre Level
11: Nanocasting: A Versatile Strategy for Creating Nanostructured Porous Materials
12: Titanate and Titania Nanotubes: Synthesis, Properties and Applications
13: Raman Spectroscopy, Fullerenes and Nanotechnology
14: Nanotechnologies in Food
15: Unravelling Single Cell Genomics: Micro and Nanotools
16: Polymer Nanocomposites by Emulsion and Suspension
17: Phage Nanobiotechnology
18: Nanotubes and Nanowires: 2nd Edition
19: Nanostructured Catalysts: Transition Metal Oxides
20: Fullerenes: Principles and Applications, 2nd Edition
21: Biological Interactions with Surface Charge Biomaterials
22: Nanoporous Gold: From an Ancient Technology to a High-Tech Material
23: Nanoparticles in Anti-Microbial Materials: Use and Characterisation
24: Manipulation of Nanoscale Materials: An Introduction to Nanoarchitectonics
25: Towards Efficient Designing of Safe Nanomaterials: Innovative Merge of Computational Approaches and Experimental Techniques

How to obtain future titles on publication:
A standing order plan is available for this series. A standing order will bring delivery of each new volume immediately on publication.

For further information please contact:
Book Sales Department, Royal Society of Chemistry, Thomas Graham House, Science Park, Milton Road, Cambridge, CB4 0WF, UK
Telephone: +44 (0)1223 420066, Fax: +44 (0)1223 420247,
Email: booksales@rsc.org
Visit our website at http://www.rsc.org/Shop/Books/

Towards Efficient Designing of Safe Nanomaterials

Innovative Merge of Computational Approaches and Experimental Techniques

Edited by

Tomasz Puzyn
Laboratory of Environmental Chemometrics, Faculty of Chemistry, University of Gdansk, POLAND

Jerzy Leszczynski
Interdisciplinary Nanotoxicity Center, Jackson State University, Jackson, Mississippi, USA
Email: jerzy@icnanotox.org

RSCPublishing

RSC Nanoscience & Nanotechnology No. 25

ISBN: 978-1-84973-453-0
ISSN: 1757-7136

A catalogue record for this book is available from the British Library

Published by The Royal Society of Chemistry,
Thomas Graham House, Science Park, Milton Road,
Cambridge CB4 0WF, UK

Registered Charity Number 207890

For further information see our web site at www.rsc.org

Printed in the United Kingdom by Henry Ling Limited, at the Dorset Press, Dorchester, DT1 1HD

Preface

Only 50 years after Richard Feynman's "*There's plenty of room at the bottom*" talk, nanotechnology has emerged at the forefront of science and technology developments. Engineered nanomaterials (ENMs) have found a wide range of applications, covering different areas of human life. The further development of new materials at the "nano" scale (100 nm or less) is highly profitable for modern chemistry, physics, medicine *etc.* There are currently over 1100 commercial products on the shells which incorporate nanomaterials and their manufacturing has become an industry worth over $1 000 000 000 000 per year. However, since at least some nanoparticles can possess a negative impact on human health and the environment, designing novel nanomaterials must be always accompanied by precaution preceded by a comprehensive risk assessment.

Unfortunately, the information available in the literature on the methods available for this purpose is fragmented. The reviews published so far refer mainly to the methodology and the results of research obtained based on the experimental work. In our opinion, there is a strong demand for a comprehensive review and discussion documenting the latest achievements not only of empirical methodologies, but also of computational techniques that might significantly support or even reduce the number of required experiments. Thus, based on our scientific experience from the last few years, we have proposed a book that for the first time brings together both points of view in one, inclusive source. We believe that really comprehensive risk assessment related to this important, fast-growing group of nanospecies is possible only when these two complementary groups of techniques are employed.

The book consists of 13 chapters. Chapter 1 briefly introduces the most important problems related to the safe use of nanomaterials and opens the volume. Since the spectrum of nanomaterials is very wide, the authors selected a

RSC Nanoscience & Nanotechnology No. 25
Towards Efficient Designing of Safe Nanomaterials: Innovative Merge of Computational Approaches and Experimental Techniques
Edited by Tomasz Puzyn and Jerzy Leszczynski
© The Royal Society of Chemistry 2012
Published by the Royal Society of Chemistry, www.rsc.org

group of nanoparticles based on graphene (its synthesis method was awarded the 2010 Nobel Prize) as the case study. The three following chapters (Chapters 2–4) provide more details related to different levels and aspects of ENM toxicity assessment by using *in vitro* and *in vivo* experimental procedures. In contrast to testing conventional chemicals, the toxicity testing of nanoparticles always requires the evaluated structure to be characterized in detail by employing various microscopic techniques. This is because the investigated structures may vary along with time of the experiment (aggregate, agglomerate *etc.*). Thus, Chapter 5 presents experimental techniques for structural characterization. As mentioned, computational (chemoinformatic) methods may serve as a significant support for experimental risk assessment. In Chapter 6 the authors discuss the potential of such techniques for exploring experimental data and discovering nano–bio interaction mechanisms. The next two chapters (Chapters 7 and 8) treat in more detail the computational modeling of interactions between (carbon- and metal-based) nanoparticles and various biological systems (*i.e.*, DNA and proteins). Since nanoparticles' toxicity and other properties are fundamentally linked to the structure, they can therefore be predicted with reliable structure–toxicity and/or structure–property relationship models. Indeed, the next three chapters present the currently developed methods for making reliable predictions by employing thermodynamic cartography and structure/property mapping techniques (Chapter 9), nano-QSAR (Chapter 10) and the structure–reactivity models (Chapter 11). Chapter 12 is devoted to exposure assessment of ENMs, which requires both experimental (analytical) work and computational modeling studies. Exposure assessment might be crucial for comprehensive assessing of risk, because even if a considered substance is toxic, the risk could be low when no one is highly exposed to this species. On the contrary, when a nanoparticle is less toxic, but the exposure level is high, the resultant risk could be relatively high. In the final chapter (Chapter 13) the authors present an example of the comprehensive risk assessment of self-decontaminating surface materials, which combines life-cycle analysis with traditional risk assessment parameters (characterization, exposure, effects *etc.*) to understand nanoparticles' exposure and effects in different environmental settings.

The scope of the book is to discuss recent progress and challenges in the risk assessment of engineered nanomaterials, performed with the use of empirical and computational techniques. We noticed that very often the main difficulty in developing computational methods for the risk assessment of ENMs is the lack of appropriately measured empirical data to calibrate the models. On the other hand, experimentalists could probably reduce costs and time spent in their labs if they knew more about the abilities of currently developed computational methods. Both issues originate from the fact that usually the efforts of empirical and computational risk assessors are not well coordinated. We hope this book would serve as a bridge that covers the gap between specialists on the experimental and computational sides.

With great pleasure, we take this opportunity to thank all authors for devoting their time and hard work enabling us to complete the current volume

"*Towards Efficient Designing of Safe Nanomaterials*". We believe that with the excellent contributions from all authors this book will provide a common platform for both the theoretician and the experimentalist – not only those involved in research on nanomaterials, but also all others who are planning to start in this area of research, and especially graduate students. We are grateful to the editors at the Royal Society of Chemistry for their excellent cooperation and to our family and friends for their support.

Tomasz Puzyn, Gdańsk, Poland
Jerzy Leszczynski, Jackson, MS, USA

Contents

RSC Nanoscience & Nanotechnology No. 25
Towards Efficient Designing of Safe Nanomaterials: Innovative Merge of Computational Approaches and Experimental Techniques
Edited by Tomasz Puzyn and Jerzy Leszczynski
© The Royal Society of Chemistry 2012
Published by the Royal Society of Chemistry, www.rsc.org

**Chapter 8   Theoretical Studies of Interaction in Nanomaterials and
             Biological Systems   148**
*H. Tzoupis, A. Avramopoulos, H. Reis, G. Leonis, S. Durdagi,
T. Mavromoustakos, G. Megariotis and M. G. Papadopoulos*

**Chapter 9   Thermodynamic Cartography and Structure-Property
             Mapping of Potential Nanohazards   186**
A. S. Barnard

Chapter 12   Modeling the Environmental Release and Exposure of Engineered Nanomaterials   284
F. Gottschalk and B. Nowack

Chapter 13   Comprehensive Environmental Assessment of Nanotechnologies: a Case Study Using Self-decontaminating Surface Materials   314
J. A. Steevens, A. Bednar, M. Chappell, K. Donohue, M. Ginsberg, K. Guy, D. Johnson, A. Kennedy, R. Moser, M. Page, A. Poda and C. Weiss Jr.

CHAPTER 1

Graphene: Properties, Biomedical Applications and Toxicity

TANDABANY C. DINADAYALANE[a],
DANUTA LESZCZYNSKA[a,b] AND
JERZY LESZCZYNSKI*[a]

[a] Interdisciplinary Center for Nanotoxicity, Department of Chemistry and Biochemistry, Jackson State University, 1400 J.R. Lynch Street, Jackson, Mississippi 39217, USA; [b] Department of Civil and Environmental Engineering, Jackson State University, 1400 J.R. Lynch Street, Jackson, Mississippi 39217, USA
*E-mail: jerzy@icnanotox.org

1.1 Introduction

Among numerous commercial endeavors, nanotechnology is regarded as the key technology of the 21st century. It provides novel products and facilitates applications of innovative techniques in medicine, pharmacy, computer technology, and sensing. Therefore, it holds promise for potential global socio-economic benefits. In 2011, there were about 1100 commercial products that include nanomaterials. It is a common belief that nanotechnology could assist in solving many global problems that society faces. These include environmental and health concerns of the fast-growing human population, as well as access to clean water and affordable energy. The quickly growing applications of nanomaterials are due to their unique properties which offer advantages over conventional materials.

The variety of various nanomaterials is too big to cover in a single chapter. Therefore, we decided to focus on one particular class of species and discuss in

RSC Nanoscience & Nanotechnology No. 25
Towards Efficient Designing of Safe Nanomaterials: Innovative Merge of Computational Approaches and Experimental Techniques
Edited by Tomasz Puzyn and Jerzy Leszczynski
© The Royal Society of Chemistry 2012
Published by the Royal Society of Chemistry, www.rsc.org

details its characteristics and applications. Since the 2010 Nobel awards validate the importance of graphene not only in basic research but also in various commercial applications, we selected this nanomaterial as the main subject of our chapter.

The demand for carbon nanostructures, particularly carbon nanotubes (CNTs) and graphene, is increasing rapidly in electrical, mechanical, and biomedical applications. This is due to their outstanding thermal, electrical, mechanical, optical and other unique properties.[1-3] Although the intense interest and continuing experimental success of graphene-based devices facilitate their various applications, the reliable production of high quality samples of graphene on a large scale is very difficult.[4] At present, great efforts have been made toward the preparation of graphene nanosheets.[5-7] Among them, the chemical reduction of exfoliated graphene oxide (EGO) is the most commonly used approach due to its low cost for large-scale production. In the case of carbon nanotubes, closely related to graphene, controlling their size and diameter is still very challenging. The availability of carbon nanotubes, both in quality and quantity, has stimulated the worldwide pursuit of carbon nanotubes for technological applications. Nevertheless, carbon nanotubes (especially single-walled carbon nanotubes (SWCNTs)) are still quite expensive.[8] To assist experimental studies, the structures, reactivities, and functionalization of defect-free and Stone–Wales defective SWCNTs have been investigated by our group, using quantum chemical calculations.[9-16]

The morphology of graphene is different from that of CNTs; for example, the length of CNTs influences their toxicity but graphene and graphene oxide (GO) do not have a "length".[17] An important similarity between these carbon nanomaterials is that both graphene/graphene oxide and carbon nanotube structures vary according to the synthetic processes employed. Such processes can also change their physical properties, including dispersity, surface functionality, and their toxicity.[18] In the materials science world, carbon nanostructures such as fullerenes, carbon nanotubes and graphene are famous for their small dimension and unique architecture, and several possible applications in diversified areas. In addition to these carbon nanostructures, scientists now produce a plethora of carbon-based nanoforms such as 'bamboo' tubes, 'herringbone' and 'bell' structures.[19] Figure 1.1 shows a "family tree" of carbon nanoforms that were obtained by applying various transformations to graphene (details of operations are provided in the figure). Though the diagram is non-exhaustive (primarily for clarity), this chart is useful to classify the nanoforms by morphology and provides a first step towards a standardized nomenclature.

The research involving graphene has grown at a spectacular pace in the last few years. Several potential applications have been proposed for graphene. These include conductive and high-strength composites, energy storage and energy conversion devices, sensors, field emission displays and radiation sources, hydrogen storage media, and nanometre-sized semiconductor devices, probes, and interconnect (Scheme 1.1). Impressive advances have been made in

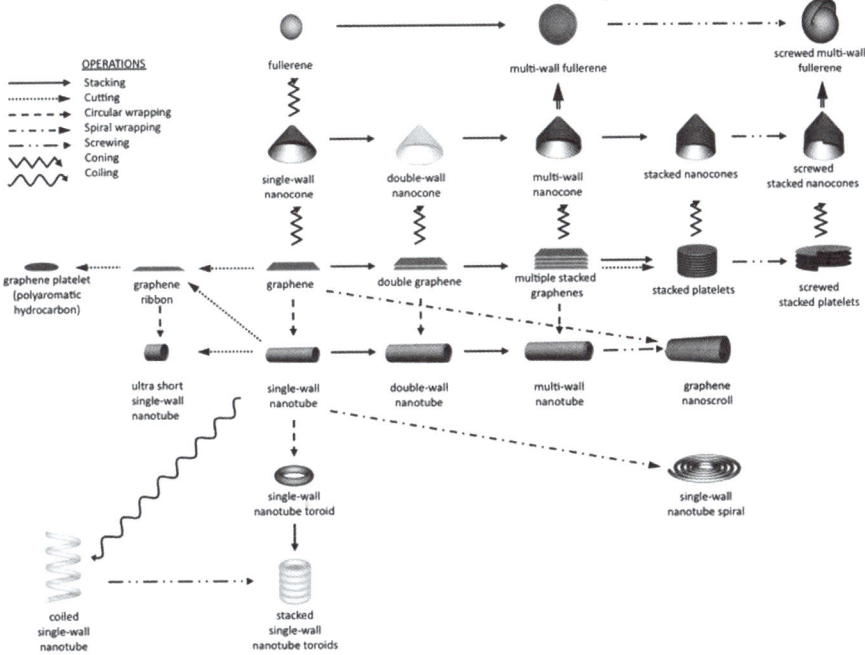

Figure 1.1 "Family tree" of primary carbon nanoforms showing the topological relationships between them. Forms which have not been identified experimentally are faded. (Reproduced from ref. 19. Copyright 2011, with permission from Elsevier Ltd.)

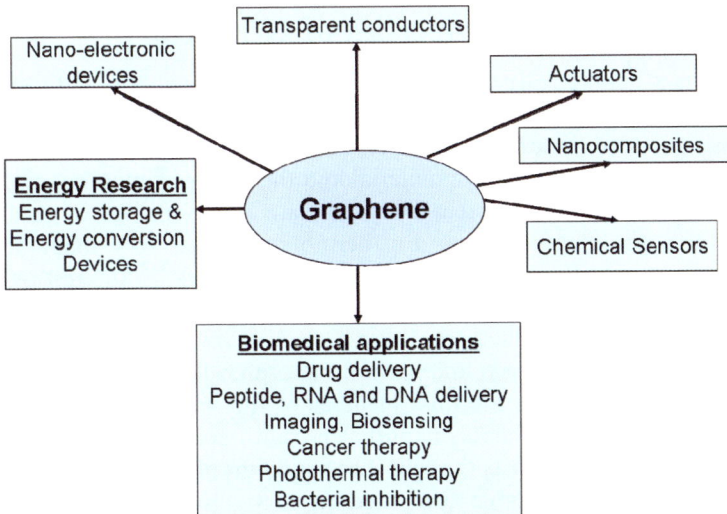

Scheme 1.1 Various applications of graphene including biomedical applications.

realizing some of the applications of carbon nanotubes and graphene.[2,3,20–27] Graphene is a possible replacement material in applications where carbon nanotubes are presently used.[20] A recent study has revealed that graphene-based liquid crystal devices (LCDs) showed an excellent performance with a high contrast ratio. Thus, LCDs might be the first realistic commercial application of graphene.[21] Like carbon nanotubes, the unique properties of graphene offer a wide range of opportunities and application potential for biology and medicine (Scheme 1.1). Bioapplications of carbon nanotubes and graphene have attracted much attention recently.[22–28]

Significant recent development and progress in the use of graphene-based materials for biosensors, drug and other delivery systems, and bioimaging have generated much excitement in the research community.[22–28] In the last few years, considerable attention has been paid to the health risks of carbon nanomaterials due to rising global production and the wide range of proposed applications.[29–31]

The lungs and skin are regarded as the two main potential exposure sites during the manufacture and handling of carbon nanotubes.[32,33] The high ratio between the length and diameter of carbon nanotubes and their low solubility in aqueous media make them potentially biopersistent and may lead to toxic effects similar to those seen with asbestos. Carbon nanotubes might induce lung cancer and mesothelioma in a similar manner to that of asbestos.[34,35] Hence, an understanding of the occupational health, public safety and environmental implications of carbon nanomaterials is required. Some recent studies have reported the *in vivo* biodistribution and toxicity of carbon nanotubes and graphene. In this chapter, we briefly present various aspects related to the toxicity of graphene-based nanomaterials.

1.2   Structure and Properties of Graphene

Carbon has a unique feature of forming a chemically stable two-dimensional (2D), one-atom thick sheet called graphene. Each carbon atom in graphene is covalently bonded to three other carbon atoms with sp^2 hybridization. Graphene is the thinnest known material and the strongest material ever to be measured. It can sustain current densities that are six orders of magnitude higher than those of copper. Furthermore, it has very high thermal conductivity and stiffness, and is impermeable to gases.[1–3] Interestingly, the honeycomb network of graphene is also the basic building block of other important carbon nanostructures. Hence, graphene may be considered as the mother of graphite, fullerene and carbon nanotubes (Scheme 1.2). Graphene can also be considered as the final member of the series of fused polycyclic aromatic hydrocarbons, such as naphthalene, anthracene and coronene.

For several years, scientists thought that planar graphene could not exist in a free state since it is unstable compared to curved structures such as soot, nanotubes and fullerenes. Surprisingly, in 2004, Novoselov *et al.* prepared graphitic sheets including a single graphene layer and studied their electronic

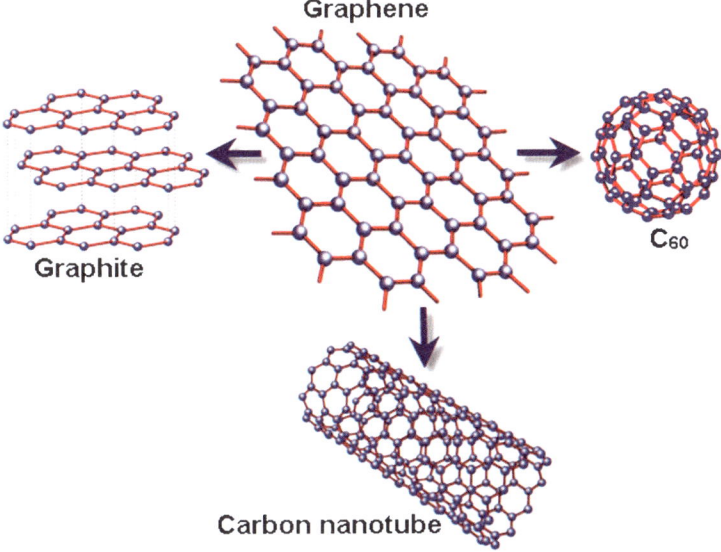

Scheme 1.2 Carbon-containing molecules (graphite, Buckminsterfullerene (C₆₀) and carbon nanotubes) derived from graphene.

properties.[36,37] Just a few years later, in 2010, Andre Geim and Konstantin Novoselov were awarded the Nobel Prize for their groundbreaking experimental isolation of single-layer graphene. The experimental realization of graphene generated vast interest in the areas of fundamental physics, materials science and engineering. Researchers have been looking for possible applications of graphene in high-speed and radio-frequency logic devices, thermally and electrically conductive reinforced composites, sensors, transparent electrodes for displays and solar cells, and biomedicine.[21–27,36–43] Numerous graphene-based biosensing devices and techniques based on various mechanisms have been developed in the past few years.[22,23] Graphene is obviously a prospective material for nanoelectronics. The electron transport in graphene was described by a Dirac-like equation.[38,39] The studies pertinent to the chemistry of graphene sheets have also been reported.[40–45] The growth and isolation of graphene have been recently reviewed.[2,3,44,45] Research in the area of graphene material has been rapidly growing due to the recent advances in technology for the growth, isolation and characterization of two-dimensional carbon sheets.

Graphene sheets need not always be perfect. Various defects such as Stone–Wales (SW), vacancies, and pore defects can occur in the graphene sheet. Like the creation of vacancies by knocking atoms out of the graphene sheet, surplus atoms can be found as ad-atoms on the graphene surface. An ad-dimer or inverse Stone–Wales (ISW) defect is characterized by two adjacent five-membered rings instead of two adjacent seven-membered rings in a Stone–Wales defect. Experimental observations of defects in graphene were reported.[46,47]

Zettl and co-workers showed the direct image of Stone–Wales defects in graphene sheets using high resolution transmission electron microscopy (HRTEM) and explored their real-time dynamics. They found that the dynamics of defects in extended, two-dimensional graphene membranes are different from in closed-shell graphenes such as nanotubes or fullerenes.[46] The results of our computational study (calculations at the B3LYP/6-31G(d) level)

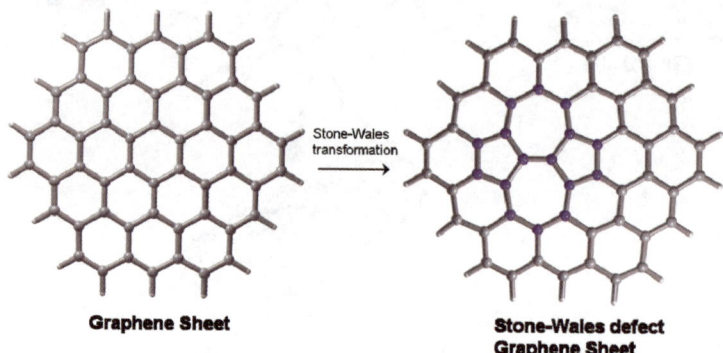

Scheme 1.3   Stone–Wales defect transformation, which is a 90° rotation of a C−C bond, in a perfect graphene sheet gives a Stone–Wales defective graphene sheet. The carbon atoms in the SW defect region are given in blue.

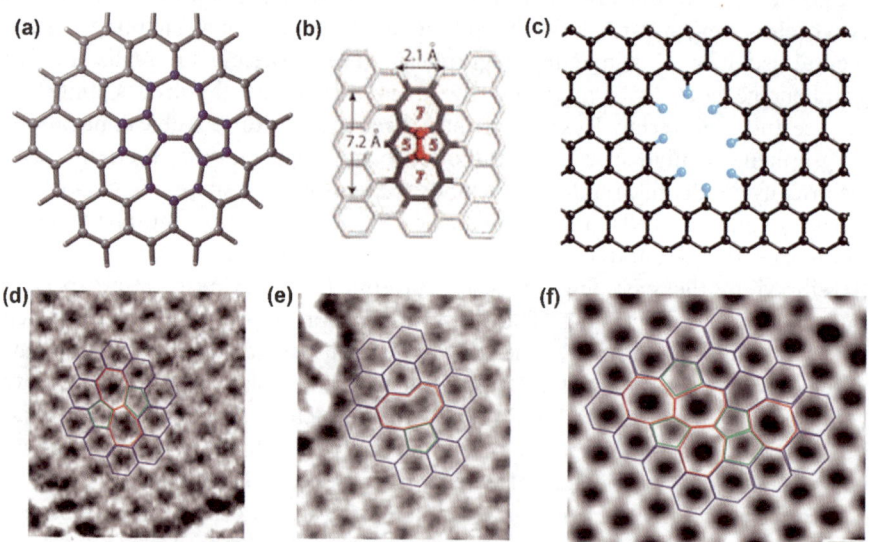

Figure 1.2   The segment of graphene containing different defects: (a) Stone–Wales (SW), (b) inverse Stone–Wales (ISW) defect, (c) all-hydrogen saturated pore in graphene, (d) HRTEM image of graphene with Stone–Wales defect, (e) image of vacancy defect, (f) defect image with four pentagons (green) and four heptagons (red). In (d)–(f), the atomic configuration was superimposed for easy recognition. (Adapted with permission from ref. 46 and 47. Copyright 2008 and 2009 American Chemical Society.)

indicate that the Stone–Wales defective graphene sheet is 83.3 kcal mol^{-1} less stable than the defect-free graphene sheet (Scheme 1.3). Figure 1.2 depicts some of the common defects and HRTEM images of defects in a graphene sheet. The characteristics of typical defects and their concentrations in graphene sheets are unclear.

1.2.1 Biomedical Applications of Graphene

Recently, graphene has showed promise similar to carbon nanotubes in various biomedical applications such as drug delivery and cancer therapy. Graphene can provide a larger specific surface area than other commonly used carbon nanomaterials and forms strong π–π interactions with the drug molecules, which can therefore act as a good candidate for drug loading.[25] Very recently, *in vivo* cancer treatment using graphene has been realized in animal experiments.[48,49] Functionalized graphene sheets have been mostly used for biomedical applications. PEGylated (PEG – polyethylene glycol) nanoscale graphene oxide (NGO-PEG), which has a high stability in physiological solutions, can be utilized for effective loading of aromatic anticancer drugs such as doxorubicin (DOX) and water-insoluble SN38. It is accomplished *via* π–π stacking (Figure 1.3a). The unique 2D shape and ultra-small size (down to 5 nm) of NGO-PEG may offer interesting *in vitro* and *in vivo* behaviors.[25,50,51]

The aromatic molecule SN38 is a camptothecin (CPT) analogue and a potent topoisomerase I inhibitor. The free SN38 was mentioned to be insoluble in water, but the NGO-PEG–SN38 complex exhibited excellent aqueous solubility.[50] The experimental study revealed no loading of SN38 on the PEG polymer in a solution free of NGO. Furthermore, the hydrophobic and π–π interactions (between SN38 and aromatic regions of the graphene oxide sheet) were attributed to the binding of SN38 onto NGO-PEG. The water-soluble NGO-PEG–SN38 exhibited high potency with IC$_{50}$ values of about 6 nM for HCT-116 cells, which is ~1000-fold more potent than CPT-11 (an FDA-approved SN38 prodrug for colon cancer treatment). Hence, NGO-PEG–SN38 has a high potential for killing cancer cells *in vitro* with a human colon cancer cell line HCT-116. The potency of NGO-PEG–SN38 was reported to be similar to that of free SN38 dissolved in DMSO. Liu *et al.* have also observed the high potency of NGO-PEG–SN38 with various other cancer cell lines in their tests.[50]

The extremely large surface area of graphene, with every atom exposed on its surface, allowed for ultra-high drug loading efficiency on NGO-PEG. The terminals of PEG chains were available for the conjugation of targeting ligands such as antibodies. This facilitated targeted drug delivery to specific types of cancer cell (Figure 1.3b).[51] Rituxan (CD20+ antibody) conjugated NGO-PEG was used to target specific cancer cells for selective cell killing. The drug delivery research revealed that approximately 40% of DOX loaded on NGO-PEG was released over 1 day in an acidic solution of pH 5.5. This phenomenon was attributed to the increased hydrophilicity and solubility of DOX at this

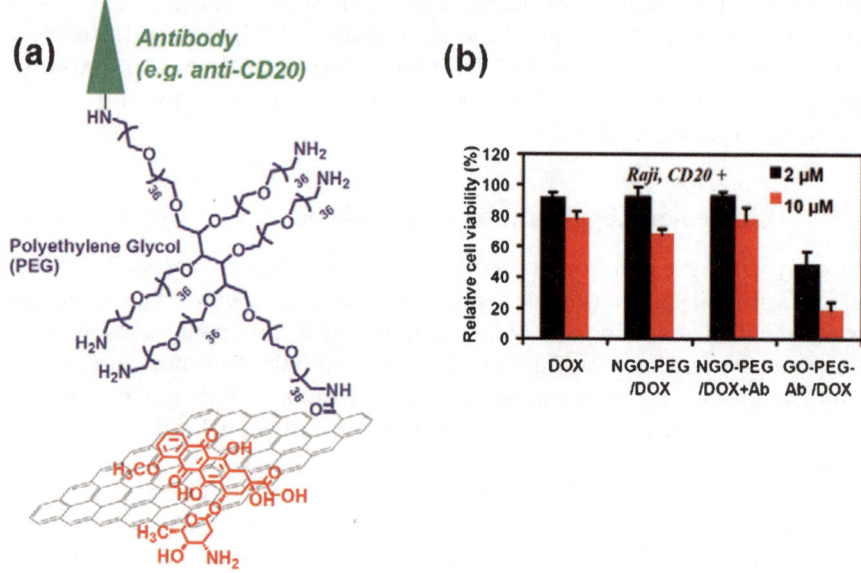

Figure 1.3   (a) A schematic illustration of doxorubicin loading onto NGO-PEG–Rituxan *via* π–π stacking. (b) *In vitro* toxicity test at 2 μM and 10 μM DOX concentrations showing that Rituxan conjugation selectively enhanced doxorubicin delivery into Raji B-cells by comparing NGO-PEG–Rituxan/DOX with free DOX, NGO-PEG/DOX, and the mixture of DOX, Rituxan and NGO-PEG. (Adapted from ref. 25. Copyright 2011, with permission from Elsevier Ltd.)

pH. Compared to single-walled carbon nanotubes for drug loading *via* π–π stacking, NGO is advantageous in terms of its low cost and ready scalability.[51]

Yang *et al.* have studied the *in vivo* behaviors of nanographene sheets (NGS) with polyethylene glycol (PEG) coating by a fluorescent labeling method.[48] They have labeled NGS-PEG with Cy7, which is a commonly used near-infrared fluorescent dye, in order to study the *in vivo* behaviors of NGS.[48] Interestingly, the majority of Cy7 dye was covalently conjugated to NGS-PEG *via* the formation of an amide bond instead of π-stacking physisorption. BALB/c mice bearing 4T1 murine breast cancer tumors, nude mice bearing KB human epidermoid carcinoma tumors, and U87MG human glioblastoma tumors were intravenously injected with NGS-PEG–Cy7. Yang *et al.* have observed a prominent uptake of NGS in the tumor with relatively low quantity in other parts of the mouse body after 24 h post-injection for all three types of tumor models.[48]

The graphical and pictorial illustrations of *in vivo* photothermal therapy study using intravenously injected NGS-PEG are given in Figure 1.4. Tumor growth curves of different groups after treatment were given. The tumor volumes were normalized to their initial sizes. There were six mice in the untreated group, ten mice in the 'laser only' group, seven mice in the 'NGS-

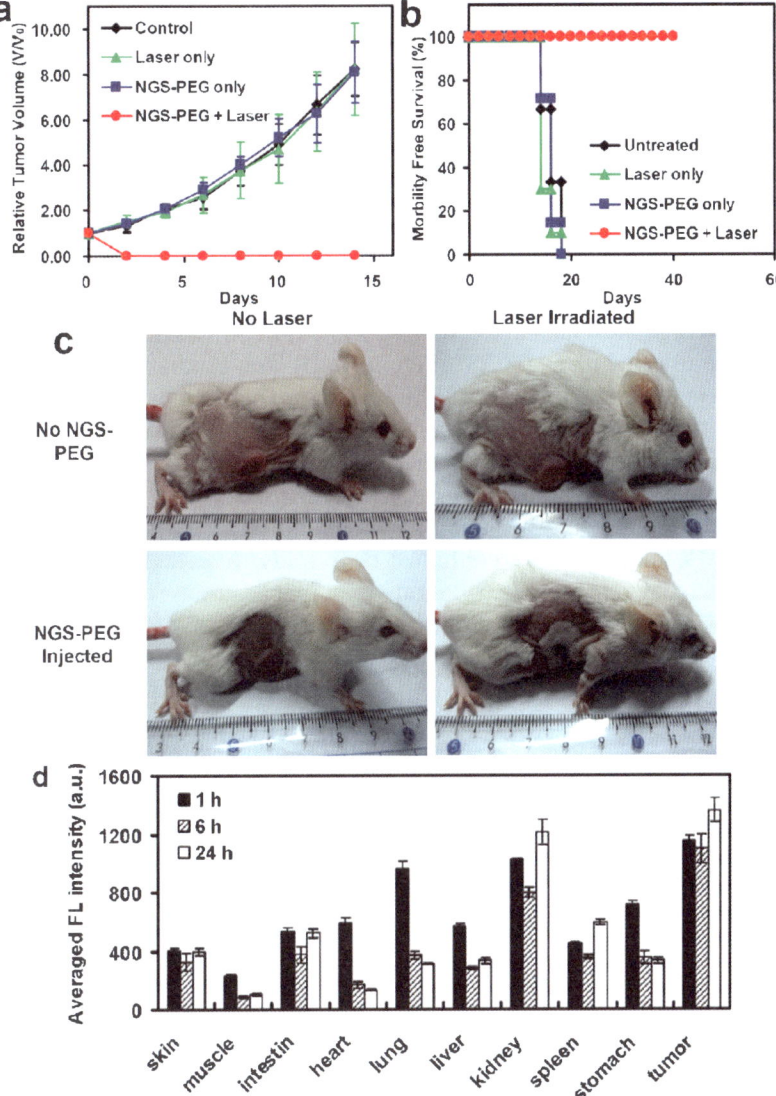

Figure 1.4 *In vivo* photothermal therapy study using intravenously injected NGS-PEG. (a) Tumor growth curves of different groups after treatment. (b) Survival curves of mice bearing the 4T1 tumor after various treatments indicated. NGS-PEG-injected mice after photothermal therapy survived over 40 days without any single death. (c) Representative photos of tumors on mice after various treatments indicated. (d) Semiquantitative biodistribution of NGS-PEG–Cy7 in mice determined by the averaged fluorescence intensity of each organ (after subtraction by the fluorescence intensity of each organ before injection). Error bars in (a) were based on standard deviations and in (d) were based on three mice per group. (Adapted with permission from ref. 48. Copyright 2010 American Chemical Society.)

PEG only', and ten mice in the 'NGS-PEG + laser' groups. While injection of NGS-PEG by itself or laser irradiation on uninjected mice did not affect tumor growth, tumors in the treated group were completely eliminated after NGS-PEG injection and followed by near-infrared (NIR) laser irradiation.

The laser-irradiated tumor on NGS-PEG injected mouse was completely destroyed. The *in vivo* study revealed that the high kidney uptake of NGS-PEG–Cy7 might indicate possible renal excretion of NGS with small sizes (Cy7 fluorescence was indeed detected in the mouse urine) but this needs further validation. Nude mice bearing KB or U87MG tumors also showed high tumor and kidney uptake of NGS (see Figure 1.4d). Mice in the three control groups showed average life spans of ∼16 days, while mice in the treated group were tumor-free after treatment (NGS injection + NIR laser irradiation) and survived over 40 days without a single death (Figure 1.4b). This finding demonstrates the excellent efficacy of NGS-based *in vivo* photothermal therapy.[48]

There are limitations of fluorescence-based *in vivo/ex vivo* imaging methods, including light absorption and scattering by tissues, the potential photo-bleaching of fluorescent dyes, and the non-quantitative nature of this method. Therefore, Yang *et al.* have continued their study to further understand the pharmacokinetics and biodistribution of graphene using ^{125}I radionuclide-labeled NGS-PEG.[49] They have measured the radioactivity levels in the blood over time after intravenous injection of ^{125}I-NGS-PEG into BALB/c mice. As shown in Figure 1.5, NGS-PEG was distributed in many different organs at 1 h post-injection, but it was mainly accumulated in the reticuloendothelial system (RES), which is an older term for the mononuclear phagocyte system, such as the liver and spleen. The less accurate result of low RES uptake of NGS-PEG from fluorescence imaging[48] was attributed to the quenching of Cy7 dyes in the liver and spleen. Thus, the radiolabeled NGS obviously provided more reliable results than the fluorescence imaging technique in application to study the *in vivo* behaviors of graphene.[49] Substantial bone uptake of NGS was reported at early stages post-injection and it is likely to be due to the macrophage uptake of nanomaterials in the bone marrow. The small amount of unpurified free ^{125}I in the injected ^{125}I-NGS-PEG sample

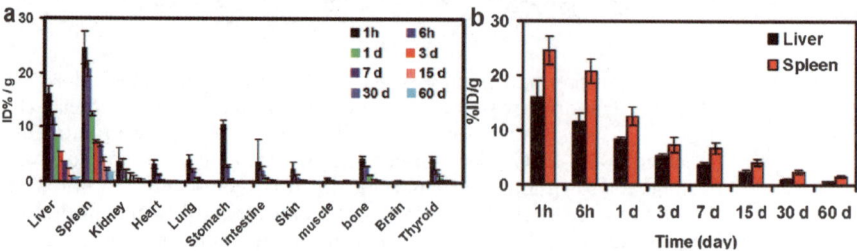

Figure 1.5 (a) Time-dependent biodistribution of ^{125}I-NGS-PEG in female BALB/c mice. (b) ^{125}I-NGS-PEG levels in the liver and spleen over time. (Reproduced with permission from ref. 49. Copyright 2011 American Chemical Society.)

caused appreciable thyroid uptake only at the earliest time points. Very low thyroid uptake was observed at later time points and was barely detectable on and after day 3.

Liu *et al.* have reported a green and facile method for the preparation of gelatin-functionalized graphene nanosheets (gelatin–GNS) by the chemical reduction of EGO. Gelatin-GNS exhibited good biocompatibility and physiological stability, and it was tested as a carrier to load anticancer drugs for cellular uptake and drug delivery.[52] Gelatin is a linear polypeptide that consists of different amounts of 18 amino acids. It is the thermally and hydrolytically denatured product of collagen. The advantage of using gelatin is that the non-polar amino acid chain of the gelatin could immobilize on the surface of graphene through hydrophobic–hydrophobic interactions. This leads to the formation of a stable dispersion of graphene. The cellular uptake of the gelatin–GNS as a drug carrier was studied by selecting R6G as the model fluorescence probe to label gelatin–GNS. As a commonly used model drug, R6G was a cationic dye that could bind with negatively charged gelatin–GNS to form R6G@gelatin-GNS complex, as a result of electrostatic attraction. The loading capacity was measured to be 8 mg of R6G to 110 mg of gelatin–GNS. This high loading capacity was attributed to the large specific surface area and the strong π–π as well as hydrogen bonding interactions between gelatin–GNS and R6G. Regardless of the pH value, R6G@gelatin-GNS exhibited sustained release properties with a relatively fast release at the initial stage. Liu *et al.* have mentioned that about 28% and 45% of the total bound R6G was released from the R6G@gelatin-GNS correspondingly at pH 4.6 and 2.0. This phenomenon is related to the partial dissociation of the hydrogen bonding interactions between R6G and gelatin–GNS under acid conditions. The investigation of the release behavior of R6G@gelatin-GNS in ethanol revealed the good release but poor uptake properties of R6G. The saturated adsorption of gelatin–GNS toward doxorubicin (DOX), which is a commonly used anticancer drug, was noted as 16 mg of DOX to 120 mg of gelatin–GNS.[52]

The combined use of two or more drugs is a widely adopted clinical practice and often displays much better therapeutic efficacy than that of a single drug. However, there are only few publications available on the controlled loading of two or more drugs onto a nanocarrier. Functional NGO was reported to be an efficient nanocarrier for the controlled loading and targeted delivery of mixed anticancer drugs. Very recently, Zhang *et al.* have investigated controlled loading of two anticancer drugs, doxorubicin (DOX) and camptothecin (CPT), onto the folic acid functionalized NGO (FA–NGO) *via* π–π stacking and hydrophobic interactions. They have chosen DOX and CPT as model drugs to demonstrate the loading of two or more drugs together on graphene with a controlled ratio of DOX to CPT.[26] These two drugs are widely used in clinics for cancer treatment and showed a much better anticancer effect when used together. Covalent binding of folic acid (FA) to the NGO allowed specifically target MCF-7 cells, human breast adenocarcinoma cell with FA receptors. (MCF-7 is the acronym of Michigan Cancer Foundation-7.)

As shown in Figure 1.6, the drug loading ratio (the weight ratio of drug loaded to FA–NGO) for DOX was remarkably higher than that of CPT (more than 400% for DOX, and only about 4.5% for CPT). More importantly, the loading ratio was linearly correlated to the concentration of drugs (Figure 1.6c), thus showing controlled loading of DOX or CPT onto the NGO sheets. Owing to the difference in the chemical structures of the two drugs, the difference in the π–π stacking and hydrophobic interactions with the NGO occurred. Hence, the two drugs exhibited different loading ability. The loading ratio of DOX or CPT was influenced by the distribution coefficient of the drug dissolved in solution and that adsorbed on the graphene carriers. Zhang *et al.* have reported that about 35% of DOX and approximately 17% of CPT loaded on FA–NGO were released after 48 h at pH 5.0, while much less DOX or CPT was released after 48 h at pH 7.0 (Figure 1.6d).[26]

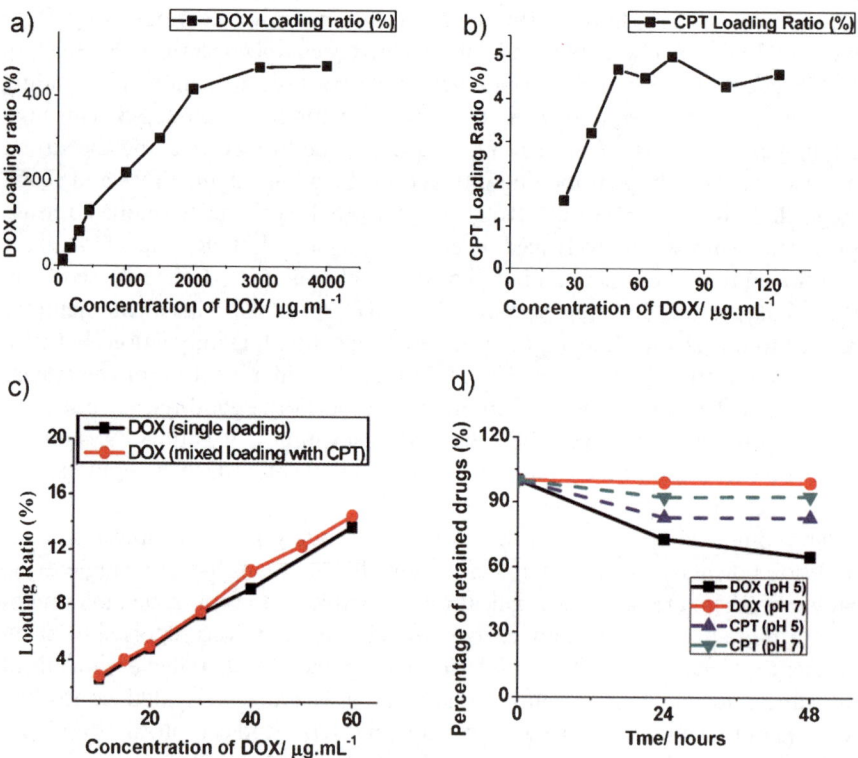

Figure 1.6   (a) Plot of loading ratio of DOX to FA–NGO *versus* the concentration of DOX; (b) plot of loading ratio of CPT to FA–NGO *versus* the concentration of CPT; (c) plot of loading ratio of DOX to FA–NGO *versus* concentration of DOX in a mixture of DOX and CPT (concentration of CPT 100 μg mL^{-1}); and (d) plot of the release of DOX and CPT from FA–NGO at pH 5 and pH 7. (Reproduced with permission from ref. 26. Copyright 2010 Wiley-VCH Verlag GmbH & Co. KGaA, Weinheim.)

Gene therapy is used to cure diseases that are difficult for traditional clinical methods. The major obstacle in this area is to develop non-viral-based safe and efficient gene delivery vehicles. Over the years, scientists have utilized cationic polymers and various nanomaterials such as carbon nanotubes, silica nanoparticles and nanodiamonds as gene delivery vehicles in their investigations.[53–57] They encounter challenges in developing non-toxic nano-vectors with high gene transfection efficiency for potential gene therapy. Among all materials used in gene delivery, polyethylenimine (PEI) has been normally recognized as the 'golden standard' cationic polymer in gene transfection,[53–57] due to its strong binding to nucleic acids, effective uptake by cells, and the excellent proton sponge effect that triggers the endosomal release of DNA or RNA. However, the high cytotoxicity and poor biocompatibility of PEI polymers, especially those with high molecular weights, have largely restricted their further applications in gene therapy. Recent studies have suggested that functionalized graphene sheets are capable of gene transfection. Negatively charged GO is non-covalently bound with cationic polymers, PEI to produce GO-PEI complexes that were reported to be stable in physiological solutions and exhibit significantly reduced cellular toxicity compared with bare PEI polymers. The positively charged GO-PEI complexes were able to further bind plasmid DNA (pDNA) for intracellular transfection of the enhanced green fluorescence protein (EGFP) gene in HeLa cells. Zhang *et al.* have reported that GO was covalently conjugated with PEI for small interfering RNA (siRNA) loading. Sequential delivery of Bcl-2 siRNA and doxorubicin into cancer cells by the GO-PEI conjugate showed significantly improved cell killing efficacy *via* a synergistic effect.[53]

Lu *et al.* have shown that functionalized nanoscale graphene oxide (NGO) can protect oligonucleotides from enzymatic cleavage and efficiently deliver oligonucleotides for gene detection and therapy.[58] They have used the model oligonucleotide of molecular beacon (MB) that was a hairpin-shaped DNA sequence of 5'-Dabcyl-CGA CGG AGA AAG GGC TGC CAC GTC G-Cy5-3'. This loop was designed to incorporate a complementary region for the survivin transcript, a target that has received significant attention due to its potential use in cancer therapeutics and diagnostics. The experiments of Lu *et al.* have demonstrated that MB is protected from DNase I cleavage after adsorption on the NGO. Furthermore, the MB/NGO complex was found to have excellent thermostability even at high temperatures. Several advantages have been reported for the use of NGO in an oligonucleotide delivery system: (1) the strong adsorption of the oligonucleotides on NGO can protect oligonucleotides from cleavage in intracellular environments. Moreover, the high stability of MB/NGO complex can prevent the release of MB during the cell transporting. (2) NGO can improve the transfection efficiency of oligonucleotides as a carrier. (3) NGO is a less cytotoxic transport agent compared with SWCNTs (SWCNTs showed obvious cytotoxicity when their concentration was 20 µg mL^{-1}). (4) The low cost and large production scale of the NGO makes it a promising biomaterial.[58]

1.2.2 Toxicity of Graphene-based Nanomaterials

The biosafety of nanomaterials has attracted significant attention from governments and scientific communities. The potential short- and long-term toxicity of nanomaterials has been the key issue that should be carefully investigated before the full implementation of such nanomaterials in a wide range of biological applications. Information on the potential health effects of nanomaterials is urgently required and should be updated regularly because of the growing industrial use of such species. Owing to the distinct and unique physicochemical properties of many nanomaterials, their possible toxicity may differ from that of the bulk material of similar chemical nature. CNTs and graphite nanofibres agglomerate and remain insoluble on inhalation exposure to the lungs. The cytotoxic effects of graphene are expected to be significantly different when compared to those of carbon nanotubes.

Norppa and co-workers have reported that treatment of the BEAS 2B cells with graphite nanofibres decreased cell viability in a dose-dependent manner after all the exposure times used (24, 48, and 72 h) when the cells were allowed to recover for 48 h post treatment (Figure 1.7). Also, both CNTs and graphite nanofibres showed similar toxicity. They were found to be genotoxic in human bronchial epithelial BEAS 2B cells. A dose-dependent increase in DNA

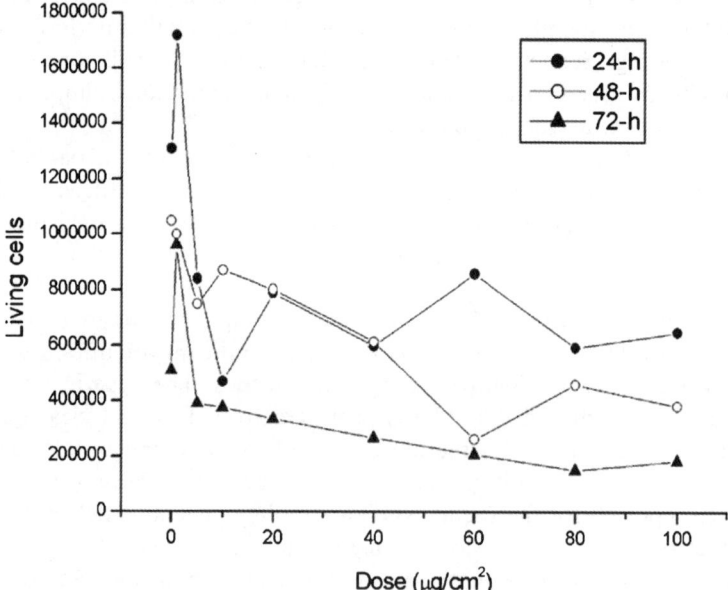

Figure 1.7   Cell viability (number of living cells) in cultured human bronchial epithelial BEAS 2B cells after 24 h, 48 h, and 72 h exposure to graphite nanofibres. The cell viability was evaluated with the Trypan blue dye exclusion assay. The symbols show mean numbers of living cells in triplicate cultures. (Reproduced from ref. 59. Copyright 2008, with permission from Elsevier Ireland Ltd.)

damage was reported and the damage effect was more significant at longer treatment times. It should also be taken into consideration that the transition metals used as catalysts in the manufacture of carbon nanomaterials may contribute to the genotoxicity.[59]

Biris and co-workers have recently investigated the cytotoxicity of highly pure graphene and compared the results with those of SWCNTs. Their study has been focused on understanding how these two different forms and shapes of nanomaterials characterized by the same chemical composition affect the cytotoxicity. Owing to different shapes of graphene and SWCNTs, their interactions with cell systems are expected to be governed by different mechanisms. Concentration and shape-dependent cytotoxic effects were induced by both graphene and SWCNT. Graphene induced a stronger metabolic activity than SWCNTs at low concentrations, while the reverse of this trend was observed at higher concentrations (Figure 1.8). At lower concentrations (0.01–10 µg mL^{-1}), graphene had no effect on the release of lactate dehydrogenase (LDH) whereas SWCNT induced a dramatic release of LDH. Therefore, lower levels of exposure (<0.01 µg mL^{-1}) to graphene could

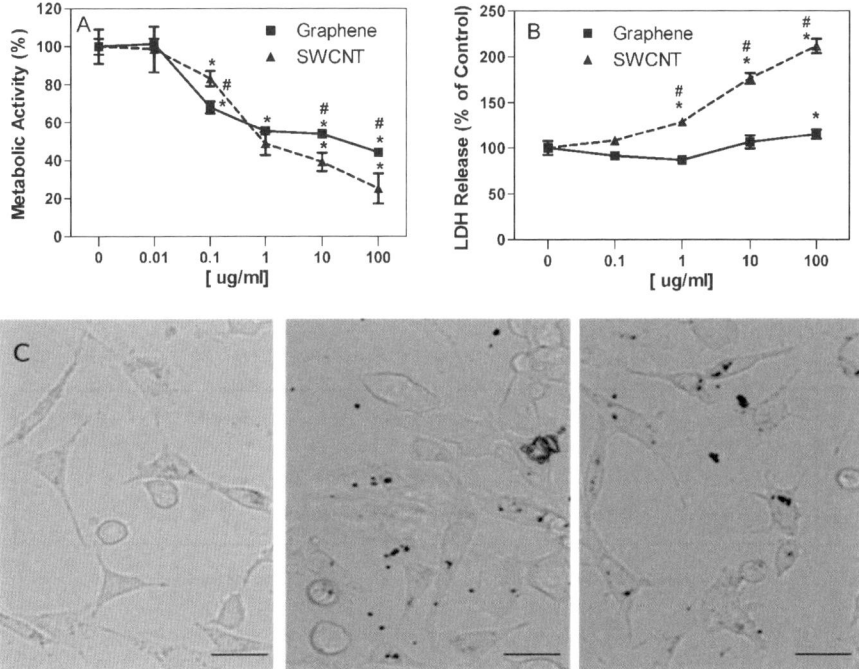

Figure 1.8 Effect of graphene or SWCNT on (A) mitochondrial toxicity and (B) LDH release (cell membrane damage marker). (C) Morphology change of PC12 cells: left, control; middle, graphene; right, SWCNT. Bar = 10 µm. (Reproduced with permission from ref. 60. Copyright 2010 American Chemical Society).

theoretically be useful in biomedical applications. Graphene was reported to have a relatively low toxicity compared to SWCNT.[60]

In the *in vitro* toxicity study, Liu *et al.* have reported no obvious toxicity of plain NGO-PEG at various concentrations without drug loading, suggesting that the PEGylated nanoscale graphene oxide sheets were not cytotoxic by themselves (Figure 1.9).[50] The relative cell viability (*versus* untreated control) data was collected for HCT-116 cells incubated with CPT-11, SN38, and NGO-PEG–SN38 at different concentrations for 72 h. Free SN38 was dissolved in DMSO and diluted in phosphate buffer saline (PBS). Water-soluble NGO-PEG–SN38 showed similar toxicity as SN38 in DMSO and far higher potency than CPT-11.[50]

In the *in vivo* toxicological studies, Yang *et al.* did not notice any obvious sign of toxic side effects for NGS-PEG-injected mice at the dose of 20 mg kg^{-1} within 40 as well as 90 days.[48,49] Neither death nor significant body weight drop was noted in the NGS-PEG + laser treated group. Yang *et al.* reported that they also did not observe any signal of organ damage from haematoxylin and eosin (H&E) stained organ slices. They have also carried out blood biochemistry and hematology analysis to understand the toxic effect of NGS-PEG on the treated mice.[49] The blood biochemistry tests were done to the liver function markers including alanine aminotransferase (ALT), aspartate aminotransferase (AST), alkaline phosphatase (ALP), and the ratio of albumin and globulin (A/G). As shown in Figure 1.10a and 1.10b, no obvious hepatic toxicity was induced by NGS-PEG treatment. The urea levels in the blood of treated mice were also normal indicating normal kidney functions. Biochemistry tests of important hematology markers of white blood cells, red blood cells, hemoglobin, mean corpuscular volume, mean corpuscular hemoglobin, mean corpuscular hemoglobin concentration, and platelet count in the NGS-PEG-treated groups at different times post-injection were reported to be normal compared with the control groups and in good agreement with the reference normal ranges.

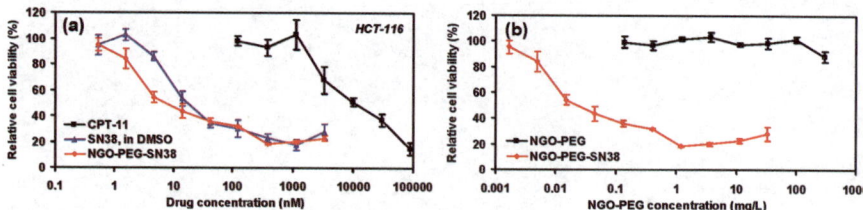

Figure 1.9   *In vitro* cell toxicity assay. (a) Relative cell viability (*versus* untreated control) data of HCT-116 cells incubated with CPT-11, SN38, and NGO-PEG–SN38 at different concentrations for 72 h. (b) Relative cell viability data of HCT-116 cells after incubation with NGO-PEG with (red) and without (black) SN38 loading. Plain NGO-PEG exhibited no obvious toxicity even at very high concentrations. Error bars were based on triplet samples. (Reproduced with permission from ref. 50. Copyright 2008 American Chemical Society).

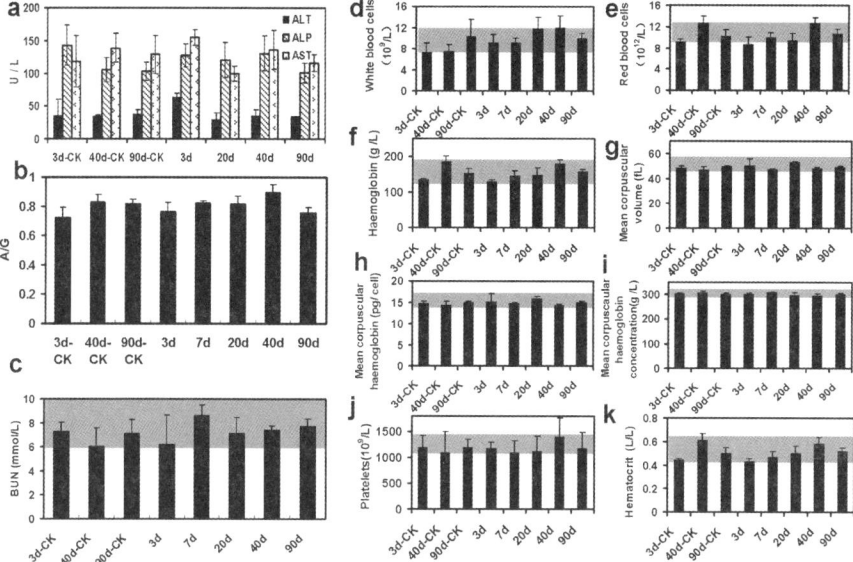

Figure 1.10 Blood biochemistry and hematology data of female BALB/c mice treated with NGS-PEG at the dose of 20 mg kg^{-1} at 3, 7, 20, 40, and 90 days post-injection. Age-matched control untreated mice were sacrificed at 3, 40, and 90 d (3d-CK, 40d-CK, and 90d-CK). (a) ALT, ALP, and AST levels in the blood at various time points after NGS-PEG treatment. (b) Time-course albumin/globulin ratios. Blood chemistry data suggested no hepatic disorder induced by NGS-PEG treatment. (c) Blood urea nitrogen (BUN) over time. (d)–(k) Time-course changes of white blood cells (d), red blood cells (e), hemoglobin (f), mean corpuscular volume (g), mean corpuscular hemoglobin (h), mean corpuscular hemoglobin concentration (i), platelets (j), and hematocrit (k) from control mice and NGS-PEG-treated mice. All blood chemistry and hematological data were all within the normal range. Statistics were based on five mice per data point. Gray areas in the figures show the normal reference ranges of hematology data of female BALB/c mice. (Reproduced with permission from ref. 49. Copyright 2011 American Chemical Society.)

Yang *et al.* have clearly showed no appreciable toxicity of NGS-PEG as evidenced from the blood biochemistry and hematological data. They also mentioned no organ damage except the color of liver and spleen turned brown due to the accumulation of NGS in these organs, which changed back to normal colors after 20 days post-injection indicating NGS-PEG clearance from these organs.[49] However, the exact clearance mechanisms of NGS-PEG, especially the size effect of graphene sheets on their *in vivo* behaviors, would be very important. More careful toxicology studies of NGS-PEG at higher doses and using different animal models are required before any potential clinical applications of graphene.

Feng *et al.* evaluated the *in vitro* cytotoxicity of different GO-PEI complexes with HeLa cells in comparison to bare PEI polymers by a cell viability assay.

They reported that PEI at a low molecular weight (1.2 kDa) was much less toxic but with a high molecular weight (at 10 kDa) appeared to be rather toxic to HeLa cells, with the half maximal inhibitory concentration (IC_{50}) value of approximately 30 mg L^{-1}.[53] They observed significantly reduced toxicity for the GO-PEI-10k complex, whose *in vitro* toxicity was reported to be one order of magnitude less than that of PEI-10k. Even at a GO-PEI-10k concentration as high as 300 mg L^{-1}, the relative cell viability was still higher than 80%. The decreased cytotoxicities of GO-PEI complexes were attributed to the rigid molecular structure that reduced the density of cationic residues interacting with cells. Feng *et al.* mentioned that GO-PEI complexes exhibited slightly higher relative cell viabilities compared to GO at high concentrations, likely owing to the improved stability of GO-PEI in physiological environments. With respect to PEI-10k, although it did offer efficient EGFP transfection at nitrogen/phosphate (N/P) ratios of 20 and 40, obvious toxicity to cells was noted at the N/P ratio of 80. In contrast, high EGFP expression together with no obvious cell toxicity was observed when using GO-PEI-10k as the transfection vector at N/P ratios from 20 to 80.[53] Further investigations are necessary to learn how the structure of graphene (*e.g.*, size and defects) would affect the gene transfection efficiency.

The toxicological studies of Zhang *et al.* suggest that the NGOs are practically non-toxic, and therefore are ideal nanocarriers for drug delivery. The cytotoxicity of FA–NGO/DOX (in terms of 20 μg mL^{-1} DOX) was much higher than that of NGO-SO$_3$H/DOX. When incubating the FA–NGO/DOX and excessive free FA with MCF-7 cells, the cytotoxicity was significantly lower, similar to that of the NGO-SO$_3$H/DOX (Figure 1.11). The study also revealed

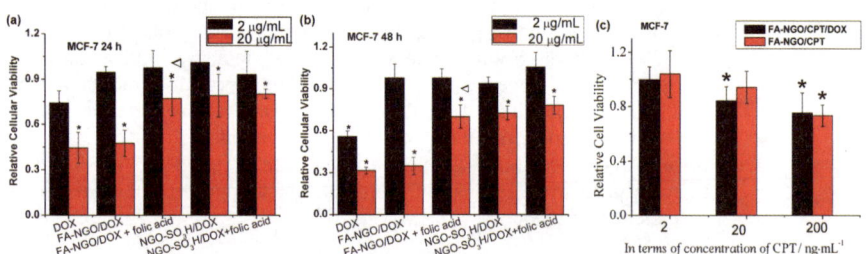

Figure 1.11   Relative cell viability of MCF-7 (a) 24 and (b) 48 h after treatment with FA–NGO/DOX, FA–NGO/DOX+FA, NGO-SO$_3$H/DOX, NGO-SO$_3$H/DOX+FA, and DOX. The asterisks indicate $P < 0.05$ *versus* normal cells and Δ indicates $P < 0.05$ *versus* FA–NGO/DOX groups at the same concentration of DOX. (c) Relative cell viability of MCF-7 treated with FA–NGO/CPT and FA–NGO/DOX/CPT. The asterisks indicate $P < 0.05$ *versus* normal cells. For (a)–(c), when the P value was less than 0.05, differences were considered statistically significant. The P value, which is associated with the statistical significance testing, is the probability of obtaining a test statistic at least as extreme as the one that was actually observed, assuming that the null hypothesis is true. (Reproduced with permission from ref. 26. Copyright 2010 Wiley-VCH Verlag GmbH & Co. KGaA, Weinheim.)

that no obvious toxicity was found for FA–NGO even at a concentration of 100 mg mL^{-1}.[26] Zhang *et al.* have also compared the cytotoxicity of FA–NGO/CPT/DOX (weight ratio of CPT:DOX was 1:0.976) with that of FA–NGO/CPT and FA–NGO/DOX. They found that the FA–NGO/DOX with a concentration of up to 2000 ng mL^{-1} showed no noticeable cytotoxicity. A similar result was also observed for FA–NGO/CPT with a CPT concentration of 2 and 20 ng mL^{-1}. However, FA–NGO/CPT/DOX (concentrations of both CPT and DOX approximately 20 ng mL^{-1}) displayed obvious cytotoxicity when CPT and DOX were loaded together onto FA–NGO (Figure 1.11c).[26]

Duch *et al.* explored strategies to improve the biocompatibility of graphene nanomaterials in the lung.[61] They found that GO is highly toxic when administered directly to the lungs of mice, causing severe and persistent lung injury. They also found that toxicity was significantly reduced by the generation of pristine graphene *via* liquid phase exfoliation and was further minimized when the graphene was dispersed with the ethylene oxide/propylene oxide block copolymer called Pluronic. The functional groups introduced during the oxidative process was noted as a major contributor to the pulmonary toxicity of GO. Duch *et al.* reported severe lung inflammation with alveolar exudates and hyaline membrane formation in mice treated with GO. Furthermore, lung inflammation following particle exposure is sufficient to induce a prothrombotic state in mice, which may contribute to the increased risk of heart attack and stroke observed in humans exposed to particulate matter air pollution. The experimental results of Duch *et al.* have revealed that pristine graphene induces minimal inflammation in the lung irrespective of how it is administered (as an aggregate or as a Pluronic dispersed material). In contrast, graphene oxide causes disruption of the alveolar-capillary barrier allowing the exudation of a protein rich fluid into the airspaces (acute lung injury) accompanied by an infiltration of inflammatory cells into the lung and the release of pro-inflammatory cytokines.[61]

Preparations of GO were reported to support redox cycling of cytochrome c and electron transport proteins in bacteria. Therefore, the oxygen moieties on the GO may accept electrons from cellular redox proteins, in the process forming highly reactive oxygen radicals.[62,63] Earlier studies have showed that aggregates of single-walled and multi-walled carbon nanotubes induce peribronchiolar lung fibrosis during the period equal or longer than 21 days after their administration.[64,65] The results of Duch *et al.* indicated that aggregates of graphene persisted in the medium or small airways of the lung and induced peribronchial inflammation and mild fibrosis, but there was no evidence of fibrosis in mice treated with dispersed graphene (Figure 1.12).[61] Although there was little evidence of lung fibrosis in mice treated with GO, persistent lung inflammation was reported 21 days after GO administration. Maintaining the nanoscale dispersion of graphene and minimizing contamination with GO were suggested to reduce the potential health consequences of workplace or environmental exposures.[61]

Very recently, Wang *et al.* have reported the effects of GO on human normal cells and mice.[66] They found that graphene oxide with dose <20 μg mL^{-1} did

Dispersed   Aggregated   Oxide

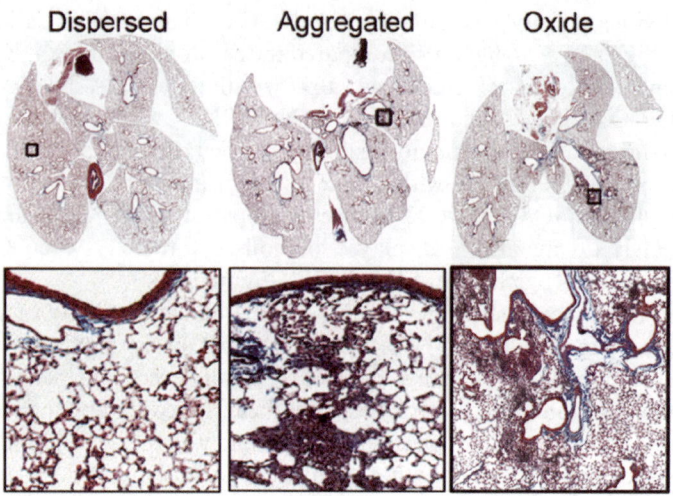

Figure 1.12   Trichrome-stained lung sections. Aggregated graphene induces patchy
fibrosis in mice. Mice were treated with highly purified and dispersed
preparations of graphene in 2% Pluronic F 108NF (dispersed),
aggregates of graphene in water (aggregated) or GO in water (oxide)
by intratracheal instillation and, 21 days later, the lungs were examined
for markers of fibrosis. (Reproduced with permission from ref. 61.
Copyright 2011 American Chemical Society.)

not exhibit toxicity to human fibroblast cells (HDF). However, the dose >50
μg mL^{-1} exhibited obvious cytotoxicity such as decreasing cell adhesion,
inducing cell apoptosis, entering into lysosomes, mitochondrion, endoplasm,
and cell nucleus. Similar phenomena were also observed in other cell lines such
as human gastric cancer MGC803, human breast cancer MCF-7, MDA-MB-
435, and liver cancer HepG2 cell lines indicating obvious toxicity of GO to
human normal cells or tumor cells. Furthermore, the amount of GO inside
HDF cells increased with increasing culture time; a lot of graphene oxides
appeared as black dots scattered in the cell cytoplasm around the cell nucleus,
a few graphene oxides were located inside the nucleus. As the cell culture time
increased, the survival rate of cells decreased, which is highly dependent on the
dose of graphene oxides (Figure 1.13a). Western blot results showed that the
expression levels of laminin, fibronectin, focal adhesion kinase (FAK), and cell
cycle protein cyclin-D3 in the HDF cells treated with GO were considerably
decreased compared with normal cells (Figure 1.13b). As shown in
Figure 1.13c, the cell adhesive ability decreased markedly with the increase
in GO concentration and culture time.[66] Graphene oxide and reduced
graphene oxide have been reported to exhibit toxicity to mice fibroblast cells
(line L929). Furthermore, the type of dispersant used to stabilize the
suspension, the type of nanomaterials, and their concentrations significantly
influence the cell cytotoxicity. Both graphene oxide and reduced graphene
oxide exhibited good cytocompatibility with concentrations between 3.125 μg
mL^{-1} and 12.5 μg mL^{-1}.[67]

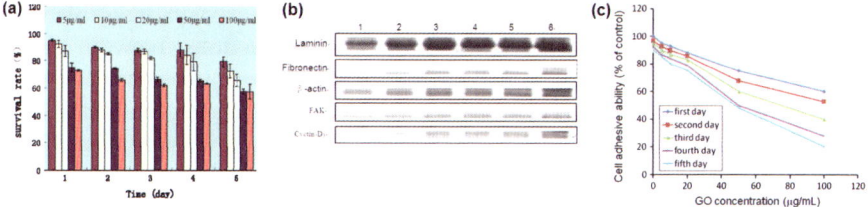

Figure 1.13   (a) Effects of GO on human fibroblast cells: the HDF survival rate at different concentrations of GO and different culture time. (b) Western blot analysis of adhesion proteins in HDF cells cultured with different concentrations of GO for 5 days. Lanes 1–6 show the expression levels of proteins in HDF cells treated with GO with the following concentrations: 100, 50, 20, 10, 5, 0 μg mL^{-1}, respectively. (c) GO-treated HDF cell adhesion ability measured by the centrifugation method. The percentage of adhesive cells decreased with the increase in GO concentration and culture time. (Reproduced with permission from ref. 66. Copyright 2010 Wang *et al.*).

In the case of mice, GO under low dose (0.1 mg) and middle dose (0.25 mg) did not exhibit obvious toxicity but under high dose (0.4 mg) exhibited chronic toxicity and lung granuloma formation, mainly located in the lung, liver, spleen, and kidney. GO could produce cytotoxicity in a dose- and time-dependent manner. Wang *et al.* suggested that exposures to GO may induce severe cytotoxicity and lung diseases. They have proposed possible mechanisms of effects of GO on human cells and mice. Graphene oxide has not been suggested as a good biocompatible material for human because of the long-term stay of GO in the kidney. Also, it is a very difficult species to be cleaned by the kidney.[66]

Chang *et al.* performed a comprehensive study on the toxicity of GO by examining the influences of GO on the morphology, viability, mortality and membrane integrity of A549 cells that are human lung carcinoma epithelial cells.[68] They also indicated that GO has pretty good biocompatibility to A549 cells compared to fullerene, CNT, and carbon nanofibres. GO was not found to be cytotoxic to A549 cells since it hardly entered cells. However, GO can cause a dose-dependent oxidative stress in cell and induce a slight loss of cell viability at high concentration. The possible mechanism for the toxicity of GO to A549 cells was proposed. Similar to the case of CNTs, GO absorbs the nutrients in culture medium and then the depletion of nutrients induces the oxidative stress and toxicity to A549 cells. Oxidative stress is a well-recognized toxicological mechanism of various nanoparticles. The oxidative stress induced by GO is considerably lower compared to fullerene and CNTs.[68]

We have noticed the inconsistency in the toxicity data of GO and such inconsistency might come from the GO synthesis/film preparation, and the testing models. Until now, the detailed information on the effect of shape of nanoparticles on the toxicity is still unknown. However, the few available results indicate that the shape affects the biological fate of nanomaterials.[69,70] Yuan *et al.* have performed a comparative study of protein profile of human hepatoma

HepG2 cells treated with graphene and single-walled carbon nanotubes to acquire a fundamental understanding of the interactions between these nanomaterials and biosystems. A distinct protein profile was observed for the cells exposed to graphene and SWCNTs. Graphene showed a moderate effect on the cellular functions compared to SWCNTs, indicating low cytotoxicity. The "snaking" effect of graphene appeared to be much weaker in comparison with SWCNTs, and this was reasoned to be due to a less toxic effect on the cells.[71]

Liao *et al.* have demonstrated that particle size, particular state, and oxygen content/surface charge of graphene have a strong impact on biological/toxicological responses to red blood cells. They have utilized two assays in assessing the cytotoxicity. The methylthiazolyldiphenyl-tetrazolium bromide (MTT) assay, a typical nanotoxicity assay, fails to assist in evaluation of the toxicity of graphene oxide and graphene toxicity especially at high doses because of the spontaneous reduction of MTT by graphene and graphene oxide. However, the application of a water-soluble tetrazolium salt (WST-8) and reactive oxygen species (ROS) assay reveal that the compacted graphene sheets are more damaging to mammalian fibroblasts than the less densely packed graphene oxide.[17] Dash and co-workers have studied the effects of graphene oxide and reduced graphene oxide on ultrastructural details of platelets. They have analyzed the interaction between graphene oxide and blood platelets, the cells responsible for acute arterial thrombotic events like ischemic heart disease and stroke. They have showed that graphene oxide can evoke strong aggregatory response in platelets in a scale comparable to that elicited by thrombin, one of the most potent physiological agonists of platelets. However, reduced graphene oxide was considerably less effective in activating platelets due to the reduced charge density on the graphene surface.[72]

Liu *et al.* have used 3-(4,5-dimethylthiazol-2-yl)-2,5-diphenyltetrazolium bromide for MTT assay to determine the cellular viabilities of DOX@gelatin-GNS with the MCF-7 cells.[52] Gelatin–GNS, even at a high concentration (200 μg mL^{-1}), was not found to be cytotoxic against MCF-7 cells. In general, the smaller size sheets are expected to be used in the drug delivery process and they showed very low cytotoxicity against MCF-7 cells. The cytotoxicity of the DOX@gelatin-GNS and free DOX enhanced gradually with increasing time. Furthermore, the DOX@gelatin-GNS exhibited a lower cytotoxicity than free DOX at the same dose. The gelatin-mediated sustained release effect of the drug, which may have potential clinical advantages pertaining to increased therapeutic efficacy, was proposed to be useful for cancer therapy in clinical applications because it could control the release rate of the drug in body. Therefore, the blood drug concentration could be maintained at a stable level for a long time.[52]

Both graphene and graphene oxide suspensions were proposed to inhibit the growth of *Escherichia coli* bacteria but with a minimal cytotoxicity.[73] The nanowalls were reported to be toxic to the bacteria. Akhavan and Ghaderi have studied the toxicity of graphene sheets, particularly the direct interaction of its extremely sharp edges with bacteria, the RNA effluxes through the damaged cell membranes of both Gram-negative *E. coli* and Gram-positive *Staphylococcus aureus* bacteria. Graphene oxide and reduced graphene oxide nanowalls were used to test the bacterial toxicity. The direct contact interaction of the bacteria with the very sharp edge of the nanowalls resulted in more damage to the cell membrane of the Gram-positive *S. aureus* bacteria, which lacks the outer membrane, as compared to the Gram-negative *E. coli* having the outer membrane.[74] The antibacterial activity of the reduced graphene nanowires is also comparable with the antibacterial activity of SWCNTs.[75]

1.3 Conclusions

In this chapter, we have reviewed the recent biomedical applications of graphene and the biocompatible systems derived from it. The key defects in graphene are highlighted. Nanoscale graphene and graphene oxide have immense potential in nanomedicine as biocompatible and supportive substrates, and as a novel tool for the delivery of therapeutic molecules. Graphene can be successfully used as a non-toxic nano-vehicle for efficient gene transfection, a novel gene delivery nano-vector with low cytotoxicity and high transfection efficiency, which is promising for future applications in non-viral-based gene therapy. Like carbon nanotubes, there are concerns about toxicity in the use of graphene. A limited number of publications have appeared on this topic. The preliminary experimental results highlighted the toxicology of graphene materials and difficulties in evaluating it. Similar to CNTs, graphitic fibers were found to be genotoxic in human bronchial epithelial BEAS 2B cells; dose-dependent DNA damage was reported with increasing effect of the damage at long treatment times. Experimental studies have demonstrated that the shape of carbonaceous nanomaterials plays an extremely important role in how they interact with cells and potentially other biological systems, such as tissues and organisms. The toxicity of graphene and graphene oxide depends on the exposure environment and mode of interaction with cells. The investigations of biomedical applications of graphene structures with different defects and their toxicity should be rewarded.

Acknowledgments

We thank the Department of Defense (DoD) for the HPCDNM project (Contract # W912HZ-09-C-0108) through the U.S. Army/Engineer Research and Development Center (Vicksburg, MS) and the National Science Foundation (NSF/CREST HRD-0833178) for the NSF-CREST center (Interdisciplinary Center for Nanotoxicity).

References

1. T. C. Dinadayalane and J. Leszczynski, *Struct. Chem.*, 2010, **21**, 1155.
2. A. K. Geim, *Science*, 2009, **324**, 1530.
3. M. J. Allen, V. C. Tung and R. B. Kaner, *Chem. Rev.*, 2010, **110**, 132.

4. R. Ruoff, *Nat. Nanotechnol.*, 2008, **3**, 10.
5. D. Li, M. B. Müller, S. Gilje, R. B. Kaner and G. G. Wallace, *Nat. Nanotechnol.*, 2008, **3**, 101.
6. A. Reina, X. Jia, J. Ho, D. Nezich, H. Son, V. Bulovic, M. S. Dresselhaus and J. Kong, *Nano Lett.*, 2009, **9**, 30.
7. L. Y. Jiao, L. Zhang, X. R. Wang, G. Diankov and H. J. Dai, *Nature*, 2009, **458**, 877.
8. F. Lu, L. Gu, M. J. Meziani, X. Wang, P. G. Luo, L. M. Veca, L. Cao and Y.-P. Sun, *Adv. Mater.*, 2009, **21**, 139.
9. T. C. Dinadayalane, J. S. Murray, M. C. Concha, P. Politzer and J. Leszczynski, *J. Chem. Theory Comput.*, 2010, **6**, 1351.
10. J. Wu, F. Hagelberg, T. C. Dinadayalane, D. Leszczynska and J. Leszczynski, *J. Phys. Chem. C*, 2011, **115**, 22232.
11. T. C. Dinadayalane and J. Leszczynski, *Chem. Phys. Lett.*, 2007, **434**, 86.
12. T. C. Dinadayalane, A. Kaczmarek, J. Òukaszewicz and J. Leszczynski, *J. Phys. Chem. C*, 2007, **111**, 7376.
13. A. Kaczmarek, T. C. Dinadayalane, J. Òukaszewicz and J. Leszczynski, *Int. J. Quantum Chem.*, 2007, **107**, 2211.
14. T. C. Dinadayalane and J. Leszczynski, in *Nanomaterials: Design and Simulation*, ed. P. B. Balbuena and J. M. Seminario, Theoretical and Computational Chemistry vol. 18, Elsevier, Amsterdam, 2007, p. 167.
15. T. C. Dinadayalane and J. Leszczynski, in *Practical Aspects of Computational Chemistry: Methods, Concepts and Applications*, ed. J. Leszczynski and M. K. Shukla, Springer, Netherlands, 2009, p. 297.
16. T. C. Dinadayalane, L. Gorb, T. Simeon and H. Dodziuk, *Int. J. Quantum Chem.*, 2007, **107**, 2204.
17. K.-H. Liao, Y.-S. Lin, C. W. Macosko and C. L. Haynes, *ACS Appl. Mater. Interfaces*, 2011, **3**, 2607.
18. M. Pumera, *Chem.–Asian J.*, 2011, **6**, 340.
19. I. Suarez-Martinez, N. Grobert and C. P. Ewels, *Carbon*, 2012, **50**, 741.
20. J. Xia, F. Chen, J. Li and N. Tao, *Nat. Nanotechnol.*, 2009, **4**, 505.
21. P. Blake, P. D. Brimicombe, R. R. Nair, T. J. Booth, D. Jiang, F. Schedin, L. A. Ponomarenko, S. V. Morozov, H. F. Gleeson, E. W. Hill, A. K. Geim and K. S. Novoselov, *Nano Lett.*, 2008, **8**, 1704.
22. B. G. Choi, H. Park, T. J. Park, M. H. Yang, J. S. Kim, S. Y. Jang, N. S. Heo, S. Y. Lee, J. Kong and W. H. Hong, *ACS Nano*, 2010, **4**, 2910.
23. J. H. Jung, D. S. Cheon, F. Liu, K. B. Lee and T. S. Seo, *Angew. Chem., Int. Ed.*, 2010, **49**, 5708.
24. C. Wan and B. Chen, *Biomed. Mater.*, 2011, **6**, 055010.
25. Z. Liu, J. T. Robinson, S. M. Tabakman, K. Yang and H. Dai, *Mater. Today*, 2011, **14**, 316.
26. L. Zhang, J. Xia, Q. Zhao, L. Liu and Z. Zhang, *Small*, 2010, **6**, 537.
27. K. Liu, J.-J. Zhang, F.-F. Cheng, T.-T. Zheng, C. Wang and J.-J. Zhu, *J. Mater. Chem.*, 2011, **21**, 12034.
28. N. Sinha and J. T.-W. Yeow, *IEEE Trans. Nanobiosci.*, 2005, **4**, 180.

29. A. D. Maynard, P. A. Baron, M. Foley, A. A. Shvedova, E. R. Kisin and V. Castranova, *J. Toxicol. Environ. Health A*, 2004, **67**, 87.
30. G. Oberdörster, E. Oberdörster and J. Oberdörster, *Environ. Health Perspect.*, 2005, **113**, 823.
31. P. P. Simeonova, *Nanomedicine*, 2009, **4**, 373.
32. A. A. Shvedova, V. Castranova, E. R. Kisin, D. Schwegler-Berry, A. R. Murray, V. Z. Gandelsman, A. Maynard and P. Baron, *J. Toxicol. Environ. Health A*, 2003, **66**, 1909.
33. S. K. Smart, A. I. Cassady, G. Q. Lu and D. J. Martin, *Carbon*, 2006, **44**, 1034.
34. K. Donaldson, R. Aitken, L. Tran, V. Stone, R. Duffin, G. Forrest and A. Alexander, *Toxicol. Sci.*, 2006, **92**, 5.
35. J. Muller, F. Huaux and D. Lison, *Carbon*, 2006, **44**, 1048.
36. K. S. Novoselov, A. K. Geim, S. V. Morozov, D. Jiang, Y. Zhang, S. V. Dubonos, I. V. Grigorieva and A. A. Firsov, *Science*, 2004, **306**, 666.
37. K. S. Novoselov, D. Jiang, F. Schedin, T. J. Booth, V. V. Khotkevich, S. V. Morozov and A. K. Geim, *Proc. Natl. Acad. Sci. U. S. A.*, 2005, **102**, 10451.
38. K. S. Novoselov, A. K. Geim, S. V. Morozov, D. Jiang, M. I. Katsnelson, I. V. Grigorieva, S. V. Dubonos and A. A. Firsov, *Nature*, 2005, **438**, 197.
39. L. A. Ponomarenko, F. Schedin, M. I. Katsnelson, R. Yang, E. W. Hill, K. S. Novoselov and A. K. Geim, *Science*, 2008, **320**, 356.
40. C. N. R. Rao, A. K. Sood, K. S. Subrahmanyam and A. Govindaraj, *Angew. Chem., Int. Ed.* 2009, **48**, 7752.
41. P. Avouris, Z. H. Chen and V. Perebeinos, *Nat. Nanotechnol.*, 2007, **2**, 605.
42. A. H. C. Neto, F. Guinea, N. M. R. Peres, K. S. Novoselov and A. K. Geim, *Rev. Mod. Phys.*, 2009, **81**, 109.
43. A. K. Geim and K. S. Novoselov, *Nat. Mater.*, 2007, **6**, 183.
44. Z. Sun, D. K. James and J. M. Tour, *J. Phys. Chem. Lett.*, 2011, **2**, 2425.
45. C. N. R. Rao, A. K. Sood, R. Voggu and K. S. Subrahmanyam, *J. Phys. Chem. Lett.*, 2010, **1**, 572.
46. J. C. Meyer, C. Kisielowski, R. Erni, M. D. Rossell, M. F. Crommine and A. Zettl, *Nano Lett.*, 2008, **8**, 3582.
47. D. Jiang, V. R. Cooper and S. Dai, *Nano Lett.*, 2009, **9**, 4019.
48. K. Yang, S. Zhang, G. Zhang, X. Sun, S.-T. Lee and Z. Liu, *Nano Lett.*, 2010, **10**, 3318.
49. K. Yang, J. Wan, S. Zhang, Y. Zhang, S.-T. Lee and Z. Liu, *ACS Nano*, 2011, **5**, 516.
50. Z. Liu, J. T. Robinson, X. Sun and H. Dai, *J. Am. Chem. Soc.*, 2008, **130**. 10876.
51. X. Sun, Z. Liu, K. Welsher, J. T. Robinson, A. Goodwin, S. Zaric and H. Dai, *Nano Res.*, 2008, **1**, 203.
52. K. Liu, J.-J. Zhang, F.-F. Cheng, T.-T. Zheng, C. Wang, J.-J. Zhu, *J. Mater. Chem.*, 2011, **21**, 12034.
53. L. Feng, S. Zhang and Z. Liu, *Nanoscale*, 2011, **3**, 1252.
54. Z. Liu, M. Winters, M. Holodniy and H. Dai, *Angew. Chem., Int. Ed.*, 2007, **46**, 2023.

55. C. Hom, J. Lu, M. Liong, H. Luo, Z. Li, J. I. Zink and F. Tamanoi, *Small*, 2010, **6**, 1185.
56. T. A. Xia, M. Kovochich,M. Liong, H. Meng, S. Kabehie, S. George, J. I. Zink and A. E. Nel, *ACS Nano*, 2009, **3**, 3273.
57. J. Zhu, A. Tang, L. P. Law, M. Feng, K. M. Ho, D. K. L. Lee, F. W. Harris and P. Li, *Bioconjugate Chem.*, 2005, **16**, 139.
58. C.-H. Lu, C.-L. Zhu, J. Li, J.-J. Liu, X. Chen and H.-H. Yang, *Chem. Commun.*, 2010, **46**, 3116.
59. H. K. Lindberg, G. C.-M. Falck, S. Suhonen, M. Vippola, E. Vanhala, J. Catalán, K. Savolainen and H. Norppa, *Toxicol. Lett.*, 2009, **186**, 166.
60. Y. Zhang, S. F. Ali, E. Dervishi, Y. Xu, Z. Li, D. Casciano and A. S. Biris, *ACS Nano*, 2010, **4**, 3181.
61. M. C. Duch, G. R. S. Budinger, Y. T. Liang, S. Soberanes, D. Urich, S. E. Chiarella, L. A. Campochiaro, A. Gonzalez, N. S. Chandel, M. C. Hersam and G. M. Mutlu, *Nano Lett.*, 2011, **11**, 5201.
62. A. J. Patil, J. L. Vickery, T. B. Scott and S. Mann, *Adv. Mater.*, 2009, **21**, 3159.
63. E. C. Salas, Z. Sun, A. Lüttge and J. M. Tour, *ACS Nano*, 2010, **4**, 4852.
64. G. M. Mutlu, G. R. S. Budinger, A. A. Green, D. Urich, S. Soberanes, S. E. Chiarella, G. F. Alheid, D. R. McCrimmon, I. Szleifer and M. C. Hersam, *Nano Lett.*, 2010, **10**, 1664.
65. A. A. Shvedova, E. R. Kisin, R. Mercer, A. R. Murray, V. J. Johnson, A. I. Potapovich, Y. Y. Tyurina, O. Gorelik, S. Arepalli, D. Schwegler-Berry, A. F. Hubbs, J. Antonini, D. E. Evans, B.-K. Ku, D. Ramsey, A. Maynard, V. E. Kagan, V. Castranova and P. Baron, *Am. J. Physiol.: Lung Cell. Mol. Physiol.*, 2005, **289**, L698.
66. K. Wang, J. Ruan, H. Song, J. Zhang, Y. Wo, S. Guo and D. Cui, *Nanoscale Res. Lett.*, 2011, **6**, 8.
67. M. Wojtoniszak, X. Chen, R. J. Kalenczuk, A. Wajda, J. Òapczuk, M. Kurzewski, M. Drozdzik, P. K. Chu and E. Borowiak-Palen, *Colloids Surf., B*, 2012, **89**, 79.
68. Y. Chang, S.-T. Yang, J.-H. Liu, E. Dong, Y. Wang, A. Cao, Y. Liu and H. Wang, *Toxicol. Lett.*, 2011, **200**, 201.
69. W. K. Oh, S. Kim, H. Yoon and J. Jang, *Small*, 2010, **6**, 872.
70. A. Simon-Deckers, S. Loo, M. Mayne-L'hermite, N. Herlin-Boime, N. Menguy, C. Reynaud, B. Gouget and M. Carrière, *Environ. Sci. Technol.*, 2009, **43**, 8423.
71. J. Yuan, H. Gao and C. B. Ching, *Toxicol. Lett.*, 2011, **207**, 213.
72. S. K. Singh, M. J. Singh, M. K. Nayak, S. Kumari, S. Shrivastava, J. J. A. Gracio and D. Dash, *ACS Nano*, 2011, **5**, 4987.
73. W. Hu, C. Peng, W. Luo, M. Lv, X. Li, D. Li, Q. Huang and C. Fan, *ACS Nano*, 2010, **4**, 4317.
74. O. Akhavan and E. Ghaderi, *ACS Nano*, 2010, **4**, 5731.
75. S. Kang, M. Herzberg, D. F. Rodrigues and M. Elimelech, *Langmuir*, 2008, **24**, 6409.

CHAPTER 2

In Vitro *Toxicity Assessment of Metallic Nanomaterials*

LAURA K. BRAYDICH-STOLLE,
NICOLE M. SCHAEUBLIN AND SABER M. HUSSAIN*

Applied Biotechnology Branch, Human Effectiveness Directorate, Air Force Research Laboratory, Wright-Patterson AFB, OH, USA
*E-mail: saber.hussain@wpafb.af.mil

2.1 Introduction

Nanotechnology involves the creation and manipulation of materials at the nanoscale level to create unique products with novel properties. Nanomaterials (NMs) such as nanotubes, nanowires, fullerene derivatives (buckyballs) and quantum dots are currently used for imaging, sensing, and targeted gene and drug delivery.[1–5] Bionanomaterials, which are by definition in the 1–100 nanometre range, have been used to create materials that have novel physical/chemical properties and functions due to their advantageous, miniscule size. In particular, nanoparticles are now used to target synthetic peptides, proteins, oligonucleotides and plasmids to specific cell types while protecting these macromolecules from enzymatic degradation.[6,7] In addition, nanoparticles have been proposed for the treatment of many diseases that need constant drug concentration in the blood or drug targeting to specific cells or organs.[8,9] In this respect, nanoencapsulated therapeutic agents such as antineoplastic drugs have been used with the aim to selectively target antitumor agents and to obtain a higher drug concentration at the tumor site.[10,11] This achievement

RSC Nanoscience & Nanotechnology No. 25
Towards Efficient Designing of Safe Nanomaterials: Innovative Merge of Computational Approaches and Experimental Techniques
Edited by Tomasz Puzyn and Jerzy Leszczynski
© The Royal Society of Chemistry 2012
Published by the Royal Society of Chemistry, www.rsc.org

appears to be important since many antineoplastic agents have several adverse side effects. Nanoparticles can be utilized to treat diseases that require a sustained presentation of the drug at several anatomical sites.[9,12] In addition, nanomaterials are of interest to defense and engineering programs because of their potential use in electronics, sensors, munitions and energetic/reactive systems involved with the advancement of propulsion technology.[13] If formulated properly with other materials, nanomaterials may provide greater stability and efficiency for propellant systems.

Despite the wide application of nanomaterials, there is a serious lack of information concerning the impact of manufactured nanomaterials on human health and the environment. Since there is a growing level of exposure to nanomaterials, it is important to answer questions regarding the toxicity of these materials. From an occupational safety standpoint three main routes of nanoparticle exposure have been identified and they are inhalation, absorption through the skin, and ingestion. Regardless of the exposure scenario, several studies have shown that nanoparticles can pass through biological membranes and thus can affect the physiology of any cell in an animal body by accumulating in tissues.[14,15] *In vitro* cultures are useful for providing a preliminary foundation for studies to assess dosing ranges, probable mechanisms of toxicity, and allow for the refinement of techniques before progressing to costly *in vivo* studies. Engineered nanomaterials can be split into two main groups: carbon-based and metal-based nanomaterials. This chapter will focus on the toxicity of the metal-based nanomaterials with regard to *in vitro* studies. The primary metal nanoparticles that have received attention are silver, gold, manganese, and titanium dioxide (Table 2.1 lists their abbreviations and uses) and this chapter will focus on them, and also discuss in lesser detail other metals such as aluminium, iron oxide, and copper. Figure 2.1 provides transmission electron microscopy (TEM) images for some of these common metallic nanoparticles.

Table 2.1 Abbreviations and uses for metallic nanoparticles.

Nanomaterial	Abbreviation	Consumer products/potential uses
Silver	Ag	Antimicrobial in food packaging, band aids, socks, *etc.*
Gold	Au	Cell imaging, drug delivery, cancer diagnostics
Titanium dioxide	TiO_2	Paint, sunscreen, cosmetics, waste water treatments
Manganese	Mn	Biosensors, coatings, magnetic data storage, imaging
Copper	Cu	Antimicrobial, circuits, nanowires
Iron oxide	FeO	Drug delivery, tissue repair, detoxification
Aluminium	Al	Fuels, munitions, coatings

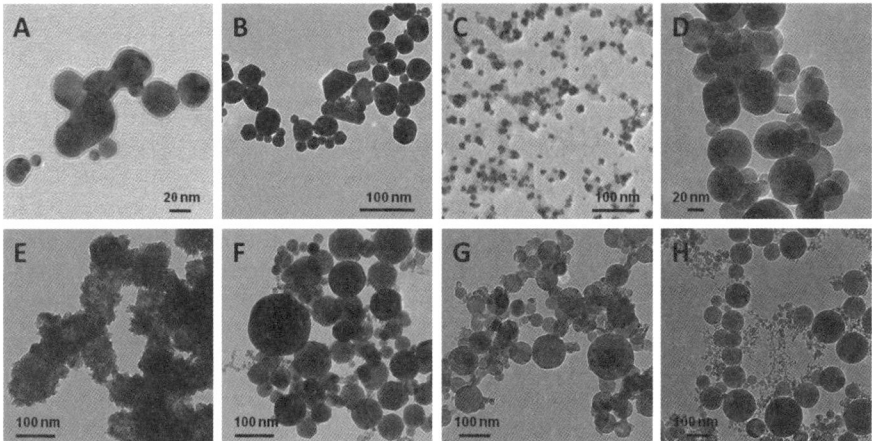

Figure 2.1 TEM images of various metallic nanoparticles. (A) Polysaccharide-coated Ag 25 nm. (B) Au 25 nm. (C) Mn 40 nm. (D) TiO_2 39 nm. (E) Cu 60 nm. (F) FeO 30 nm. (G) Al 50 nm. (H) Al_2O_3 40 nm.

2.2 Silver Nanomaterials

Typically scientists turn to gold nanoparticles to detect target DNA, but recent evidence is showing silver may actually be better than its gold counterpart.[16] Silver has a greater extinction coefficient; therefore, enhanced visual and absorption analyses can be performed using lower concentrations. Silver nanoparticles have been synthesized and shown to be effective antimicrobial agents due to their ability to bind to proteins and interfere with bacterial and viral processes.[17,18] Perhaps one of the most interesting examples is the ability of silver nanoparticles to bind to HIV-1 and prevent the virus from infecting host cells.[19] Therefore, nano-sized silver has found antimicrobial uses in bandages, coatings on clothing and other surfaces, and in paints.[9,20–22] However, silver nanoparticles have been shown to produce a size- and concentration-dependent toxicity effect.[23] In the liver cells, iron(II,III) oxide (Fe_3O_4 30, 47 nm), aluminium (Al 30 nm, 103 nm), manganese oxide (MnO_2 1–2 µm), and W (27 µm) displayed little or no toxicity, while molybdenum oxide (MoO_3 30, 150 nm) was moderately toxic, and silver (Ag 15, 100 nm) was highly toxic, and oxidative stress was identified as the participating mechanism of Ag toxicity.[24] Furthermore, the mouse germ cell stem cell line C18-4 was shown to be more sensitive to Ag nanoparticles than the rat liver cells, demonstrating a toxicity as low as 10 µg ml^{-1}.[25]

As stated earlier, silver nanoparticles have been shown to bind to HIV-1 and prevent the virus from entering the cell. Further studies by Elechiguerra *et al.* have shown this interaction is dependent on the size of the silver nanoparticles.[26] Silver within the size range of 1–10 nm has been shown to bind to the glycoprotein gp 120 knobs, which inhibits the virus from binding to the host cells. Not only is size important for the antibacterial properties of silver nanoparticles, but new studies are showing that shape plays a role in the antimicrobial abilities of silver.[27] A study by Pal *et al.* showed silver nanoparticles interact with *Escherichia coli* in a shape-dependent manner. The shape is important, because it is the number of {111}-facets that are displayed that have higher densities and therefore greater reactivity.[27] Furthermore, a study reported by Speshock *et al.* described that silver nanoparticles were capable of inhibiting the replication of a prototype arenavirus at non-toxic concentrations when administered prior to viral infection or early after initial virus exposure, suggesting that viral neutralization occurs during the early phases of viral replication.[28] In addition, several reports have identified that low levels of silver nanoparticle exposure were not toxic but interfered with key cell functions such as signaling pathways and immune responses.[29–31] These *in vitro* studies have illustrated that, while silver nanomaterials have great potential for antimicrobial effects, exposure to these materials needs to be further explored since toxicity as well as altered cell function in the absence of toxicity can occur.

2.3 Gold Nanomaterials

Gold nanoparticles have been used for cell imaging, targeted drug delivery, cancer diagnostics and medical therapeutic applications.[32–38] They serve as a model system for nanoparticle studies and products due to their large size range (can easily be synthesized from 1–100 nm) and large shape range with a 1:1 or 1:5 aspect ratio.[39] Furthermore, these particles are easily characterized using a variety of characterization techniques which include UV-Vis, inductively coupled plasma mass spectroscopy (ICP-MS), and transmission electron microscopy (TEM). A study by Schulz *et al.* determined that the particles appeared to be entering the alveolar macrophages and the lung epidermal cells by endocytotic pathways.[40] As stated earlier, nanoparticle uptake can be changed by surface modification. Gold nanoparticles have been modified with a combination of cell-penetrating peptides (CPPs) and a peptide acting as a nuclear localization signal to target the nucleus.[41] Nativo *et al.* showed this concept with 16 nm gold particles, which were surface-modified with CPPs and a nuclear localization target and brought directly to the nucleus, skipping the normal method of nanomaterial uptake.

Traditionally, these nanoparticles have been considered non-toxic;[39,42,43] however, recent studies have illustrated that these nanoparticles do induce some toxicity. For example, Pan and colleagues evaluated Au nanoparticle (NP) exposure in four different cell types with the NP size ranging from 0.8–15 nm. They found that there was size-dependent toxicity with the 1.4 nm being 60-fold more toxic than the 15 nm.[44] Furthermore, another study illustrated that 20 nm Au NPs repressed cell proliferation in lung fibroblasts after 72 h and this inhibition was a result of oxidative damage and downregulation of cell cycle genes.[45] Another nano-property that could impact toxicity is charge. Anionic Au NPs have been found to be non-toxic while cationic Au NPs are moderately toxic.[46] In contrast, another study demonstrated that 1.5 nm gold nanoparticles with differently charged ligands illustrated that the neutral particles were the least toxic, and the charged nanoparticles were toxic regardless of the charge (Figure 2.2).[47] Figure 2.2 illustrates the morphology of the human keratinocytes when treated with the gold nanoparticles. In Figure 2.2A, when the cells are untreated they form a nice adherent monolayer with very few disruptions. A similar morphology is seen after treatment with the neutral Au 1.5 nm-MEEE (MEEE – 2,2-mercaptoethoxyethoxyethanol) particles, but treatment with the positively charged Au 1.5 nm-TMAT (TMAT – *N,N,N*-trimethylammoniumethanethiol) and the negatively charged Au 1.5 nm-MES (MES – 2-mercaptoethanesulfonate) causes severe disruption in the monolayer. The microscopy data correlated with the viability data with the neutral nanoparticles causing a lesser decline in cell viability compared to the positively and negatively charged nanoparticles.[47] Furthermore, the negatively charged Au 1.5 nm nanoparticles cause a massive downregulation of DNA repair genes and, while the neutral and positive nanoparticles also down-regulated DNA repair genes, there was also some upregulation to compensate.[47] Furthermore, a study evaluating toxicological effects in human keratinocytes treated with Au nanospheres and Au nanorods with biocompatible surface chemistries showed that the shape of the nanomaterial plays a critical role in the cellular response. Data from the viability, mitochondrial stress, gene, and protein studies illustrated that the Au nanorods had considerable detrimental effects on the cells, while exposure to the Au nanospheres was non-toxic.[48] Interestingly, a report by Comfort *et al.* has demonstrated that at low levels of Au NM exposure where no toxicity was reported, the Au NM interfere with cell signaling pathways that are critical for cell growth and survival.[31] Therefore, while initial assessments indicated that gold nanoparticles were non-toxic, studies are beginning to show that gold nanoparticles display toxicity based on factors such as size, charge, surface modifications, and shape. Similarly to the Ag exposures, these Au NM studies have demonstrated the importance of *in vitro* evaluations for nanomaterials prior to their widespread implementation into consumer products.

2.4   Titanium Dioxide Nanomaterials

A key nanomaterial that has been immersed in the market place is titanium dioxide (TiO_2). Nano-sized TiO_2 has properties of high transparency to visible light and high UV absorption. Some components of visible light are reflected and refracted differentially which leads to an iridescent quality. These properties have generated widespread use of TiO_2 in pearlescent or "metallic" paint formulations, cosmetics, hard coatings, plastics, and self-cleaning additives for porcelain, ceramics, and specialty coatings.[49] Other TiO_2 applications include filters which exhibit strong germicidal properties and remove odors; TiO_2 has been used in conjunction with silver as an antimicrobial agent.[50] Furthermore, due to its photocatalytic activity, TiO_2

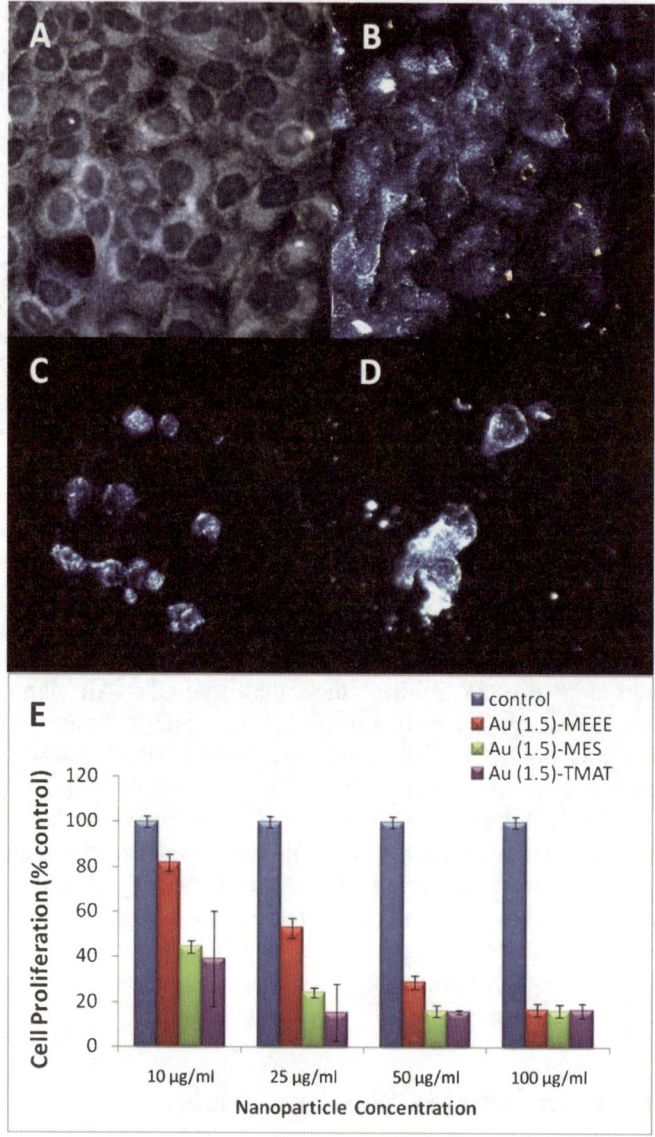

Figure 2.2   The role of charge in gold nanotoxicity. Cytoviva images of human keratinocytes treated with 1.5 nm gold nanoparticles with varying charges. (A) Untreated cells. (B) Treated with 25 µg ml^{-1} Au 1.5 nm-MEEE. (C) Treated with 25 µg ml^{-1} Au 1.5 nm-TMAT. (D) Treated with 25 µg ml^{-1} Au 1.5 nm-MES. (E) Cell viability when the human keratinocytes are treated with 1.5 nm gold nanoparticles with varying charges. The MEEE neutral nanoparticles have the highest viability and the negatively and positively charged Au nanoparticles demonstrated a greater reduction in cell viability; however, there was no significant difference between the positively and negatively charged nanoparticles.

has been used in waste water treatments.[51] This photocatalytic activity is influenced by the crystal structure (anatase and/or rutile), surface area, size distribution, porosity, surface, and hydroxyl group density.[51]

Nanotoxicity studies examining the effects of TiO_2 have shown the induction of inflammatory responses, cytotoxicity, and reactive oxygen species (ROS) formation in a variety of cell types and tissues.[52–57] However, the findings from these initial studies have been contradictory as to which nano-property correlates to the toxicity of this metal. Ultrafine TiO_2 particles (29 nm) increased inflammation and altered macrophage chemotactic responses in rat lungs when compared to larger TiO_2 particles (250 nm).[51] In addition, TiO_2 particles induce oxidative damage in a human bronchial epithelial cell line in a size-dependent manner.[55] In contrast, other studies have shown that exposure to nanoscale TiO_2 rods/dots produced inflammatory responses that were not different from pulmonary effects of larger TiO_2 particles in rats,[52] and that the composition of nanoscale titania correlates to cytotoxicity, with the anatase TiO_2 being more toxic than the rutile TiO_2.[54] However, it is important to note that none of these studies examined size while controlling for the crystal structure composition or examined crystal structure while controlling the size. In the study by Renwick *et al.* the crystal structure composition of the two different types of TiO_2 particles is not identified[51] and in the study by Gurr *et al.* three of the particles are anatase while the fourth is rutile.[55] Warheit *et al.* contradicted the size-dependent study: 300 nm rutile nanoparticles were compared to TiO_2 rods that were 200 nm × 35 nm and 10 nm anatase dots.[52] Additionally, the study by Sayes *et al.* did not consider size based on the formation of aggregates,[54] yet Nel and colleagues have identified aggregates as one of the biokinetic properties of nanomaterials that must be taken into consideration.[57] Another study recently published by Braydich-Stolle *et al.* has illustrated that size and crystal structure both impact TiO_2 nanotoxicity, with crystal structure mediating the mechanism of cell death. While many of these studies illustrate the toxic properties of TiO_2 nanoparticles, other studies have demonstrated that TiO_2 nanoparticles are not capable of penetrating the skin and therefore do not pose a health risk since there is no internal exposure.[58–61]

2.5 Manganese Nanomaterials

Manganese nanoparticles have potential uses in biomedical imaging and gene delivery. One study has shown that manganese zinc ferrite magnetic nanoparticles are able to induce gene expression using heat generated by an alternating magnetic current and that these nanoparticles help control the *in vivo* temperature so that safe and effective heat induced transgene expression can occur.[62] Furthermore, manganese oxide nanoparticles have showed promise as a positive contrast agent in MR imaging.[63] However, high doses of MnO have been found to be toxic[63] and manganese-containing silica nanoparticles can generate ROS production with up to an 8-fold increase in ROS when compared to controls.[64] In addition, manganese has been shown to induce apoptotic mitochondrial-induced signaling in neurons. Similarly, one study using the PC-12 cell line showed that Mn 40 nm particles induced dose-dependent dopamine depletion as well as the depletion of dopamine metabolites,[65] and further studies need to examine if dopamine depletion occurs *in vivo*.

2.6 Copper Nanomaterials

Copper nanomaterials are being incorporated into textiles, polymers, fuel cells, and composites due to their potential antimicrobial, electrical, magnetic, optical, imaging, and catalytic properties. In addition to causing toxicity in PC-12 cells, varying sizes of copper nanoparticles (40, 60, and 80 nm) have shown significant depletion in dopamine production at concentrations as low as 2.5 μg ml^{-1}. An acellular ROS assay determined that the copper nanoparticles themselves produced high levels of ROS in solution and this resulted in high levels of ROS production from the PC-12 cells. Furthermore, rat alveolar macrophages exposed to copper nanoparticles (40, 60, and 80 nm) exhibited toxicity in a concentration-dependent manner but not a size-dependent manner. In the macrophages, reactive oxygen species production was low; however, exposure to copper nanoparticles increased cytokine release.[66]

2.7 Iron Oxide Nanomaterials

Iron oxide nanoparticles are being used for magnetic resonance imaging contrast, detoxification, drug delivery, and tissue repair.[67] Currently, there are mixed reviews regarding whether these nanoparticles are harmful to cells and tissues. One study evaluated the toxic effects of iron oxide nanoparticle complexes in human mesenchymal stem cells and HeLa cells, and demonstrated that in these cell types there were no short- or long-term negative effects.[68] Similarly, iron oxide nanoparticles were injected into rat brains for MRI imaging and showed no change in myelin or neuron pathology over time.[69] In contrast, these nanoparticles have demonstrated toxicity in rat liver cells[70] and decreased viability and inhibition of neurite extension in the presence of nerve growth factor in a dose-dependent manner in PC-12 cells.[25] Furthermore, low levels of iron oxide exposure demonstrated an inhibition of EGF-induced (epidermal growth factor) gene expression indicating that, even in the absence of toxicity, cellular responses can be impacted by iron oxide nanoparticles.[31]

2.8 Aluminium Nanomaterials

Nanoenergetic aluminium has potential military, medical, and industrial applications, but there is limited information concerning the health effects of

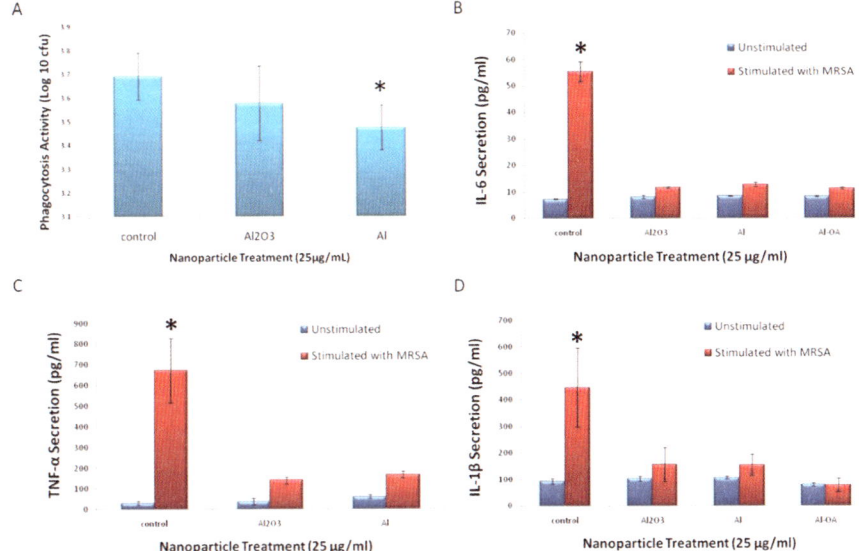

Figure 2.3 Alteration of macrophage function after treatment with a non-toxic concentration of aluminium nanoparticles. (A) Phagocytic activity is decreased in the presence of aluminium nanoparticles but not aluminium oxide nanoparticles. (B)–(D) Following a 24 h treatment with 25 µg ml^{-1} of aluminium nanoparticles, the macrophages were unable to secrete inflammatory cytokines in the presence of bacteria. (B) IL-6. (C) TNF-α. (D) IL-1β.

aluminium nanomaterial exposure and, since bulk aluminium is known to cause respiratory and neurological toxicity, these studies are imperative. Aluminium nanoparticles exhibit toxicity and significantly diminish the phagocytotic ability of macrophages in comparison to aluminium oxide nanoparticles which are not toxic in both rat and human lung models.[71,72] Furthermore, aluminium nanoparticles alter the macrophage's ability to initiate an inflammatory response to bacterial pathogens regardless of coating. Figure 2.3 illustrates that after treatment with Al nanoparticles and then exposure to bacteria, IL-6, TNF-α, and IL-1β secretion are altered. These initial studies with various nanometals have demonstrated the need to pursue not only more mechanistic *in vitro* studies but also further evaluations in *in vivo* models in order for these materials to be declared "biocompatible" and "safe".

2.9 Biocompatibility of Nanomaterials

Several studies have shown that altering the surface chemistry of nanoparticles has been effective in preventing toxicity from the core nanomaterials.[73–78] One study by Derfus and colleagues demonstrated that ZnS "capping" of CdSe quantum dots resulted in decreased toxicity to rat primary hepatocytes, while another study examining the effect of coating on quantum dot toxicity in human epidermal kerotinocytes found neutral, polyethylene glycol-coated quantum dots to be non-toxic, while amine surfaces were cytotoxic, and carboxylic surfaces were both cytotoxic and inflammatory.[78] Furthermore, iron oxide nanoparticles coated with amino-polyvinylalcohol proved to be more biocompatible, potentially making them candidates for imaging agents.[79] However, other studies by Schrand *et al.* and Braydich-Stolle *et al.* have demonstrated that while these coatings may initially reduce the toxic potential of a nanoparticle that the cellular environment impacts the properties of these coatings and therefore the stability of the coating must also be evaluated

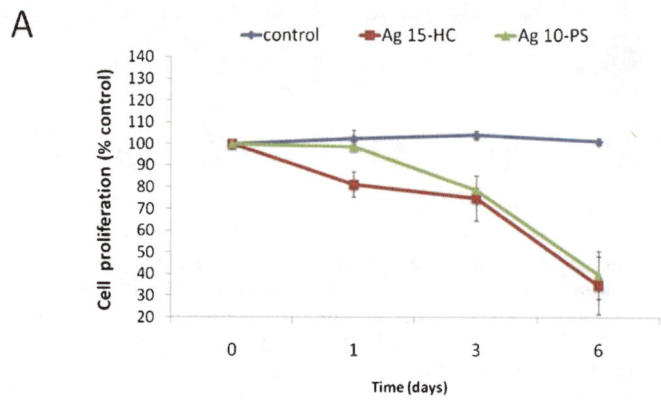

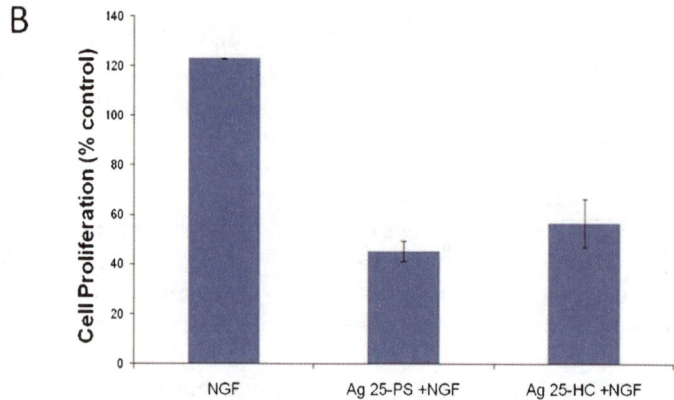

Figure 2.4   The effect of coatings on silver nanoparticle toxicity. (A) A spermato-gonial stem cell line (C18-4 cells) was treated with a non-toxic concentration of silver nanoparticles (10 µg ml^{-1}) for six days and cell proliferation was assessed daily. Initially, the silver nanoparticles coated with the polysaccharide were more biocompatible than the non-coated silver nanoparticles. However, after three days in culture there was no difference in biocompatibility between the two nanoparticle types. (B) Graph indicating that coating stability is another factor to consider when working with nanoparticles.

(Figure 2.4). For example, spermatogonial stem cells were treated with two different types of Ag NPs, one had a hydrocarbon surface while the other was coated with a polysaccharide (PS). Initially, the PS-coated Ag was biocompatible but when the cells were exposed for one week after three days in culture the biocompatibility of these nanoparticles was lost (Figure 2.4A). Additionally, Schrand *et al.* demonstrated that the PS-coated Ag was more biocompatibile in neuroblastoma cells; however, when nerve growth factor (NGF) was added to the cultures to induce proliferation of the cells, the presence of both types of Ag prevented the NGF from inducing a response (Figure 2.4B).

2.10 Conclusions

Nanotechnology is emerging as one of the most innovative technologies of the 21st century and promises to yield personal, social, and commercial benefits by being incorporated into medical, energy, materials, and microelectronics applications. However, the potential risks of nanomaterials must be evaluated prior to incorporation of these materials. Preliminary studies have shown that the nanoparticle type, size, crystal structure, coating, and charge are all properties that contribute to the cytotoxicity of the nanoparticle. Additionally, gold nanoparticles serve as an excellent warning about quick assessments of toxicity and should not be taken lightly. While earlier studies suggested gold nanoparticles to be non-toxic, emerging studies are finding that different nano-properties can contribute to making gold nanoparticles toxic. Furthermore, these studies illustrate that there is still much that is not known about how these materials behave in a cellular environment and how the cellular environment impacts the nanomaterial.

References

1. M. De Wild, S. Berner, H. Suzuki, L. Ramoino, A. Baratoff and T. A. Jung, Molecular assembly and self-assembly: molecular nanoscience for future technologies. *Ann. N. Y. Acad. Sci.*, 2003, **1006**, 291–305.
2. D. Kim, Y. Zhang, W. Voit, K. Rao, J. Kehr, B. Bjelke and M. Muhammed, Superparamagnetic Iron Oxide Nanoparticles for Bio-Medical Applications, *Scr. Mater.*, 2001, **44**, 1713–1717.
3. R. Murugan and S. Ramakrishna, Development of Nanocomposites for Bone Grafting, *Compos. Sci. Technol.*, 2005, **65**, 2385–2406.
4. V. Sinha and A. Trehan, Biodegradable Microspheres for Protein Delivery, *J. Controlled Release*, 2003, **90**, 261–280.
5. Y. Wu, W. Yang, C. Wang, J. Hu and S. Fu, Chitosan Nanoparticles as a Novel Delivery System for Ammonium Glycyrrhizinate, *Int. J. Pharm.*, 2005, **295**, 235–245.
6. C. Chavany, T. Saison-Behmoaras, D. T. Le, F. Puisieux, P. Couvreur and C. Helene, Adsorption of oligonucleotides onto polyisohexylcyanoacrylate

nanoparticles protects them against nucleases and increases their cellular uptake, *Pharm. Res.*, 1994, **11**, 1370–1378.

7. K. A. Janes, P. Calvo and M. J. Alonso, Polysaccharide colloidal particles as delivery systems for macromolecules, *Adv. Drug Delivery Rev.*, 2001, **47**, 83–97.

8. S. M. Moghimi, A. C. Hunter and J. C. Murray, Long-circulating and target-specific nanoparticles: theory to practice, *Pharmacol. Rev.*, 2001, **53**, 283–318.

9. J. Panyam and V. Labhasetwar, Biodegradable nanoparticles for drug and gene delivery to cells and tissue, *Adv. Drug Delivery Rev.*, 2003, **55**, 329–347.

10. J. S. Chawla and M. M. Amiji, Biodegradable poly(ε-caprolactone) nanoparticles for tumor-targeted delivery of tamoxifen, *Int. J. Pharm.*, 2002, **249**, 127–138.

11. S. K. Sahoo, W. Ma and V. Labhasetwar, Efficacy of transferrin-conjugated paclitaxel-loaded nanoparticles in a murine model of prostate cancer. *Int. J. Cancer*, 2004, **112**, 335–340.

12. Z. Z. Li, L. X. Wen, L. Shao and J. F. Chen, Fabrication of porous hollow silica nanoparticles and their applications in drug release control, *J. Controlled Release*, 2004, **98**, 245–254.

13. S. P. Ringer and K. R. Ratinac, On the role of characterization in the design of interfaces in nanoscale materials technology, *Microsc. Microanal.*, 2004, **10**, 324–335.

14. Y. Chen, Z. Xue, D. Zheng, K. Xia, Y. Zhao, T. Liu, Z. Long and J. Xia, Sodium chloride modified silica nanoparticles as a non-viral vector with a high efficiency of DNA transfer into cells, *Curr. Gene Ther.*, 2003, **3**, 273–279.

15. P. J. Borm and W. Kreyling, Toxicological hazards of inhaled nanoparticles–potential implications for drug delivery, *J. Nanosci. Nanotechnol.*, 2004, **4**, 521–531.

16. D. G. Thompson, A. Enright, K. Faulds, W. E. Smith and D. Graham, Ultra-Sensitive DNA Detection Using Oligonucleotide-Silver Nanoparticle Conjugates, *Anal. Chem.*, 2008, **80**, 2805–2810.

17. M. Karhanek, J. T. Kemp, N. Pourmand, R. W. Davis and C. D. Webb, *Nano Lett.*, 2005, **5**, 403–407.

18. T. A. Taton, G. Lu and C. A. Mirkin, *J. Am. Chem. Soc.*, 2001, **123**, 5164–4165.

19. I. Medintz, A. R. Clapp, J. S. Melinger, J. R. Deschamps and H. Mattoussi, *Adv. Mater.*, 2005, **17**, 2450–2455.

20. P. Yang, X. Sun, J. Chiu, H. Sun and Q. He, *Bioconjugate Chem.*, 2005, **16**, 494–496.

21. N. Kohler, C. Sun, J. Wang and M. Zhang, *Langmuir*, 2005, **21**, 8858–8864.

22. S. Shukla, A. Priscilla, M. Banerjee, R. R. Bonde, J. Ghatak, P. V. Satyam and M. Sastry, *Chem. Mater.*, 2005, **17**, 5000–5005.

23. C. Carlson, S. Hussain, A. Schrand, L. Braydich-Stolle, K. Hess, R. Jones and J. Schlager, Unique Cellular Interaction of Silver Nanoparticles: Size

Dependent Generation of Reactive Oxygen Species, *J. Phys. Chem. B*, 2008, **112**, 13608–13619.

24. A. Hoshino, F. Kujioka, T. Oku, M. Suga, Y. F. Sasaki, T. Ohta, M. Yasuhara, K. Suzuki and K. Yamamoto, *Nano Lett.*, 2004, **4**, 2163–2169.

25. L. Braydich-Stolle, S. M. Hussain, J. Schlager and M.-C. Hofmann, In vitro cytotoxicity of nanoparticles in mammalian germ-line stem cells, *Toxicol. Sci.*, 2005, **88**, 412–419.

26. J. L. Elechiguerra, J. L. Burt, J. R. Morones, A. Camacho Bragado, X. Gao, H. H. Lara and M. J. Yacaman, Interaction of silver nanoparticles with HIV-1, *J. Nanobiotechnol.*, 2005, **3**, 6–16.

27. S. Pal, K. Tak and J. M. Song, Does the Antibacterial Activity of Silver Nanoparticles Depend on the Shape of the Silver Nanoparticle? A Study of the Gram-negative bacterial *Escherichia coli*, *Appl. Environ. Microbiol.*, 2007, **73**, 1712–1720.

28. J. L. Speshock, R. C. Murdock, L. K. Braydich-Stolle, A. M. Schrand and S. M. Hussain, Interaction of silver nanoparticles with Tacaribe virus, *J. Nanobiotechnol.*, 2010, **8**, 19.

29. L. K. Braydich-Stolle, B. Lucas, A. M. Schrand, R. C. Murdock, T. Lee, J. J. Schlager, S. M. Hussain and M. C. Hofmann, Silver nanoparticles disrupt GDNF/Fyn kinase signaling in spermatogonial stem cells, *Toxicol. Sci.*, 2010, **116**(2), 577–589.

30. J. L. Speshock, L. K. Braydich-Stolle, E. R. Szymanski and S. M. Hussain, Silver and Gold Nanoparticles Alter Cathepsin Activity In vitro, *Nanoscale Res. Lett.*, 2011, **6**, 17.

31. K. K. Comfort, E. I. Maurer, L. K. Braydich-Stolle and S. M. Hussain, Interference of Silver, Gold, and Iron Oxide Nanoparticles on Epidermal Growth Factor Signal Transduction in Epithelial Cells, *ACS Nano*, 2011, **5**(12), 10000–10008.

32. J. C. Riboh, A. J. Haes, A. D. McFarland, C. R. Yonzon and R. P. Van Duyne, An Optical Biosensor: Real-Time Immunoassay in Physiological Buffer Enabled by Improved Nanoparticle Adhesion, *J. Phys. Chem. B*, 2003, **107**(8), 1772–1780.

33. P. Ghosh, G. Han, M. De, C. K. Kim and V. M. Rotello, Gold nanoparticles in delivery applications, *Adv. Drug Delivery Rev.*, 2008, **60**(11), 1307–1315.

34. D. Pissuwan, S. M. Valenzuela and M. B. Cortie, Therapeutic possibilities of plasmonically heated gold nanoparticles, *Trends Biotech.*, 2006, **24**(2), 62–67.

35. Y. Wang, X. Xie, X. Wang, G. Ku, K. L. Gill, D. P. O'Neal, G. Stoica and L. V. Wang, Photoacoustic Tomography of a Nanoshell Contrast Agent in the *in vivo* Rat Brain, *Nano Lett.*, 2004, **4**(9), 1689–1692.

36. D. Yelin, D. Oron, S. Thiberge, E. Moses and Y. Silberberg, Multiphoton plasmon-resonance microscopy, *Opt. Express*, 2003, **11**(12), 1385–1391.

37. T. B. Huff, M. N. Hansen, Y. Zhao, J. X. Cheng and A. Wei, Controlling the Cellular Uptake of Gold Nanorods, *Langmuir*, 2007, **23**(4), 1596–1599.

38. D. P. O'Neal, L. R. Hirsch, N. J. Halas, J. D. Payne and J. L. West, Photo-thermal tumor ablation in mice using near infrared-absorbing nanoparticles, *Cancer Lett.*, 2004, **209**(2), 171–176.

39. I. H. El-Sayed, X. Huang and M. A. El-Sayed, Surface Plasmon Resonance Scattering and Absorption of anti-EGFR Antibody Conjugated Gold Nanoparticles in Cancer Diagnostics: Applications in Oral Cancer, *Nano Lett.*, 2005, **5**(5), 829–834.

40. S. Takenaka, E. Karg, W. G. Kreyling, B. Lentner, W. Moller, M. Behnke-Semmler, L. Jennen, A.Walch, B. Michalke, P. Schramel, J. Heyder and H. Schulz, Distribution Pattern of Ultra-fine gold nanoparticles in the rat lung, *Inhal. Toxicol.*, 2006, **18**, 733–740.

41. P. Nativo, I. A. Prior and M. Brust, Uptake and Intracellular Fate of Surface Modified Gold Nanoparticles, *ACS Nano*, 2008, **2**, 1639–1644.

42. Y. Pan, S. Neuss, A. Leifert, M. Fischler, F. Wen, U. Simon, G. Schmid, W. Brandau and W. Jahnen-Dechent, Size-Dependent Cytotoxicity of Gold Nanoparticles, *Small*, 2007, **3**, 1941–1949.

43. J. J. Li, L. Zou, D. Hartano, C. N. Ong, B. H. Bay, L. Yung, Gold Nanoparticles Induce Oxidative Damage in Lung Fibroblasts In Vitro, *Adv. Mater.*, 2008, **20**, 138–142.

44. C. M. Goodman, C. D. McCusker, T. Yilmaz and V. M. Rotello, Toxicity of gold nanoparticles functionalized with cationic and anionic side chains, *Bioconjugate Chem.*, 2004, **15**(4), 897–900.

45. N. M. Schaeublin, L. K. Braydich-Stolle, A. M. Schrand, J. M. Miller, J. Hutchison, J. J. Schlager and S. M. Hussain, Surface charge of gold nanoparticles mediates mechanism of toxicity, *Nanoscale*, 2011, **3**, 410.

46. N. Schaeublin, L. K. Braydich-Stolle, E. I. Maurer, K. Park, R. I. Maccuspie, A. N. Afrooz, N. B. Saleh, R. A. Vaia and S. M. Hussain, Does Shape Matter? Bioeffects of Gold Nanomaterials in a Human Skin Cell Model, *Langmuir*, 2012, **28**, 3248–3258.

47. A. Wold, Photocatalytic Properties of TiO_2, *Chem. Mater.*, 1993, **5**, 280–283.

48. D. K. Ellsworth, D. Verhurst, T. M. Spitler and B. J. Sabacky, Titanium nanoparticles move to the marketplace, *Chem. Innovation*, 2000, **30**(12), 30–35.

49. L. C. Renwick, D. Brown, A. Clouter and K. Donaldson, Increased inflammation and altered macrophage chemotactic responses caused by two ultrafine particle types, *Occup. Environ. Med.*, 2004, **5**, 442–447.

50. D. B. Warheit, T. R. Webb, C. M. Sayes, V. L. Colvin and K. L. Reed, Pulmonary instillation studies with nanoscale TiO_2 rods and dots in rats: toxicity is not dependent upon particle size and surface area, *Toxicol. Sci.*, 2006, **1**, 227–236.

51. V. H. Grassian, P. T. O'Shaughnessy, A. Adamcakova-Dodd, J. M. Pettibone and P. S. Thorne, Inhalation exposure study of titanium dioxide nanoparticles with a primary particle size of 2 to 5 nm, *Environ. Health Perspect.*, 2007, **115**(3), 397–402.

52. C. M. Sayes, R. Wahi, P. A. Kurian, Y. Liu, J. L. West, K. D. Ausman, D. B. Warheit and V. L. Colvin, Correlating nanoscale titania structure with toxicity: a cytotoxicity and inflammatory response study with human dermal fibroblasts and human lung epithelial cells, *Toxicol. Sci.*, 2006, **1**, 174–85.

53. J. R. Gurr, A. S. Wang, C. H. Chen and K. Y. Jan, Ultrafine titanium dioxide particles in the absence of photoactivation can induce oxidative damage to human bronchial epithelial cells, *Toxicology*, 2005, **1–2**, 66–73.

54. T. C. Long, N. Saleh, R. D. Tilton, G. V. Lowry and B. Veronesi, Titanium dioxide (P25) produces reactive oxygen species in immortalized brain microglia (BV2): implications for nanoparticle neurotoxicity, *Environ. Sci. Technol.*, 2006, **40**(14), 4346–4352.

55. A. Nel, T. Xia, L. Mädler and N. Li, Toxic potential of materials at the nanolevel, *Science*, 2006, **5761**, 622–627.

56. A. Gamer, E. Leibold and B. van Ravenzway, The in vitro absorption of microfine ZnO and TiO_2 through porcine skin, *Toxicol. In Vitro*, 2006, **20**, 301–307.

57. J. Lademann, H. Weigmann, C. Rickmeyer, H. Barthelmes, H. Schaefer, G. Mueller and W. Sterry, Penetration of titanium dioxide microparticles in a sunscreen formulation into the horny layer and the follicular orifice, *Skin Pharmacol. Appl.*, 1999, **12**, 247–256.

58. F. Pflücker, V. Wendel, H. Hohenberg, E. Gärtner, T. Will, S. Pfeiffer, R. Wepf and H. Gers-Barlag, The human stratum corneum layer: an effective barrier against dermal uptake of different forms of topically applied micronised titanium dioxide, *Skin Pharmacol. Appl. Skin Physiol.*, 2001, **14**(Suppl. 1), 92–97.

59. J. Schulz, H. Hohenberg, F. Pflücker, E. Gärtner, T. Will, S. Pfeiffer, R. Wepf, V. Wendel, H. Gers-Barlag and K. P. Wittern, Distribution of sunscreens on skin, *Adv. Drug Delivery Rev.*, 2002, **54**(Suppl. 1), S157–S163.

60. M. Kreilgaard, Influence of microemulsions on cutaneous drug delivery, *Adv. Drug Delivery Rev.*, 2002, **54**, S77–S98.

61. E. G. de Jalon, M. J. Blanco-Prieto, P. Ygartua and S. Santoyo, PLGA microparticles: possible vehicles for topical drug delivery, *Int. J. Pharm.*, 2001, **226**, 181–184.

62. J. Lademann, N. Otberg, H. Richter, H. J. Weigmann, U. Lindemann, H. Schaefer and W. Sterry, Investigation of follicular penetration of topically applied substances, *Skin Pharmacol. Appl. Skin Physiol.*, 2001, **14**, 17–22.

63. J. Wang, M. F. Rahman, H. M. Duhart, G. D. Newport, T. A. Patterson, R. C. Murdock, S. M. Hussain, J. J. Schlager and F. Ali, Expression Changes of Dopaminergic System-Related Genes in PC12 Cells Induced by Manganese, Silver, or Copper Nanoparticles, *NeuroToxicology*, 2009, **30**(6), 926–933.

64. A. K. Gupta and M. Gupta, Synthesis and surface engineering of iron oxide nanoparticles for biomedical application, *Biomaterials*, 2005, **26**, 3995–4021.

65. T. J. Brunner, P. Wick, P. Manser, P. Spohn, R. N. Grass, L. K. Limbach, A. Bruinink and W. J. Stark, In vitro cytotoxicity of oxide nanoparticles: comparison to asbestos, silica, and the effect of particle solubility, *Environ. Sci. Technol.*, 2006, **40**, 4374.

66. T. R. Pisanic 2nd, J. D. Blackwell, V. I. Shubayev, R. R. Finones and S. Jin, Nanotoxicity of iron oxide nanoparticle internalization in growing neurons, *Biomaterials*, 2007, **28**, 2572.

67. S. M. Hussain, K. L. Hess, J. M. Gearhart, K. T. Geiss and J. J. Schlager, In vitro toxicity of nanoparticles in BRL 3A rat liver cells, *Toxicol. In Vitro*, 2005, **19**, 975–983.

68. A. J. Wagner, C. A. Bleckmann, R. C. Murdock, A. M. Schrand, J. J. Schlager and S. M. Hussain, Cellular interaction of aluminum and aluminum oxide (Al_2O_3) nanoparticles, *J. Phys. Chem. B*, 2007, **111**, 7353–7359.

69. L. K. Braydich-Stolle, J. L. Speshock, A. B. Castle, M. Smith, R. C. Murdock and S. M. Hussain, Nanosized aluminum altered immune function, *ACS Nano*, 2010, **4**(7), 3661–3670.

70. C. Wilhelm, C. Billotey, J. Roger, J. N. Pons, J.-C. Bacri and F. Gazeau, *Biomaterials*, 2003, **24**, 1001–1011.

71. C. Chen, G. Xing, J. Wang, Y. Zhao, B. Li, J. Tang, G. Jian, T. Wang, J. Sun, L. Xing, H. Yuan, Y. Gao, H. Meng, Z. Chen, F. Zhao, Z. Chai and X. Fang, *Nano Lett.*, 2005, **5**, 2050–2057.

72. W. Lesniak, A. U. Bielinska, K. Sun, K. W. Janczak, X. Shi, J. R. Baker Jr. and L. P. Balogh, *Nano Lett.*, 2005, **5**, 2123–2130.

73. H. Dumortier, S. Lacotte, G. Pastorin, R. Marega, W. Wu, D. Bonifazi, J.-P. Brian, M. Prato, S. Muller and A. Bianco, *Nano Lett.*, 2006, **6**, 1522–1528.

74. J. P. Ryman-Rasmussen, J. E. Riviere and N. A. Monteiro-Riviere, Surface coatings determine cytotoxicity and irritation potential of quantum dot nanoparticles in epidermal keratinocytes, *J. Invest. Dermatol.*, 2007, **127**, 143–153.

75. F. Cengelli, D. Maysinger, F. Tschudi-Monnet, X. Montet, C. Corot, A. Petri-Fink, H. Hofmann and L. Juillerat-Jeanneret, The Journal of pharmacology and experimental therapeutics, ASPET, 2006, **318**(1), 108–116.

76. A. M. Schrand, L. K. Braydich-Stolle, J. J. Schlager, L. Dai and S. M. Hussain, Can Silver Nanoparticles be Useful as Potential Biological Labels?, *Nanotechnology*, 2008, **19**, 235104.

77. L. K. Braydich-Stolle, N. M. Schaeublin, R. C. Murdock, J. Jiang, P. Biswas, J. J. Schlager and S. M. Hussain, Crystal structure mediates mechanism of cell death in TiO2 nanotoxicity. *J Nanopart Res.*, 2008, **11**(6), 1361–1374.

78. A. Derfus, W. Chan and S. Bhatia, *Nano Lett.*, 2004, **4**, 11–18.

79. L. K. Braydich-Stolle, B. Lucas, A. M. Schrand, R. C. Murdock, T. Lee, J. J. Schlager, S. M. Hussain and M. C. Hofmann, Silver nanoparticles disrupt GDNF/Fyn kinase signaling in spermatogonial stem cells. *Tox Sci*, 2010, **116**(2), 577–589.

CHAPTER 3

In Vivo *Testing of Nanomaterials*

SEISHIRO HIRANO

Environmental Nanotoxicology Project, RCER, National Institute for
Environmental Studies, Tsukuba, Ibaraki, Japan
E-mail: seishiro@nies.go.jp

3.1 Administration Methods

There are several routes for exposure to nanomaterials and the absorption, distribution, metabolism, excretion, and toxicity (ADMET) of nanomaterials largely depends on the exposure route like many other toxicants. As most of nanomaterials occur in particulate substances, exposure *via* the airways is the most important route by which we are exposed to nanomaterials. Dermal exposure is also important because silver, titania, and zinc oxide nanoparticles are reported to have been used as deodorant sprays and sunscreens. For *in vivo* testing of nanomaterials small laboratory animals such as rats, mice, hamster, and guinea pigs are commonly used. However, minipigs are also used for dermal exposure tests because the anatomical structures of minipigs and humans are more similar to each other than when compared to the other small rodents. It should be emphasized that the animal experiments including the anesthesia procedure have to comply with the Institutional Animal Care and Experimentation Ethics Act. In this section exposure to nanomaterials, technical aspects of *in vivo* toxicity testing, and toxicological outcomes are described.

RSC Nanoscience & Nanotechnology No. 25
Towards Efficient Designing of Safe Nanomaterials: Innovative Merge of Computational Approaches and Experimental Techniques
Edited by Tomasz Puzyn and Jerzy Leszczynski
© The Royal Society of Chemistry 2012
Published by the Royal Society of Chemistry, www.rsc.org

3.1.1 *Via* Airways

There is no doubt that inhalation is the more realistic exposure method for evaluating pulmonary effects of particulate toxicants including nanomaterials in *in vivo* testing. However, the inhalation setup is expensive and is not always available in most laboratories. As a surrogate exposure method for inhalation exposure, the instillation of a test liquid or suspension samples directly into the upper and thoracic airways (nose, larynx, and trachea, main bronchus) is commonly used.

3.1.1.1 *Inhalation*

There are two typical inhalation methods for the exposure of small laboratory animals to toxicants. One is a nose-only (head-only) and the other is whole-body exposure (Figure 3.1). The whole-body exposure is suitable for long-term exposure studies and when a large amount of testing materials is available. The nose or head exposure is used for short-term exposure studies and when only a small amount of test material is available. The inner space of the exposure unit is much smaller than that of the whole-body inhalation chambers. The animals need to be kept in an animal holder with tight sealing around the nose or neck for the nose or head exposure. Dusting fluffy fibrous particles such as carbon nanotubes without serious agglomeration is difficult and a proper dispersing apparatus is required for the generation of inhalable or respirable dusts. Several methods have been proposed to form aerosols of carbon nanotubes for inhalation studies using vibrational generators and mechanical separators such as cyclones.[1–3]

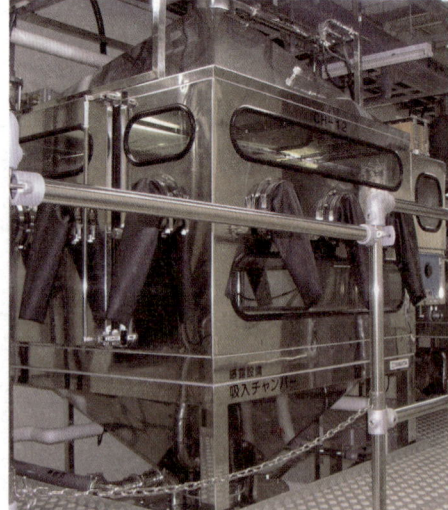

Figure 3.1   A nose-exposure chamber (left) and a whole-body exposure chamber (right).

3.1.1.2   Intratracheal or Intranasal Instillation

Usually anesthesia is required for intratracheal instillation to avoid the laryngeal reflex, and to administer a correct amount to the animals. The larger the amount of test sample instilled, the more evenly the administered materials spread in the respiratory tract. However, the volume to be intratracheally instilled should be less than 1–1.5 mL kg^{-1} body weight to avoid the apnea caused by instillation shock. An overload of particulate substances results in non-specific lung dysfunction. It has been shown that the toxicological outcomes of inhalation exposure to respirable single-walled carbon nanotubes (SWCNT) are very similar to those seen after pharyngeal aspiration.[4]

3.1.2   Dermal Exposure

In contrast to the concerns that nanosized titanium dioxide particles used as the sunscreen may penetrate the stratum corneum and epidermis and cause damage to the dermis, the related experimental data suggest that those nanosized particles are less likely to be able to permeate through the epidermis. The repeated topical application of nanosized titanium dioxide particles accumulated in the epidermis but did not translocate to the lymph and liver significantly in minipigs regardless of whether the nanoparticles are coated (hydrophobic) or not coated (hydrophilic) with polymers.[5] The follicular openings are compatible with nanoparticulate dimensions and topically applied nanoparticles have been shown to reach major organs such as the heart and liver and cause oxidative stress in hairless mice.[6]

3.1.3   Oral and Intravenous Routes

Although oral exposure is the most common route for many chemicals and most *in vivo* ADMET data are available for orally administered test materials, this is not always the case for nanomaterials. Even though there is a good chance of taking in nanomaterials through food and drinking water, nanomaterials may not be harmful. Asbestos (a natural nanofiber) is known to cause mesothelioma and lung cancer when deposited in the lung. However, there is no clear evidence that ingested asbestos causes serious diseases. Nanoparticles may permeate through the epithelium by diffusion when they are dispersed well. However, it is less likely that nanoparticles move freely in the intestinal contents and are absorbed from the intestinal wall. Thus, ingested particles are probably excreted directly without intestinal absorption and the bioavailability of nanomaterials must be very low when they are taken up orally. Acute oral administration (by gavage using a cannula) or chronic oral administration (through drinking water or dietary mixture) to laboratory animals is still required in case nanomaterials affect the gastrointestinal tracts or are absorbed after being dissolved in the gastrointestinal tract. The pH value of the gastric fluid is about 2.0 and nanomaterials are dissolved much faster than bulk ones because the larger surface area facilitates the solubilization of nanomaterials such as metals and metal oxides.

Intravenously administered particulate substances are taken up by reticuloendothelial phagocytic cells such as Kupffer cells in the liver and macrophages in the spleen. However, when the particle size is small enough they are not well recognized by those phagocytes. If drugs, such as anticancer ones, are tagged to nanoparticles, the drug can be delivered to the target cells efficiently before the nanoparticles are taken up by the phagocytes. This drug delivery system (DDS) has been gaining great attention from pharmaceutical scientists and medical staff for this reason and this manipulation is now called "Nano-DDS." However, a poor dispersion of nanoparticles should result in the removal of the drug by the phagocytic cells in the liver and spleen.

3.1.4   Other Routes

Nanomaterials can be administered into peritoneal and thoracic cavities or subcutaneously in animal experiments, although humans are seldom exposed in these ways. Intraperitoneal administration is often used in *in vivo* toxicity testing, because technically it is easier than other administration methods. Anesthesia is not necessary for intraperitoneal administration, and it mimics intravenous injection when the test materials can be dissolved. Moreover, a large amount of test material can be administered in the peritoneal cavity. When biopersistent materials such as carbon nanotubes are injected intrapenitoneally, they remain in the cavity for a long time and cause inflammation and other mesothelial or fibrotic lesions. Thus, this administration route may be beneficial for *in vivo* screening of a large number of nanomaterials. Intrathoracic injection is similar to intraperitoneal injection. This administration route may be used to investigate the effects on the thoracic mesothelium assuming that the nanomaterials may translocate to the thoracic cavity from the alveolar space like asbestos does. Similarly, subcutaneous injection can be used like intraperitoneal injection to evaluate the potency of nanomaterials.

3.2   Kinetics, Dynamics and the Translocation of Nanoparticles

There seems to be a critical size for immediate translocation and clearance of particles from the alveolar space. The quantum dots which are smaller than 34 nm (hydrodynamic diameter) can penetrate the lung tissue and reach to the mediastinal lymph nodes quickly unless the particles are positively charged. The particles less than 6 nm are excreted rapidly into urine.[7] When gold nanoparticles (2, 40, and 100nm) are instilled intratracheally into mice, only 2 nm particles were found to translocate to the liver and 40 nm gold nanoparticles seem to have a potential to do the same.[8] Similarly only 20 nm polystyrene latex nanospheres appear to translocate to the thymus quickly

(<1 day), while 100 nm and 1000 nm particles do not, when the particles are administered by pharyngeal aspiration in rats.[9] Those results clearly indicate that the permeability depends on the size and charge of the particles and the critical size for the rapid clearance of particles from the lung is somewhere around 30 nm.

Both multi-walled carbon nanotubes (MWCNT) and SWCNT seem to be excreted into urine after intravenous administration in rats[10] and mice[11] respectively, although the mechanism of urinary excretion of nano-fibrous particles *via* renal glomerular filtration is not known.[11] Inhalational exposure to carbon nanoparticles causes fibrinogen deposition and platelet accumulation in the hepatic microvasculature without pulmonary and systemic inflammation in mice, suggesting that inhaled carbon nanoparticles may migrate into the circulation from the alveolar space.[12] Although inhaled asbestos is known to penetrate the lung tissue and reach the pleural cavity, it is not clear whether inhaled carbon nanotubes reach to the pleural cavity. Inhaled carbon nanotubes are found in the subpleural tissue in mice,[13] suggesting the possibility of carbon nanotubes migrating to the pleural cavity. If the fiber is still long enough, the carbon nanotube would cause pleural mesothelioma like asbestos.[14]

There are two portals of entry for nanoparticles to the brain and central neural system: the nasal olfactory pathway and passing through the blood–brain barrier (BBB).[15,16] If the nanoparticles migrate through the BBB, they can probably migrate through the blood–placenta barrier. In that case

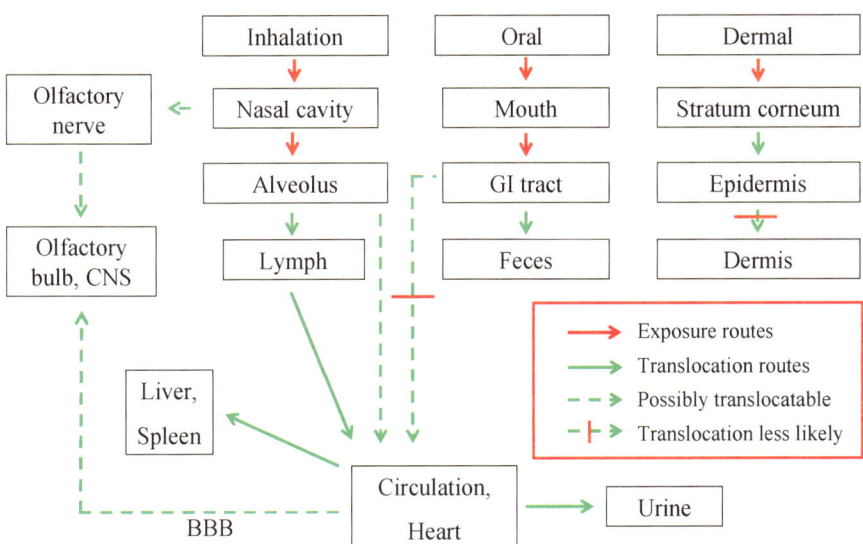

Figure 3.2 Exposure and translocation routes of nanoparticles and nanomaterials (CNS central nervous system; BBB, blood–brain barrier; GI tract, gastrointestinal tract).

comprehensive genotoxicity studies are required. Physiologically-based pharmacokinetic (PBPK) or toxicokinetic (PBTK) models would be helpful to
understand the ADMET of nanomaterials and facilitate the screening of
nanomaterials for safety usage. The possible exposure routes, translocation,
and behavior of nanomaterials are summarized in Figure 3.2.

3.3   Toxicity Outcome of Nanomaterials

There are various types of nanomaterials and accordingly the toxicity
outcomes are different among different types of nanomaterials. In this chapter
on *in vivo* toxicity, nanomaterials are categorized into three types depending on
their chemical characteristics: carbons, metals and metal oxides, and ceramics
and other materials. However, some of the nanomaterials are in the shape of
rods and fibers. The toxicity mechanisms of those irregular particles are
different from spherical particles even though the chemical characteristics are
the same[17] and this issue is discussed separately.

3.3.1   Carbons

There are five allotropes in the carbons: amorphous, graphite, diamond,
fullerene, carbon nanotubes. Of those carbons, fullerenes and carbon
nanotubes (and in some cases carbon black (amorphous) particles) are
considered to be target materials for nanotoxicology. Carbons are ubiquitous
and present as the cores of atmospheric particulate substances like diesel
exhaust particles. However, we do not know well whether carbons are
'nuisance' or health-threatening substances when they are in nanosized
particles or nanofibers.

3.3.1.1   Fullerenes

There are several types in the fullerenes and of those the C_{60} buckyball is the
most common. Hydroxylated or carboxylated fullerenes are soluble or water-
miscible and can be used as a DDS,[18,19] although they may cause reactive
oxygen species (ROS)-independent apoptosis. In contrast the pristine fullerene
is highly hydrophobic and soluble in toluene, generating a clear violet solution.
Tetrahydrofuran (THF) has been used to prepare stable aqueous dispersions
of fullerene.[20] Those fullerenes reportedly cause oxidative stress and are highly
toxic for bacteria[21] and fish,[22] although there is a possibility that reactive by-
products may be produced from THF in the preparation of water-miscible
fullerene and these impurities may cause significant effects. Controversially it
has been reported that intratracheal instillation of the pristine fullerene
prepared from THF causes as little effect on the rat lung as its water-soluble
derivatives.[23]

3.3.1.2 Single-Walled Carbon Nanotubes (SWCNTs)

The fiber width of a SWCNT is *ca.* 1nm and what makes SWCNTs different from MWCNTs is that a SWCNT can be solubilized in aqueous solution or biological fluid, while a MWCNT is insoluble in those conditions. It has been shown that peroxidase oxidizes the grapheme structure of SWCNT but not MWCNT in the presence of hydrogen peroxide.[24] It has been shown that SWCNTs are also degraded by neutrophil myeloperoxidase and the degraded SWCNTs induce less pulmonary inflammation.[25] Pharyngeally aspirated SWCNT has been reported to cause severe inflammatory and fibrotic responses in the lung and NADPH-derived ROS are implicated in the course of pulmonary inflammatory responses.[26]

3.3.1.3 Multi-Walled Carbon Nanotubes (MWCNTs)

Intratracheal instillation of surfactant-dispersed MWCNTs at a final dose of 6.25 µg per mouse (single and repeated administration) increases the production of keratinocyte chemoattractant (KC), macrophage, neutrophil and eosinophil cell numbers in bronchoalveolar lavage fluid, and causes inflammatory responses such as collagen deposition in BALB/c mice.[27] Several lines of evidence suggest that MWCNTs are carcinogenic and cause mesothelioma like asbestos.[28] However, it should be noted that short fibers (<5 µm) are much less reactive than long fibers (>5 µm).[29,30]

3.3.1.4 Carbon Black and Carbon Particles

The health effects of fine and ultrafine carbonaceous particles have been investigated intensely from the standpoint of material science and atmospheric science, since epidemiological studies indicated a close relationship between concentrations of fine ambient particulate substances and respiratory or cardiovascular mortality.[31] Intratracheal instillation of different types of carbon black particles suggested that the surface area is a good metric for the inflammatory potency of carbon black particles. There seems to be a threshold value for the surface area of carbon particles at 20 cm^2 (Brunauer–Emmett–Teller, BET area) per mouse, when acute inflammatory lung responses such as neutrophil infiltration and an increase in IL-1β and MIP-2 concentrations in bronchoalveolar lavage fluid are used for indices of adverse effects.[32]

3.3.2 Metals and Metal Oxides

Silver and gold are precious metals and they may not dissolve in biological fluid or tissues. On the other hand zinc and iron and their oxide forms can be solubilized and may cause chemico-biological effects as metal ions rather than physico-biological stress as particulate substances, if their surfaces are not coated. After being dissolved, the toxicity outcome of nano-metals would not be much different from those of metals or heavy metals. The large number of atoms on the surface of nanomaterials may confer biological reactivity to the nanomaterials compared to the bulk samples. Silver nanoparticles at a concentration of 100 µg m^{-3}, which is a limit of silver dusts proposed by American Conference of Governmental Industrial Hygienists (ACGIH), seems not to cause health effects; however, a higher concentration of silver nanoparticles causes lung and liver abnormalities.[33]

3.3.3 Ceramics and Other Materials

Ultrafine polystyrene particles (64 nm) cause more severe pulmonary inflammation than larger polystyrene particles (202 and 535 nm) after intratracheal instillation in rats and ultrafine polystyrene particles generate a larger amount of ROS and increase the intracellular calcium concentration by more than larger particles.[34] The pulmonary clearance of ultrafine titanium dioxide particles (20 nm) is slower ($t_{1/2}$ = 501 days) than that of large particles (250 nm, $t_{1/2}$ = 174 days) after subchronic inhalation exposure (*ca.* 23 mg m^{-3}, 6h day^{-1}, 5 days per week for up to 12 weeks) in rats.[35] The burden of ultrafine TiO$_2$ particles in the interstitial tissues and lymph nodes was higher than that of fine particles and the impairment of clearance is accompanied by pulmonary inflammation.[36] It should be noted that the size-dependent toxicity of silica may depend on cell types and that nano-silicate particles of 30 nm seem not to be cytotoxic to RAW264.7 (macrophage) cells, while they are toxic to HEK293 (kidney) cells.[37]

3.3.4 Nanofibers

As described in sections 3.3.1.2 and 3.3.1.3 (the SWCNT and MWCNT sections), fibrous particles with a nanosized width are harmful to cells and tissues. The size of macrophages are somewhere in the region of 10–20 µm and a macrophage loaded with long nanofibers shows 'frustrated phagocytosis' which may result in a loss of membrane integrity and toxicity.[38] The carcinogenicity of long fibers (> 8 µm) has long been known as "Stanton's theory".[39] As nanotoxicology has emerged with serious concerns about commercial use of carbon nanotubes and its health effects, the comprehensive studies of Stanton *et al.* should be re-visited.

3.4 Summary and Implications

The toxicity evaluations of nanomaterials are still controversial. The differences between *in vivo* toxicological outcomes of the same type of nanomaterials among experiments worldwide are probably brought about by differences in sample preparation and administration procedure. There is no standard and well-confirmed method for the *in vivo* toxicity testing of nanomaterials. In general the following are the most important factors to qualify as a reliable *in vivo* testing standard.

- Good dispersion in aqueous solution. Even though single particle suspension may not be achieved, avoid the administration of large agglomerations.
- Characterization of primary particles together with test suspensions. Particle size distribution, ζ-potential, and the stability of suspended particles in the test sample are important for other toxicologists to understand nanotoxicological aspects of the studies.
- Measurement of impurity and dissolved substances. A parallel experiment after the removal of particulate substances from the suspension and administration of the cleared solution may be helpful.
- Removal of endotoxin. Biological substances such as lipopolysaccharides may be adsorbed on the surface of nanomaterials. A trace amount of endotoxin may change the inflammatory responses. Heating the sample at 250 °C for 2 h and suspending the particulate sample in pyrogen-free media is recommended to avoid the endotoxin effects if the sample is heat-resistant.

References

1. P. A. Baron, G. J. Deye, B. T. Chen, D. E. Schwegler-Berry, A. A. Shvedova and V. Castranova, *Inhalation Toxicol.*, 2008, **20**, 751.
2. Y. Fujitani, A. Furuyama and S. Hirano, *Aerosol Sci. Technol.*, 2009, **43**, 881.
3. L. Ma-Hock, S. Treumann, V. Strauss, S. Brill, F. Luizi, M. Mertler, K. Wiench, A. O. Gamer, B. van Ravenzwaay and R. Landsiedel, *Toxicol. Sci.*, 2009, **112**, 468.
4. A. A. Shvedova, E. Kisin, A. R. Murray, V. J. Johnson, O. Gorelik, S. Arepalli, A. F. Hubbs, R. R. Mercer, P. Keohavong, N. Sussman, J. Jin, J. Yin, S. Stone, B. T. Chen, G. Deye, A. Maynard, V. Castranova, P. A. Baron and V. E. Kagan, *Am. J. Physiol.: Lung Cell. Mol. Physiol.*, 2008, **295**, L552.
5. N. Sadrieh, A. M. Wokovich, N. V. Gopee, J. Zheng, D. Haines, D. Parmiter, P. H. Siitonen, C. R. Cozart, A. K. Patri, S. E. McNeil, P. C. Howard, W. H. Doub and L. F. Buhse, *Toxicol. Sci.*, 2010, **115**, 156.
6. J. Wu, W. Liu, C. Xue, S. Zhou, F. Lan, L. Bi, H. Xu, X. Yang and F. D. Zeng, *Toxicol. Lett.*, 2009, **191**, 1.
7. H. S. Choi, Y. Ashitate, J. H. Lee, S. H. Kim, A. Matsui, N. Insin, M. G. Bawendi, M. Semmler-Behnke, J. V. Frangioni and A. Tsuda, *Nat. Biotechnol.*, 2010, **28**, 1300.
8. E. Sadauskas, N. R. Jacobsen, G. Danscher, M. Stoltenberg, U. Vogel, A. Larsen, W. Kreyling and H. Wallin, *Chem. Cent. J.*, 2009, **3**, 16.
9. K. Sarlo, K. L. Blackburn, E. D. Clark, J. Grothaus, J. Chaney, S. Neu, J. Flood, D. Abbott, C. Bohne, K. Casey, C. Fryer and M. Kuhn, *Toxicology*, 2009, **263**, 117.
10. D. Georgin, B. Czarny, M. Botquin, M. Mayne-L'hermite, M. Pinault, B. Bouchet-Fabre, M. Carriere, J. L. Poncy, Q. Chau, R. Maximilien, V. Dive and F. Taran, *J. Am. Chem. Soc.*, 2009, **131**, 14658.

11. R. Singh, D. Pantarotto, L. Lacerda, G. Pastorin, C. Klumpp, M. Prato, A. Bianco and K. Kostarelos, *Proc. Natl. Acad. Sci. U. S. A.*, 2006, **103**, 3357.

12. A. Khandoga, T. Stoeger, A. G. Khandoga, P. Bihari, E. Karg, D. Ettehadieh, S. Lakatos, J. Fent, H. Schulz and F. Krombach, *J. Thromb. Haemostasis*, 2010, **8**, 1632.

13. J. P. Ryman-Rasmussen, M. F. Cesta, A. R. Brody, J. K. Shipley-Phillips, J. I. Everitt, E. W. Tewksbury, O. R. Moss, B. A. Wong, D. E. Dodd, M. E. Andersen and J. C. Bonner, *Nat. Nanotechnol.*, 2009, **4**, 747.

14. K. Donaldson and C. A. Poland, *Nat. Nanotechnol.*, 2009, **4**, 708.

15. M. L. Block and L. Calderon-Garciduenas, *Trends Neurosci.*, 2009, **32**, 506.

16. M. Simko and M. O. Mattsson, *Part. Fibre Toxicol.*, 2010, **7**, 42.

17. S. Hirano, C. D. Anuradha and S. Kanno, *Am. J. Respir. Cell Mol. Biol.*, 2000, **23**, 313.

18. S. Yamago, H. Tokuyama, E. Nakamura, K. Kikuchi, S. Kananishi, K. Sueki, H. Nakahara, S. Enomoto and F. Ambe, *Chem. Biol.*, 1995, **2**, 385.

19. S. Foley, C. Crowley, M. Smaihi, C. Bonfils, B. F. Erlanger, P. Seta and C. Larroque, *Biochem. Biophys. Res. Commun.*, 2002, **294**, 116.

20. S. Deguchi, R. G. Alargova and K. Tsujii, *Langmuir*, 2001, **17**, 6013.

21. J. D. Fortner, D. Y. Lyon, C. M. Sayes, A. M. Boyd, J. C. Falkner, E. M. Hotze, L. B. Alemany, Y. J. Tao, W. Guo, K. D. Ausman, V. L. Colvin and J. B. Hughes, *Environ. Sci. Technol.*, 2005, **39**, 4307.

22. E. Oberdorster, *Environ. Health Perspect.*, 2004, **112**, 1058.

23. C. M. Sayes, A. A. Marchione, K. L. Reed and D. B. Warheit, *Nano Lett.*, 2007, **7**, 2399.

24. B. L. Allen, P. D. Kichambare, P. Gou, Vlasova, II, A. A. Kapralov, N. Konduru, V. E. Kagan and A. Star, *Nano Lett.*, 2008, **8**, 3899.

25. V. E. Kagan, N. V. Konduru, W. H. Feng, B. L. Allen, J. Conroy, Y. Volkov, I. I. Vlasova, N. A. Belikova, N. Yanamala, A. Kapralov, Y. Y. Tyurina, J. W. Shi, E. R. Kisin, A. R. Murray, J. Franks, D. Stolz, P. P. Gou, J. Klein-Seetharaman, B. Fadeel, A. Star and A. A. Shvedova, *Nat. Nanotechnol.*, 2010, **5**, 354.

26. A. A. Shvedova, E. R. Kisin, A. R. Murray, C. Kommineni, V. Castranova, B. Fadeel and V. E. Kagan, *Toxicol. Appl. Pharmacol.*, 2008, **231**, 235.

27. C. Ronzani, C. Spiegelhalter, J. L. Vonesch, L. Lebeau and F. Pons, *Arch. Toxicol.*, 2012, **86**, 137.

28. A. Takagi, A. Hirose, T. Nishimura, N. Fukumori, A. Ogata, N. Ohashi, S. Kitajima and J. Kanno, *J. Toxicol. Sci.*, 2008, **33**, 105.

29. J. Muller, M. Delos, N. Panin, V. Rabolli, F. Huaux and D. Lison, *Toxicol. Sci.*, 2009, **110**, 442.

30. C. A. Poland, R. Duffin, I. Kinloch, A. Maynard, W. A. Wallace, A. Seaton, V. Stone, S. Brown, W. Macnee and K. Donaldson, *Nat. Nanotechnol.*, 2008, **3**, 423.

31. D. W. Dockery, C. A. Pope, , *N. Engl. J. Med.*, 1993, **329**, 1753.

32. T. Stoeger, C. Reinhard, S. Takenaka, A. Schroeppel, E. Karg, B. Ritter, J. Heyder and H. Schulz, *Environ. Health Perspect.*, 2006, **114**, 328.
33. J. H. Sung, J. H. Ji, J. D. Park, J. U. Yoon, D. S. Kim, K. S. Jeon, M. Y. Song, J. Jeong, B. S. Han, J. H. Han, Y. H. Chung, H. K. Chang, J. H. Lee, M. H. Cho, B. J. Kelman and I. J. Yu, *Toxicol. Sci.*, 2009, **108**, 452.
34. D. M. Brown, M. R. Wilson, W. MacNee, V. Stone and K. Donaldson, *Toxicol. Appl. Pharmacol.*, 2001, **175**, 191.
35. J. Ferin, G. Oberdorster and D. P. Penney, *Am. J. Respir. Cell Mol. Biol.*, 1992, **6**, 535.
36. G. Oberdorster, J. Ferin and B. E. Lehnert, *Environ. Health Perspect.*, 1994, **102**(Suppl. 5), 173.
37. A. Petushkov, J. Intra, J. B. Graham, S. C. Larsen and A. K. Salem, *Chem. Res. Toxicol.*, 2009, **22**, 1359.
38. S. Hirano, S. Kanno and A. Furuyama, *Toxicol. Appl. Pharmacol.*, 2008, **232**, 244.
39. M. F. Stanton, M. Layard, A. Tegeris, E. Miller, M. May, E. Morgan and A. Smith, *J. Natl. Cancer Inst.*, 1981, **67**, 965.

CHAPTER 4

Nanotoxicity: Are We Confident for Modelling? – An Experimentalist's Point of View

DEBORAH BERHANU[a] AND
EUGENIA VALSAMI-JONES*[a,b]

[a] Natural History Museum, Department of Mineralogy, Cromwell Road, London, SW7 5BD, UK; [b] School of Geography, Earth and Environmental Sciences (GEES), University of Birmingham, Edgbaston, Birmingham, B15 2TT, UK
*E-mail: e.valsamijones@bham.ac.uk

4.1 Introduction

Nanotoxicity, the potential of nanosized objects to cause harm to an organism, is a complex discipline, requiring the close collaboration between life and material sciences. In this context, it has become crucial to identify key parameters in the physico-chemical properties of nanomaterials that can affect toxicity, to make it possible for toxicology experiments to accommodate the unique ways in which nanomaterials behave in standard toxicological media. Furthermore, in order to understand their impact on the environment and human health, the behaviour of nanomaterials has to be understood generically as well as in the context of toxicity experiments[1] and, based on that, further predictive modelling has to be developed for a comprehensive regulation to be established and for the safe development of nanotechnology.[2]

RSC Nanoscience & Nanotechnology No. 25
Towards Efficient Designing of Safe Nanomaterials: Innovative Merge of Computational Approaches and Experimental Techniques
Edited by Tomasz Puzyn and Jerzy Leszczynski
© The Royal Society of Chemistry 2012
Published by the Royal Society of Chemistry, www.rsc.org

It is clear that individual testing of all nanomaterials currently available, industrially or in the lab, will never be possible and models will be crucial to establish a generic understanding, in order to safeguard safety and to support regulation.

Nanotoxicity data are becoming rapidly available (Figure 4.1), yet a lot of gaps persist and comparisons between studies need precaution at this stage. Keeping this in mind, we here take specific examples from on-going projects and the literature to show how data can be used to understand the behaviour of nanoparticles (NPs) for the development of predictive modelling. The existing data is scattered between several disciplines and requires truly interdisciplinary collaborations to assemble them for further understanding. However, this has to be preceded by characterisation methods and datasets evaluation to allow inter-study comparisons. It is anticipated that these efforts will feed back into a sustainable nanotechnology. This chapter will not present a model but an experimentalist's point of view on the existing data that could be linked in order to progress current models.

A further consideration for the successful implementation of nanotechnology is that of sustainability. A technology may be considered sustainable if it uses small amounts of limited resources, consumes minimal energy, causes minimum pollution and generates materials that can be recycled/reused. Although in this chapter we do not consider the term sustainable in its full meaning, we anticipate that nanotechnology can be developed successfully only if safe, green and cost-effective manufacturing lines are in place that allow one to obtain high quality, safe nanoproducts. If novel nanomaterials are not designed following these minimum requirements, they cannot be considered sustainable, and as a result their successful commercialisation may be compromised.

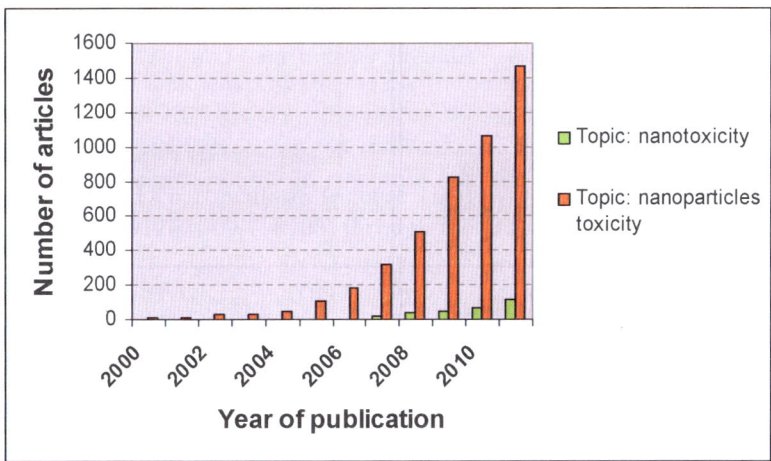

Figure 4.1 Number of publications per year indexed on Web of Science topically related to nanoparticles' toxicity.

Although nanomaterials are already present in many commercial products, the full potential of nanotechnology has yet to be realised. While progress in synthesis towards greener routes can be observed, the overall sustainability still requires improvement. For example, if the synthesis of quantum dots is performed using capping agents as the solvent, the actual content in NPs can be very low (<20 wt%). The development of greener synthesis protocols has helped in this way.[3] From pyrophoric precursors[4] to stable salts[5] and single-source precursors,[6] the use of less toxic capping and the reduction of waste generated[7] during quantum dot synthesis was a breakthrough in the 1990s. Without overcoming these challenges, it would have been difficult to promote any development of semiconducting nanocrystals. Hence, once the properties were optimised and the potential applications of quantum dots established, key concerns in the fabrication processes were altered. The use of quantum dots was envisaged for green energy sources but how could this be promoted when the synthesis of the particles was not green or user-friendly? A feedback loop (Figure 4.2) therefore allowed investigating user-friendly precursors and the reduction of wastes. This fits overall in the sustainability of nanotechnology.

We are now at the step where the toxicity has to feed back into the synthesis procedures in order to improve the next generation of nanomaterials. Not only

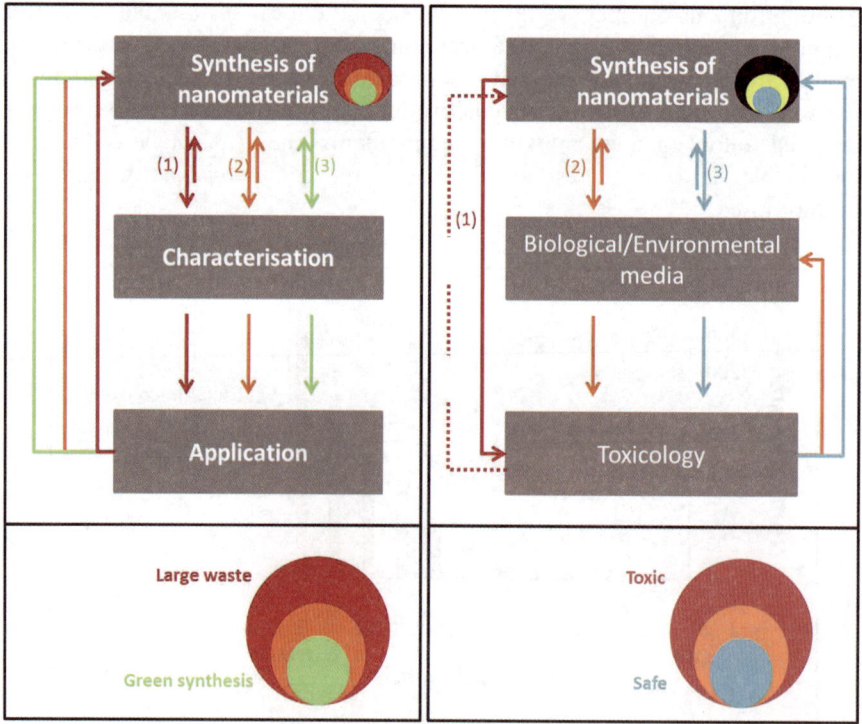

Figure 4.2   Diagrams featuring the feedback loops necessary to the development of green and safe nanotechnology.

will the particles have to be made using green synthesis[8] but the end product also needs to be safer (Figure 4.2). While toxicity data on many nanoparticle samples are available, rare are those that can be used in order to understand which physico-chemical properties can affect toxicity. In the schematic global representation (Figure 4.2), this context can be visualised as gaps in the feedback information. This chapter will present a comparison between sample specific toxicity and physico-chemical-based toxicity to underline the importance of the latter in the comprehensive picture.

Another important step which needs understanding is the behaviour of nanoparticles in environmental and biological media.[9] The initial nanoparticles (primary nanoparticles) can be drastically changed in the biological system (secondary nanoparticles) which in the context of biomedical application can affect the properties that were initially desired. In the context of toxicity, this step is a major complication to the understanding of mechanisms. While the toxicity of primary particles is what needs to be understood, the toxicity responses observed will only be those from the secondary particles. One way of addressing this issue as described in this chapter is the use of simulated fluids in order to assess the property changes prior to biological experiments. COST (European Cooperation in Science and Technology) organised a successful meeting (3–6 April 2011) gathering world-wide representatives from the nanosafety research community in order to bridge the existing gaps and encourage computational approaches (http://www.cost.eu/events/qntr). Learning from neighbouring fields (*e.g.* surface science) and developing existing statistical approaches towards the prediction of nanomaterials' toxicity, reducing costly and extensive experiments, is a general aim. However, many available data in the current literature are not candidates for statistical analysis which highlights the importance of well-designed nanotoxicity studies.

Designing a global experimental set-up, from synthesis to toxicity, is another way to avoid misinterpretations and ensure that relevant properties can be followed. The example of isotopic labelling will here be used to emphasise the need for more specific designs of experimental set-ups.

Finally, inputs from various related research areas that could help the current understanding of nanotoxicity and encourage modelling will be discussed.

4.2 The Complexity of Nano Compared to Bulk

4.2.1 From One Material to Hundreds of Different Nanoparticles

Prior to nanotechnology, a material had a defined set of properties. Each material could therefore be tested for toxicity in its one unique form. The knowledge built up for centuries was then used to understand the novel modifications in materials that occurred with advanced technologies in the past century. From one material, we can now produce thousands of nano-objects which exhibit different properties, simply by changing the particle size

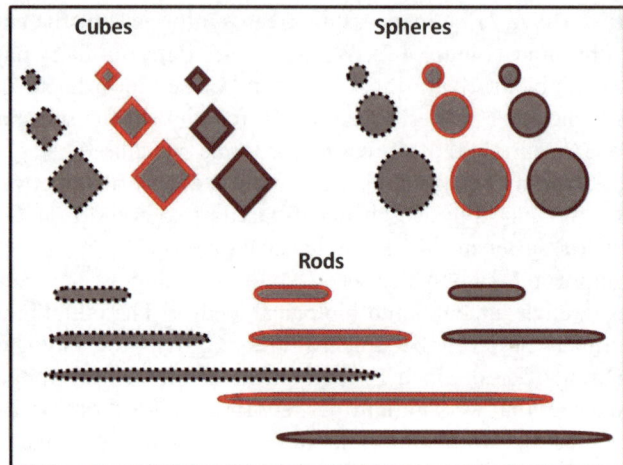

Figure 4.3 Schematic representation of 27 types of nanoparticles that can be obtained from a single material by varying only 3 parameters: size, shape and capping agent.

or shape, or modifying the surface, producing a series of materials which may considerably and unpredictably vary in terms of reactivity and toxicity (Figure 4.3). For example, let us define a nanoparticle's structure by its size, shape and capping agent only. Considering only 3 variables for these 3 parameters (3 sizes, 3 shapes and 3 capping agents), from one material we obtain 27 nanoparticles with distinct properties; for 5 variables, 125 nanoparticles. The number of parameters and variables can easily be increased further, if we consider impurities, doping, crystal structure, exposed crystal faces... If each nanoparticle type is considered separately, thousands of sample specific datasets can be generated. At this stage, it is easy to intuitively understand the role that statistical modelling can have in this area. However, one of the prerequisites for modelling is the ability to classify datasets rigorously. This is a major challenge that nanotoxicology is currently facing.

4.2.2 From Hundreds of Sample-specific Datasets to Physico-chemical Properties-based Toxicity

Most of the available nanotoxicity data are sample-specific; the specialised characterisation of the nanoparticles (only the properties considered in the study are characterised in most cases), followed by sample preparation techniques that are not always extensively described reduce the ability one has to compare data that is available in the literature. Hence, inter-study comparisons are difficult, as in most cases one type of nanoparticle has been used in one toxicity experiment using a specific protocol. Following the example in the paragraph above, if each of the 27 nanomaterials is given to 27 different laboratories to perform tests, the amount of data that is generated will be tremendous. However, if each sample is characterised specifically (size only for example), 27 sample-specific toxicity datasets will be obtained. Each set of data will be specific to the considered sample and the 27 sets of toxicity data obtained will not be fully comparable. If each toxicity data is accompanied with size, shape and capping-agent characterisation, then these data can be compared more fully and useful conclusions deduced, but one will still have to consider the preparation protocols and the possible transformations in the biological media. This route will allow a link between several specific physico-chemical properties and toxicity data which, in turn, may offer the possibility of modifying any specific property linked to toxicity and therefore developing sustainable nanomaterials.

While the purpose of such toxicological studies is clear and their designs, at least in principle, well anticipated, it is still very challenging to perform them. The underlying technical facilities and expertise required can only be overcome by inter-disciplinary groups.

4.2.3 Poorly Produced Nanoparticles *vs.* Well-defined Samples

A major challenge in linking specific parameters to toxicity is the quality of the samples; an important requirement, for example, for toxicity studies is that nanoparticles used in these studies have to be monodispersed in shape and size. This is important so that we can assign a specific toxicity response to a well-defined size and shape. Such high quality nanoparticles are usually tailor made in the lab, whereas industrially produced nanoparticles are usually less well controlled in size. However, it is worth recognising that polydispersed particles still have a place in this field. For biological tests that require large amounts of nanoparticles, polydispersed samples may be the only option at this stage as bulk production of high quality nanoparticles is expensive and time-consuming. While this data on its own will not allow mechanistic understanding, it will allow at least a generic link between a nanoparticle composition and toxicity. Also, since such types of samples are usually derived from industrial/commercial sources, they bring some real-life scenarios to nanotoxicity studies.

Commercially available and/or low-cost/large-quantity samples are convenient for starting a project; the design of experimental set-ups has proven to be difficult in nanotoxicity. Hence, the use of low-cost polydispersed samples can help design procedures involved and establish the challenging steps, prior to more detailed mechanistic studies. The authors of this chapter organised a study where, in order to design multiple sets of ecotoxicity experiments, several labs were given large quantities of a well-characterised CuO nanomaterial (Figure 4.3) from the same batch.[10] Made by plasma synthesis, the particles were polydispersed in size and shape but disperse easily in aqueous media and were amenable to detailed characterisation. Beside the interesting results obtained, this sample allowed a number of scientists involved from diverse backgrounds to establish a good collaboration, avoiding the frustration of wastage of "expensive" samples.[11] Note that expensive applies in both monetary terms but also in terms of time and effort required to be produced.

4.3   How to Design a Toxicity Experiment

4.3.1   Comparative Nanotoxicity Studies

Many nanotoxicology studies performed using poorly characterised samples have produced datasets that cannot be used further. To clarify this point, it is important that one understands the complexities of the nanoscale (explained above) but also considers the different types of characterisation that can be provided. Hence, even in the case where a single sample of nanoparticles is considered, the bulk and ionic counterparts should be included.

It is difficult to specify what would constitute a full characterisation as there are numerous techniques available which can be expensive to carry out but also require skilled personnel and maintenance. In this context, a best-practice suggestion would be a pragmatic approach, depending on sample complexity, experimental design (for the toxicological experiments) and technique availability. A recommended approach would be to begin by generating robust data for the comparison of the specific physico-chemical property under investigation. For example, if a study is to compare the size effect of a given material, it has to include nanoparticles of different sizes that are "identical" in all other aspects. The dry particle size will have to be characterised using robust and well-established techniques, which include electron microscopy (EM) or atomic force microscopy (AFM). The synthesis route employed for different sizes has to be very similar to reduce by-product effects or the use of different capping agents. If we now consider the comparison of shapes complications arise. It is difficult to obtain a variety of different shapes using one capping agent. How can these challenges be overcome? This section will concentrate on finding experimental solutions that will ultimately promote understanding and enable modelling of the generated datasets.

The most common comparative studies consider one type of nanoparticle, the corresponding ionic form and a bulk (usually micron-size) particulate form. Once in biological medium, the ionic form complexes with the counter-ions or organic molecules present. Similarly, the nanoparticles and micro-particles will undergo changes at their surfaces. The exposed surface areas are much higher for the nanoparticles than their micron-sized counterparts. The nanoparticles will therefore undergo changes very rapidly, in the form of dissolution, aggregation and/or interactions with organic molecules. In most cases, this form of comparison will mainly allow us to observe whether or not the nanoparticles behave like the complexed ions. If so, dissolution of the nanoparticles can be suspected.

However, in many cases, an intermediate effect of the nanoparticles is observed which is likely to be due to a combined effect of the particulate and soluble fraction, as suggested in some recent studies.[12,13] In order to establish the effect of dissolution on the nanoparticles, it is essential to carry out solubility studies in parallel with the toxicity experiments, and in relevant media/duration.

4.3.2 Property-based Nanotoxicity Studies

Many studies have focussed on property-based toxicity assessment. Examples include the effect of nanoparticle sizes or dissolution on various biological endpoints. Combined physico-chemical properties and their effect on toxicity are now taking place. Peng *et al.* investigated the effect of shape on dissolution and toxicity.[14] Such experimental approaches should be favoured as they provide links between the evolution of a specific behaviour (*e.g.* dissolution) and toxicity.

Nanoparticles shapes and reactivity: several studies have shown that the crystallographic faces exposed to the environment can be different when the size of the particles decreases.[15] This can be easily demonstrated (Figure 4.4) using cubes that are truncated to expose different faces. While the corner-faces (*e.g.* {111} planes) are a minor component of the truncated cube A, they become prevalent in the cuboctahedron C. Different planes have different atom–atom interactions which lead to different properties and the reactivity of different faces can be drastically different. In the well-known case of Au, while the bulk material is considered inert, Au nanoparticles can catalyse several reactions. This phenomenon is partly attributed to the increase of corner atoms, *i.e.* low-coordinated sites.[15,16] However, by growing different shapes, different faces can be considered as particle/media interfaces.

The use of capping agents can modify the rate of exposure of different crystallographic faces. The ability of thiol ligands to preferentially attach Au {111} faces and inhibit their linear assembly is well known.[17] The introduction

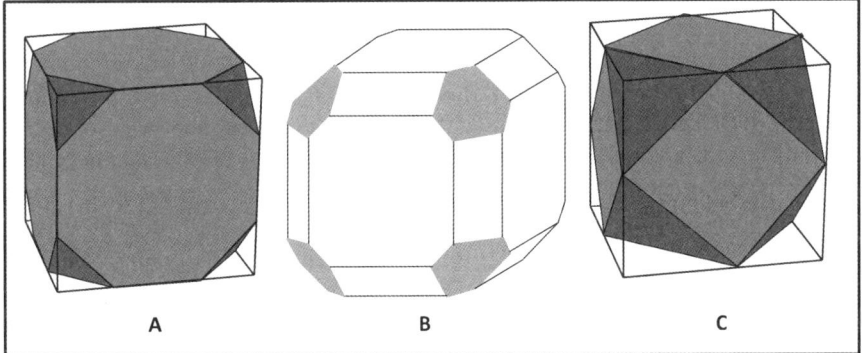

Figure 4.4 Truncated cubes exposing different crystallographic faces to the environment. (A) Truncated cube in a cube. (B) Truncated cube that exposes corners and edges. (C) Cuboctahedron in a cube.

of two types of gold nanoparticles, exposing different faces, in the same medium could therefore interact in different manners with biomolecules.

The case of soluble metal oxides: an example of a soluble metal oxide is ZnO, as nanoparticles, which are well known to dissolve in aqueous media with an enhancement of the dissolution rate as the pH decreases from the point of zero charge. While the actual effects of particulate ZnO seemed negligible, the underlying toxicity of Zn^{2+} ions accumulating in the media cannot be ignored. Lately, to avoid the loss of the dissolving particulates or enhance other properties, Fe-doped ZnO nanoparticles are being produced. Nano-ZnO has been considered in several commercial applications (e.g. cosmetics) but its fast dissolution has been problematic for its efficient function. The incorporation of Fe atoms in the nanoparticles has been proven to reduce their dissolution and consequently affect their toxicity.[18,19] If these results are confirmed, it will then be necessary to investigate the long-term accumulation effects in the organisms and the environment in order to assess its full sustainability. Furthermore, it is important to establish the dissolution kinetics, as this will show whether the nanoparticles dissolve early and potentially outside the organism, thus leaving only soluble Zn as the toxin in the media, or late, in which case the dissolution may occur after the nanoparticle has been delivered inside the organism or cell. This is a specific example where nanotoxicology results have influenced the synthesis in order to produce more sustainable nanoparticles, showing the importance of feedback mechanisms (Figure 4.1).

4.4 Remaining Challenges of Nanoparticles' Characterisation

It is now well accepted that the characterisation of NPs in the considered toxicity context is essential in order to interpret biological responses. Challenges remain due to the complexity of environmental and biological media. The behaviour of NPs in the environment is a crucial step in order to assess the changes in physico-chemical properties but also helps determine which compartment is at risk. NPs in environmental fluids commonly agglomerating and/or aggregating are most likely to settle in the sediment compartment. However, current characterisation methodologies lack the capability to characterise sediments at the nanoscale and trace the form and behaviour of nanoparticulate matter. Characterisation in fluids is more accessible and more data is available.

4.4.1 Can a Minimum Set of Suitable Techniques be Established?

Another major challenge is the fact that different characterisation data are available in different studies, which does not allow clear inter-study comparisons. In other words, if two characterisation groups (separate institutions) were given an identical nanoparticle sample and asked to characterise it, it is possible that the results they will produce will be different; without even considering reproducibility, techniques available vary widely. Hence, consequent toxicity studies on these samples are unlikely to be comparable, unless it is known to both toxicological ends that the sample is the same. For similar reasons, toxicity studies of the same materials may deliver different answers. It is therefore essential that any nanotoxicity data published contains multiple samples and tests, to produce more statistically robust datasets. Although most groups characterise particles extensively, the presented data is generally in direct correlation with the biological finding, which also limits the possibilities of inter-study comparisons. However, adding the data that was acquired along but not necessary for publication in supplementary information should be encouraged in order for others to use the data for future interpretation and comparisons. Furthermore, a number of nanosafety-related round robins are currently taking place across Europe (see for example http://www.qnano-ri.eu/networking/na-2.html) which, in the medium term, will harmonise characterisation and improve inter-laboratory comparisons.

The above discussion also highlights technical availabilities: can we overcome these? Nanotoxicology is evolving in parallel to nanotechnology, which complicates the picture. The improvement and adaptation of existing methods to toxicity is one way to go. The tracing of nanoparticles benefits uptake mechanisms to bioaccumulation studies. One way to overcome this issue is to label the nanoparticles. Three main ways to label nanoparticles are currently available: incorporation of fluorescent dyes on nanoparticle surfaces or manipulation of isotopic composition by incorporation of either stable or radioactive isotopes. The main challenge in fluorescent labelling, especially when the nanoparticles are used in an environmental context, is the detachment of the fluorophore (Figure 4.5). Hence this technique requires an extensive knowledge of the media and the surface chemistry of the nanoparticles in the considered environment. Furthermore, by modifying the surface of the nanoparticle with

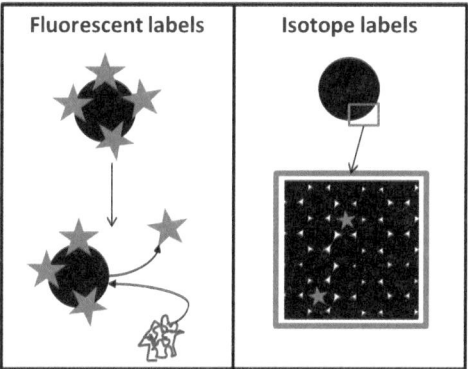

Figure 4.5 Nanoparticle labelling. Fluorescent molecules can be attached to their surfaces or their atomic structures by substituting some atoms with different isotopes, radioactive or natural.

the label, its properties may also be altered. Isotope labelling is an adequate alternative as the "tag" is contained in the structure (Figure 4.5). The challenges for using radioactive labelling are more technical: purpose-built laboratory and equipment are required. Stable isotope labelling is mainly challenged by complexities in synthesis, due to the limited availability of modified precursors and also by the potential high costs of the precursors. Once the chemistry is overcome, stable isotope labelling has shown to be an efficient way to distinguish nanoparticles' contribution from the background.[20,21]

4.4.2   The Intermediate State: Nanoparticles in Media

A scheme that links nanoparticles directly to toxicity is in reality much more complex as an intermediate step between the initial nanoparticles and the toxicity tests has to be considered: the reactivity of nanoparticles in environmental and biological media. This step will commonly favour agglomeration/aggregation which could reduce the enhanced properties of the nanoscale that are being evaluated. Decreased toxicity responses could therefore be observed. It is important to stress that while this is the realistic scenario and such experiments should be reported, it is important to understand the different steps involved in the inhibition of toxicity for mechanistic understanding; if decreased toxicity is observed, can it be directly linked to aggregation/agglomeration or are the particles somehow pacified by interactions with biological components? This challenge is even more obvious for bionanotechnology. One of the first challenges in using nanoparticles for biological application was their hydrophobicity. "Water-soluble" nanoparticle synthesis was therefore investigated in order to obtain particles that exhibited the desired property in the biological medium considered. Similarly, the nanoparticles that are received by toxicologists (primary nanoparticles) are not the particles that are being tested for toxicity. The cells will be exposed to an altered version (secondary nanoparticles) that is the result of the interaction between the exposure medium and the primary nanoparticles. If we consider primary particles that are capped, the secondary particles could be uncapped or capped with a different (media-derived, biological) molecule. An affinity competition occurs when the nanoparticles are introduced in the exposure medium. While suspension stability can be relatively easily assessed (using ζ-potential measurements or UV-Vis spectrometry), mechanistic understanding of the interaction between NPs and organic matter remains challenging.

Nanoparticle–protein interactions are being closely investigated and a mechanistic understanding is arising. In practice, however, most toxicity studies cannot be followed with such detailed experiments on nanoparticle–protein interactions. It is therefore important to develop models that would allow a generic understanding of the behaviour of nanoparticles in biological media.

The behaviour of nanoparticles in environmental media is even less known. The likelihood is that in such environments, nanoparticles will interact with a common molecule, humic acid (HA), for which we have many fewer mechanistic studies. Unlike proteins (repetitive structure, made of sequential amino acids of known structures), the structure of HA is more variable and less known. The Suwannee River HA (SRHA) is a model HA of known structure but in the natural environment many types of HA molecules exist which expose different functional groups. However, when one wants to apply this understanding to the interaction of proteins with nanoparticles, the complex structures of proteins do not allow direct uses of the acquired knowledge. A protein is not a simple molecule but a macromolecule which possesses a 3D conformation; a given protein can structurally evolve due to a modification of the medium. The interactions between different parts of this macromolecule give an almost mechanical property to the protein. Hence, the conformational structure of the protein may favour specific interactions that could have been neglected if the macromolecule was "flat". The 3D conformation can "hide" important interaction groups leaving the nanoparticles to interact with less predictable sites. On the other hand, the possible modification of proteins that interact with nanoparticles and their consequences on health has to be considered. The irreversible changes of human transferrin, a common blood protein, interacting with supramagnetic iron oxide nanoparticles were demonstrated.[22]

In this context, one way of generating data is to perform abiotic reactivity studies prior to biological tests, in order to evaluate the evolutions of the physico-chemical properties of the primary particles. While simulated fluids will allow an understanding of specific mechanisms involved in the evolution of nanoparticles in contact with common environmental and biological components, nature is a more complex matrix. While some knowledge can be transferred from abiotic reactivity other will have to be searched for in natural environments. The study of natural colloids is the obvious answer but modelling may also help in this case.

4.5 Integration of Datasets in Models: How Can We Contribute?

Can we make sense out of the nanotoxicity data that is being generated? And are we ready to use these results to generate models? Initial data is difficult to use due to the lack of characterisation and discrepancies. Currently, particles are characterised but different studies can present different types of data creating gaps in datasets for statistical analysis. Modelling using data specifically provided will be compared to modelling using literature results. "Intuitively" bridging the gaps could be one way to go, using chemistry or models that have developed the understanding of nanoparticles' properties. This section will highlight the important role that other fields can play.

4.5.1 Data Assessment for Literature Data Modelling

For modelling to take place, experimentalists have to "clean up" the data that is currently available. One way to pursue is to evaluate different synthesis and characterisation techniques. It is important to consider that this evaluation is in the only concern of advancing nanotoxicity. Synthesis protocols that may not be best suitable for nanotoxicity (linking properties to toxicity) should not be neglected, if they have significant industrial development potential. Is it possible to develop a classification scheme that would allow modellers to use the data with more or less precaution depending on the nanotoxicological assessment quality of the nanoparticles, characterisation or toxicity data? This "grade" would not be based on safety and would exclusively be used for toxicological assessments.

Statistical measurement of dispersions (SMD) (*e.g.* standard deviation, variance) is commonly used to process data in various fields and could help build predictive nanotoxicity models. While simple SMDs are used for production line quality control (which can directly be applied for synthesis techniques), more sophisticated ones (*e.g.* volatility) are used for complex systems that require multiple SMDs. Volatility is used in finance to estimate and predict the variations of price. Similarly, the assessment of different synthesis and characterisation protocols, followed by an evaluation of the toxicological tests and their variations would allow modellers to construct reliable nanotoxicity predictions.

4.5.2 Bridging the Gaps with the Knowledge Acquired in Other Fields

The understanding of nanoparticle behaviour in natural media is a prerequisite for important challenges to be overcome in nanotoxicity. In the current state, nanotoxicology has not generated enough data for statistical analysis. Other related fields may therefore be considered as data sources to enhance the current level of knowledge. Surface chemistry is an obvious candidate for such exercise; the interaction of biological entities with materials surfaces is a heavy field of research that provides clues if not qualitative data on nanoparticles' surface interactions. The models developed in such fields (*e.g.* catalysis) could be adapted to the needs in nanotoxicity. Organometallic chemistry and related toxicity and modelling data can provide qualitative understanding or at least platforms to build on.

Traditional theoretical studies of nanostructures were modelled using quantum mechanical calculation and molecular dynamics simulations but their use for understanding nanoparticles involvement in toxicity is scarce. The development of QNTR[23] (quantitative nanostructure–toxicity relationships) or nano-QSAR[24] (quantitative structure–activity relationships) models, built on statistical analysis have shown their importance in this context. While these models need to be extended, their ultimate potential to avoid lengthy and costly experiments is obvious. The expansion of these models requires an active involvement of experimentalists. Data generated in current projects are not all accessible for statistical analysis. The creation of databases should help gather results from various studies and this activity should be encouraged among experimentalists to populate models with data and make them more reliable.

4.6 Conclusions

The current fast development of nanotechnology requires that nanotoxicology advances in parallel to allow regulation for the sake of public safety. Although there is not enough time to produce long and specialised experiments, well-constructed and targeted experiments that can feed results into models could be the optimum approach.

A very wide range of nanoparticle synthesis protocols have been developed in the past decades. Yet, when toxicity is the envisaged application of the nanoparticles, a selection has to be made assuring that the toxicity responses can be correctly interpreted. Adapting the same idea, addressing the application needs at the synthesis stage, labelling of nanoparticles can be an important tool in understanding their behaviour and their tracing. In general, the overall designs of experiments from dedicated synthesis to biological experiments will enhance the quality of the generated data. When this happens, the acquired knowledge and consequently the models built will be more reliable.

Acknowledgements

This work was funded by the European Commission's Seventh Framework Programme (FP7/2007-2013) under grant agreement no. 214478, project NanoReTox.

References

1. A. Elsaesser and C. V. Howard, Biological Interactions of Nanoparticles, *Adv. Drug Delivery Rev.*, 2012, **64**(2), 129–137.
2. E. Burello and A. P. Worth, *Wiley Interdiscip. Rev.: Nanomed. Nanobiotechnol.*, 2011, **3**(3), 298–306.
3. A. Bayer, D. S. Boyle, M. R. Heinrich, P. O'Brien, D. J. Otway and O. Robbe, *Green Chem.*, 2000, **2**(2), 79–86.
4. C. B. Murray, D. J. Norris and M. G. Bawendi, *J. Am. Chem. Soc.*, 1993, **115**(19), 8706–8715.
5. Z. A. Peng and X. Peng, *J. Am. Chem. Soc.*, 2000, **123**(1), 183–184.
6. M. A. Malik, N. Revaprasadu and P. O'Brien, *Chem. Mater.*, 2001, **13**(3), 913–920.
7. J.-H. Liu, J.-B. Fan, Z. Gu, J. Cui, X.-B. Xu, Z.-W. Liang, S.-L. Luo and M.-Q. Zhu, *Langmuir*, 2008, **24**(10), 5241–5244.
8. P. Raveendran, J. Fu and S. L. Wallen, *J. Am. Chem. Soc.*, 2003, **125**(46), 13940–13941.
9. I. Lynch, A. Salvati and K. A. Dawson, *Nat. Nanotechnol.*, 2009, **4**(9), 546–547.
10. NanoReTox, Nano Risks to the Environment & Human Health, FP7 framework programme, www.nanoretox.eu

11. P. E. Buffet, O. F. Tankoua, J. F. Pan, D. Berhanu, C. Herrenknecht, L. Poirier, C. Amiard-Triquet, J. C. Amiard, J. B. Bérard, C. Risso, M. Guibbolini, M. Roméo, P. Reip, E. Valsami-Jones and C. Mouneyrac, *Chemosphere*, 2011, **84**(1), 166–174.

12. C. Pang, H. Selck, S. K. Misra, D. Berhanu, A. Dybowska, E. Valsami-Jones and V. E. Forbes, *Aquatic Toxicol.*, 2012, **106**–**107**(0), 114–122.

13. Y. Cong, G. T. Banta, H. Selck, D. Berhanu, E. Valsami-Jones and V. E. Forbes, *Aquatic Toxicol.*, 2011, **105**(3–4), 403–411.

14. Peng, X., S. Palma, N. S. Fisher, and S. S. Wong, *Aquatic Toxicol*, 2011, **102**(3–4), 186–196.

15. Z. L. Wang, M. B. Mohamed, S. Link and M. A. El-Sayed, *Surf. Sci.*, 1999, **440**(1–2), L809–L814.

16. B. Hvolbaek, T. V. W. Janssens, B. S. Clausen, H. Falsig, C. H. Christensen and J. K. Nørskov, *Nano Today*, 2007, **2**(4), 14–18.

17. K. G. Thomas, S. Barazzouk, B. I. Ipe, S. T. S. Joseph and P. V. Kamat, *J. Phys. Chem. B*, 2004, **108**(35), 13066–13068.

18. M. Li, S. Pokhrel, X. Jin, L. Mädler, R. Damoiseaux and E. M. V. Hoek, *Environ. Sci. Technol.*, 2010, **45**(2), 755–761.

19. T. Xia, Y. Zhao, T. Sager, S. George, S. Pokhrel, N. Li, D. Schoenfeld, H. Meng, S. Lin, X. Wang, M. Wang, Z. Ji, J. I. Zink, L. Mädler, V. Castranova, S. Lin and A. E. Nel, *ACS Nano*, 2011, **5**(2), 1223–1235.

20. A. D. Dybowska, M.-N. Croteau, S. K. Misra, D. Berhanu, S. N. Luoma, P. Christian, P. O'Brien and E. Valsami-Jones, *Environ. Pollut.*, 2011, **159**(1), 266–273.

21. S. K. Misra, A. Dybowska, D. Berhanu, M. N. Croteau, S. N. Luoma, A. R. Boccaccini and E. Valsami-Jones, *Environ. Sci. Technol.*, 2011, **46**(2), 1216–1222.

22. M. Mahmoudi, M. A. Shokrgozar, S. Sardari, M. K. Moghadam, H. Vali, S. Laurent and P. Stroeve, *Nanoscale*, 2011, **3**(3), 1127–1138.

23. D. Fourches, D. Pu, C. Tassa, R. Weissleder, S. Y. Shaw, R. J. Mumper and A. Tropsha, *ACS Nano*, 2010, **4**(10), 5703–5712.

24. T. Puzyn, B. Rasulev, A. Gajewicz, X. Hu, T. P. Dasari, A. Michalkova, H.-M. Hwang, A. Toropov, D. Leszczynska and J. Leszczynski, *Nat. Nanotechnol.*, 2011, **6**(3), 175–178.

CHAPTER 5

Experimental Approach to the Structure and Properties of Nanoparticles

KRZYSZTOF J. KURZYDLOWSKI*[a],
MALGORZATA LEWANDOWSKA[a] AND
MICHAL J. WOZNIAK[b]

[a] Faculty of Materials Science and Engineering, Warsaw University of Technology, Warsaw, Poland; [b] University Research Centre – Functional Materials, Warsaw University of Technology, Warsaw, Poland
*E-mail: kjk@inmat.pw.edu.pl

5.1 Introduction

Nanoparticles attract the attention of scientists and engineers because they generally exhibit properties far different from their large-sized counterparts. Generally, the special behaviors of nanoparticles can be explained in two complementary ways:

- as a consequence of their size being comparable or smaller than some dimensions specific to the phenomena of interest (for example, the length of the light waves interacting on them or the size of the pores in membrane-like structures *via* which they may permeate); and
- in terms of the distinct structure distinguished by a large fraction of surface atoms.

RSC Nanoscience & Nanotechnology No. 25
Towards Efficient Designing of Safe Nanomaterials: Innovative Merge of Computational Approaches and Experimental Techniques
Edited by Tomasz Puzyn and Jerzy Leszczynski
© The Royal Society of Chemistry 2012
Published by the Royal Society of Chemistry, www.rsc.org

Thus an experimental approach to nanoparticles must undoubtedly address these two specificities – size and structure of surfaces – which are discussed in this chapter. This discussion starts with the size of nanoparticles, more generally with their geometrical features, which in turn determine to a large degree their surface properties.

5.2 Imaging Nanoparticles

Imaging plays an important role in the development and control of shape/size-sensitive forms of materials such as powders and their aggregates. This in particular refers to nanoparticles, the definition of which is based on strict size specifications. It should be noted at this point that making nanoparticles visible to human beings requires magnifications in the range of 10^6. This is far beyond capacity of light microscopy, which offers magnifications in the range of 10^3. In the past imaging at the nanoscale was possible with relatively user-unfriendly electron microscopes and required special skills in sample preparation. It became relatively widely available only recently thanks to the development of atomic force microscopes allowing a very high magnification in the standard laboratory environment. Also, major progress has been made in simplifying the operation of high-resolution electron microscopes, and new techniques have been developed for sample preparation.

It is a well-known adage that a picture is worth a thousand words. On the other hand, it is also clear that we infer viable information only if we understand how a given image has been obtained. Again, the understanding of the underlying principles is of prime importance if the image is formed by illuminating the object with an electron beam (not light) a phenomenon not experienced by human beings. Also, although atomic force microscopy reminds us to some degree of checking the topography with one finger, it provides far more than the one "sense". Thus, before we discuss some more details it is necessary to provide general information about the techniques currently used for imaging nanoparticles.

5.2.1 Electron Microscopy

Historically, the first type of microscope allowing for imaging at the nanoscale was a transmission electron microscope. The principle of transmission electron microscopy (TEM) is basically the same as in conventional light microscopy. However, a sample is illuminated by a beam with much shorter wavelength which substantially improves the resolution. For the accelerating voltage of 100 kV (typical accelerating voltage in these microscopes), the wavelength is 0.0037 nm which gives a resolution of 0.2 nm. This is enough for imaging objects of the size of nanometres. On the other hand, one should also observe that the length of the electron waves is less than the distance between atoms in solids. Thus the electron beam may, and usually does, interact with the imaged object, also providing information on its atomic structure. Retrieving this information requires a good understanding of physics. However, there is also another challenge to overcome – sample preparation.

The first challenge in sample preparation is related to the fact that electron imaging implies passing an electric current (the electron beam) and non-conductive objects must be coated with conductive layers. The next major difficulty in TEM imaging is the requirement of "transparency" which translates into the use of very thin samples. In fact electrons can pass through films or foils of thickness typically less than 100 nm. In the case of bulk materials, it requires special methods of thinning down to such a thickness. In the case of nanopowders, the thickness of the particles is generally expected to be below the transmission limit. However, this is the case only if the nanoparticles do not agglomerate – if they do, then they need to be de-agglomerated, which is a problem by itself. In practice, the preparation of nanopowder samples includes the following steps:

- preparation of the nanopowder suspension;
- de-agglomeration, *e.g.* using ultrasonic techniques;
- deposition of a suspension droplet on the special support (amorphous carbon film on a copper mesh);
- drying; and
- deposition of conductive layer (in the case of non-conductive materials).

A TEM image of nanoparticles prepared in such a way is shown in Figure 5.1A. In this case, the agglomeration was efficiently prevented and the images obtained adopting this approach provide information on the size and shape of nanoparticles, as detailed in section 5.3. TEM observations of crystalline particles allow one to obtain a diffraction image, as illustrated in Figure 5.1B. Such images are valuable source of information about crystal symmetry and inter-atomic distances, thus enabling phase identification – an

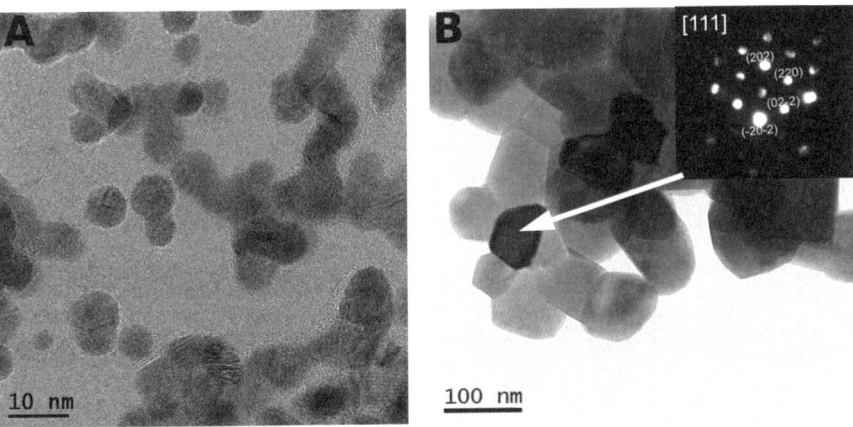

Figure 5.1 TEM image of Pd-Au (A) and $LiMn_2O_4$ (B) nanoparticles. Diffraction pattern (insert in A). (Images provided by courtesy of Dr Justyna Grzonka (A) and Dr Mariusz Andrzejczuk (B).)

exercise crucial for polymorphism determination and any assessment of their toxicity.

Modern transmission electron microscopes are equipped with better electron guns and, more frequently, various image correctors. As a result they offer the possibility of high-resolution imaging revealing the atomic structure of nanoparticles, as illustrated in Figure 5.2. From such an image one can determine not only size and shape but also interplanar distances and the crystallographic structure of nanoparticles. There is also a possibility of imaging defects inside the particles and evaluating their homogeneity. The capacity of the modern electron microscopes is demonstrated by Figure 5.2A which reveals the internal structure of a nanoparticle with five-fold twins formed during synthesis. On the other hand Figure 5.2B shows a thin layer on a nanoparticle's surface. These are organic substances developed during powder synthesis which prevents particle agglomeration. However, the meaning of this image is more general as nanoparticles may adsorb unwanted chemicals on their surface – the process known as contamination. In fact, contamination of nanoparticles may take place during nearly all stages of sample preparation and in some extreme situations also during TEM observations. (It should be remembered at this point that illumination of the particles with an electron beam implies that they are heated with electron beam current.)

In general TEM imaging enables precise characterization of size, shape and structural features of individual nanoparticles. However, it should be noted that nanoparticles prepared for TEM observation by de-agglomeration may not reflect their true geometry and chemistry. The toxicity of nanoparticles may be different for de-agglomerated nanoparticles and their clusters. Therefore, it is of prime importance to characterize them in the same form as that in which they exist in the environment of reference. This in part is

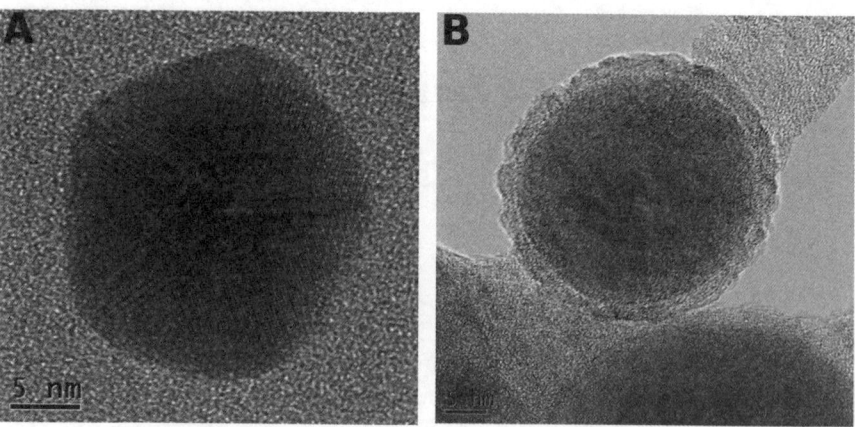

Figure 5.2   (A) High-resolution image of a Au nanoparticle with defects (five-fold twins) and (B) Ni nanoparticle with oxide surface layer. (Image provided by courtesy of Dr Justyna Grzonka.)

possible using scanning electron microscopy (SEM). This instrument does not require any special sample preparation, except in some cases the deposition of a conductive layer. It should also be noted that in standard microscopes only dried samples can be observed (not suspensions) since the specimen chamber is under high vacuum. However, there are also microscopes with so-called low vacuum and in extreme cases wet samples can also be investigated.

In SEM, the sample is scanned (point by point) by the focused electron beam. At the same time, special detectors collect signals emitted from each "point". The image is formed "point by point", in the form of maps revealing the intensity contrast. The magnification obtained is thus a ratio of the dimensions of the raster on the screen to those of the sample. Assuming that the display screen has a fixed size, a higher magnification results from reducing the size of the raster on the specimen, *i.e.* the size of the electron beam. This in turn depends on the accelerating voltage, which is usually smaller than in TEM (typically in the range 0.2 to 30 kV); however, modern SEMs equipped with a detector of transmitted electrons (so-called scanning transmission electron microscopes) may have higher accelerating voltages (*e.g.* 200 kV).

In general, one may use a number of signals emitted from the surface scanned with an electron beam. The signals more frequently used for imaging in SEM include:

- secondary electrons (SE),
- back-scattered electrons (BSE), and
- transmitted electrons (TE).

Low-energy (<50 eV) secondary electrons are ejected from the *k*-orbitals of the specimen atoms by inelastic interactions with beam electrons. As low-energy electrons, they originate from a few nanometres below the surface layer. As a result, the resolution of a SE image is similar to the size of the electron beam and is the best in SEM. SE images are very sensitive to surface topography and can be used to image individual nanoparticles as well as clusters in which individual particles can be distinguished, as illustrated in Figure 5.3.

Back-scattered electrons are high-energy ones which were reflected or back-scattered out of the specimen due to elastic interactions with specimen atoms. It should be noted that heavy elements (high atomic number) back-scatter electrons more strongly than light elements (low atomic number). As a result the solids with heavy elements appear brighter in the SEM images. Thus BSE imaging allows one to detect contrast between areas with different chemical compositions. As high-energy electrons, BSE originate from a much larger volume of a specimen, thus the resolution of BSE imaging is much lower.

A major advantage of SEM is the possibility to equip the microscopes with a number of analytical spectrometers enabling the mapping of nanostructures. The most common is an energy-dispersive X-ray spectrometer which collects characteristic X-rays emitted from the specimen due to the interaction with the incident beam. In this method, simultaneous analysis of the elements with

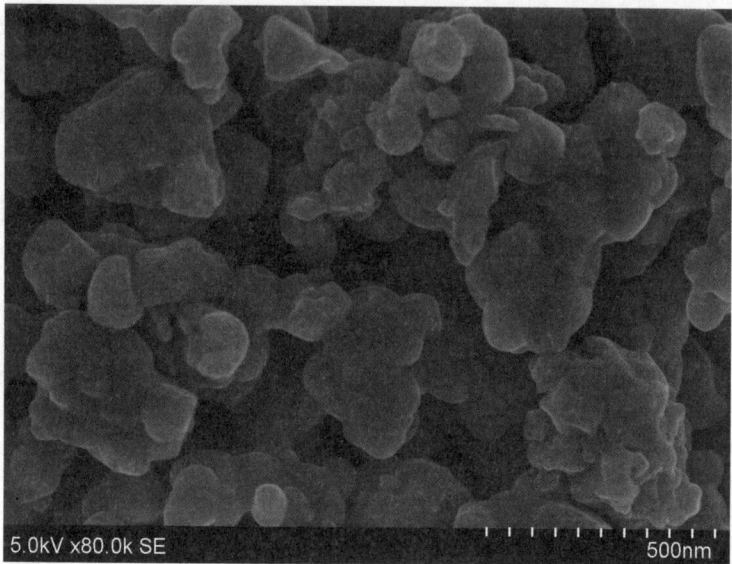

5.0kV x80.0k SE   500nm

Figure 5.3   SEM image of clusters of nanoparticles. (Image provided by courtesy of
Dr Mariusz Andrzejczuk.)

atomic number higher than 5 is possible creating maps showing the chemical
composition of the object in question.

It should be noted that the resolution of the maps obtained by energy-
dispersive X-ray spectrometry (EDS) is relatively poor – worse than in BSE
imaging. On the other hand, modern scanning electron microscopes are also
equipped with other detectors which allow the mapping of chemical
composition at the nanoscale. Much better resolution (even atomic resolution)
is in particular available in electron energy loss spectroscopy (EELS). In this
case, the electrons which undergo inelastic scattering are detected. The amount
of energy loss can be measured using special spectrometer and interpreted in
terms of the origin of such an effect. It should be noted that the EELS method
is particularly suitable in the case of detecting light atoms, *i.e.* He, Li, O, N and
C, which are difficult to reveal by EDS.

5.2.2   Scanning Probe Microscopy

Scanning probe microscopy (SPM) has emerged as a powerful tool in the
imaging of materials, probing of properties and manipulation of objects at the
nano scale. The first SPM microscope was a scanning tunneling microscope
developed by Gerd Binnig and Heinrich Rohrer at IBM laboratories in Zurich
in 1981. The most popular type of SPM is atomic force microscopy (AFM)
invented by Binnig, Gerber and Quate in 1986, which over the last 20 years has
evolved into various types and a whole range of SPM techniques, described
below.

Scanning tunneling microscopy (STM) is based on the quantum tunneling effect. The conducting probe is placed in very near contact to the examined sample. A bias voltage is applied between the probe and sample surface which allows electrons to tunnel through the small gap between them. The resulting tunneling current is a function of local density of states of the sample and changes with probe position and applied voltage.

The STM microscope can be operated in two basic modes: (1) topography, when the tunneling current is kept constant by a feedback loop, and (2) current imaging, when the height is constant during imaging. For STM a good resolution is considered to be 0.1 nm in the xy plane and 0.01 nm in the z-direction. This allows one to reveal the nanostructure of investigated samples. Importantly, STM can also be used in vacuum, ambient or liquid environments and over a wide temperature range. On the other hand the limitation of STM is the fact that it can basically only be used for conductive samples.

The other SPM techniques, *e.g.* AFM, are based on interaction forces between the probe and the investigated sample. These forces can be most generally divided into the following: van der Waals, electrostatic, magnetic, capillary, ionic repulsion and frictional forces. The typical SPM is composed of the probe (tip placed at the free end of elastic cantilever), the deflection of the cantilever monitoring system (laser diode + photodiode), the scanning system (based on a piezoelectric scanner) and the electronic controller plus computer (Figure 5.4).

The three main factors which limit SPM microscopes resolutions are: (1) probe, (2) pixel resolution, and (3) environmental conditions.

(1) Probe. During scanning the AFM probe always convolutes with the sample details. It means that the collected images are a superposition of the shapes of tip and probe. This tip–probe convolution effect is critical in the case of nanomaterials. It makes apparent images of nanoparticles a few times bigger than in reality. To obtain the real particle size a

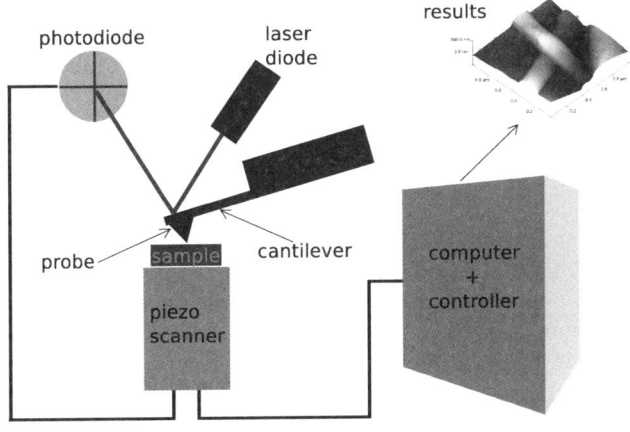

Figure 5.4   Schema of a scanning probe microscope.

deconvolution process is needed. Most commercially available probes usually have the tip radius of curvature in the range of 5–20 nm. However, very often the real tip radius is bigger than the one in technical specification. The real tip radius and shape can be estimated by observation of the probe with electron microscopy. A reliable method is scanning with the probe a special calibration sample of well-defined topography. After establishing the real diameter and shape of the probe, by use of some mathematical models, the real topography of the sample can be restored.

(2) Pixel resolution. The SPM microscopes operate the scanning mode. The probes move on samples by raster pattern *e.g.* a probe "draws" 256 lines on the sample and on each single line there are 256 points. When the area of 10 × 10 μm is scanned the pixel size is equal to 10 μm/256 lines or 1 point = ∼0.04 μm (40 nm). Thus the smallest object which can be resolved in the image is a single 40 nm pixel. When the smaller area and bigger scan lines/points number is used, *e.g.* 2 μm and 512 lines, the pixel resolution is equal to about 4 nm. This clearly demonstrates how the pixel size impacts the investigation of nanomaterials.

(3) Environmental conditions. SPM microscopes are extremely sensitive to acoustic noise, vibrations and electromagnetic noise. All of these factors can affect the process of image acquisition. They influence the image quality when smaller objects are scanned. The microscopes should thus be protected from electromagnetic noise by the use of special metal covers and from vibration and noise by using hoods and passive or active anti-vibration tables. A good practice is to place the microscope in lowest floor of the building or preferably underground. A compact construction of SPM equipment is also an advantage.

The interaction forces cause the probe cantilever to bend. During scanning over the sample surface the cantilever bending is monitored to form the image. By operating AFM in air or liquid, atomic resolution can be obtained in the z-direction without a problem. However, in the xy plane such a resolution is only possible for a few kinds of samples *e.g.* highly ordered pyrolytic graphite (HOPG) or mica. To collect images with atomic resolution in xyz, a special pretreatment of the sample is usually needed and microscopes working in ultra-high vacuum (UHV) and at ultra-low temperatures (ULT) are necessary.

AFM microscopes can operate in three principal modes based on the van der Waals forces: contact (CM), non-contact (NCM), and semi-contact mode (SCM). However, other forces, like capillary or frictional, can influence probe–sample interactions. In the CM, during the scanning process the probe is in the delicate physical contact with the sample surface. The images are collected in two modes, constant force and constant height, depending on feedback. In the CM the topographical images are generated. However, together with topography it is possible to obtain a map of frictional forces when the probe scans the sample at 90°. In scanning at this angle, the probe cantilever is very sensitive for lateral bending. The technique based on lateral scanning is called the frictional or lateral force mode. The contact of the probe with the sample surface can provide other useful information regarding the mechanical properties of sample (elasticity, elastic modulus) and adhesion forces between the probe and the sample surface. The mode is called force spectroscopy.

In the NCM, the probe cantilever is oscillating with a frequency near to the free resonant one, very close to the sample surface but without a physical contact with it. The amplitude of the cantilever free-end oscillations usually ranges from 2 up to 5 nm. The NCM is very useful in the case of delicate samples which can be damaged or moved because of the physical contact with the probe. However, the image can be influenced by a thin water layer which develops on the sample in the ambient atmosphere. To overcome this difficulty the SCM was developed.

In this mode the oscillating probe gently taps the sample due to higher amplitudes of oscillations of the free end of the cantilever, in the range of 5–100 nm. Such oscillations allow the water layer to be penetrated, giving the possibility of receiving the real image of the sample. In the SCM it is possible to visualize very delicate samples without causing damage. It is also possible to collect more sophisticated information about the sample. This is due to the fact that the local sample mechanical properties, adhesion and crystallographic orientation influence the probe–sample interaction, causing the phase shift in cantilever oscillation. This generates a new kind of contrast, called phase contrast.

There are also two-pass techniques, which allow one to obtain an image of topography and map of forces. For the same area a first profile of the topography is acquired in the CM, NCM or SCM. Afterwards the probe is lifted is some specified distance above the sample surface and the scan is repeated line by line. During this second scan information related to other interaction forces between the sample and the probe is recorded.

When the sample exhibits permanent magnetic behavior it is necessary to use a probe covered with a layer having different magnetic properties *e.g.* made of hard or soft magnetic materials. This allows one to obtain magnetic images which would not be recorded with an improperly chosen probe. Before the measurements, the probe should be magnetized to align the magnetic domains in its coat. Sometimes it is also necessary to magnetize the sample. Usually the best magnetic contrast is generated if the second pass is taken between 30 and 300 nm above the surface. The technique explained above is called magnetic force microscopy (MFM). In a similar way it is possible to map electric interactions. In such a case, it is necessary to use a conductive or semi-conductive probe. During the second scan, usually taken in the range 5–100 nm above the surface, the probe is attracted or repulsed by Coulomb forces generated by electrostatic charge from the sample. This permits one to obtain a map of the charge distribution. This technique is called electric force microscopy (EFM). There is also a very similar mode – Kelvin probe microscopy or Kelvin force microscopy (KPM or KFM). During the second pass the voltage is applied to the probe to set the voltage difference between the sample and the probe to zero. The quantitative information (sample–probe voltage difference) is obtained in this way, while in EFM only qualitative information is available. After microscope calibration the map of the real surface potential distribution can be obtained. Closely related to the EFM and KFM is scanning capacitance microscopy based on the tip–surface capacitance measurements.

The technique which allows one to measure sample conductivity is called conductive atomic force microscopy (C-AFM). During one scan with a conductive probe, the topographical profile and profile of conductivity of the sample is recorded. It is quite similar to STM; however, there is one important advantage – non-conductive samples can be investigated with this technique.

In the case of C-AFM, the topographical profile can be recorded even for non-conductive areas of the sample, which is impossible with STM. The other technique derivative of C-AFM is scanning spreading resistance microscopy (SSRM), which allows one to measure the conductivity and the resistance of the sample.

5.3   Measuring the Size, Size Distribution and Shape of Nanoparticles

The size of nanoparticles can be determined by a number of methods which can be generally grouped into direct and indirect. Direct methods are based on images obtained using various microscopic techniques, such as electron, atomic force and more recently X-ray microscopy. Examples of indirect methods include small angle scattering (neutron, X-ray, for example), and diffraction peak broadening. The direct methods offer a clear advantage by providing images which can be used to quantify not only the size of the nanoparticles but also their shape. These images also contain information on the structure and chemistry of the individual particles. However, the price we pay in this case is difficulty in obtaining statistically meaningful characteristics. Such characteristics are relatively easy to obtain with indirect methods which, on the other hand, may require *a priori* knowledge of the particles' shape. Thus it is highly recommended that geometrical features of the nanoparticles are experimentally approached by a combination of direct and indirect methods to obtain in-depth understanding of both their geometry and statistical characteristics.

In the following section of this chapter some direct and indirect methods currently available for sizing nanoparticles are described. Beside this, some general considerations are provided regarding the description of the size and shape of nanoparticles.

5.3.1   X-ray Diffraction

X-ray diffraction techniques are widely applicable to investigations of materials in the form of crystalline powders. First of all, they provide qualitative as well as quantitative information on phase content. Furthermore, they can be also used to retrieve information on the size of crystallites in powders. An example of a diffraction intensity obtained in the case of γ-Al_2O_3

nanoparticles is shown in Figure 5.5. There are three parameters which can be determined from the diffraction intensity:

- the position of peaks, which allow indentifying phases in powders,
- peak area, which is used in assessing the percentage content of components (quantitative analyses of the phase composition), and
- the width of the peaks, which depends on the average size of crystallites.

It should be noted, however, that the term "crystallites" in the context of X-ray measurements has a specific meaning. It must be understood as referring to the parts of the powder particles which coherently scatter the X-ray. In the case of perfect mono-crystalline particles, the size of the crystallites determined from X-ray diffraction is the same as the size of the particles imaged in microscopic observation. A different situation happens if particles consist of two or more parts of different, even only slightly, crystal orientations. In this case, the size of crystallites measured by X-ray is determined by the size of the particle building blocks and much smaller than the size in the image of the particles. This applies in particular to the particles of a mosaic structure as shown in Figure 5.2A. On the other hand, if these two estimates of size are available, from X-ray and from direct imaging, additional information can be retrieved regarding the structure of the particles in question.

The precision of the size measurements by X-ray diffraction strongly depends on the quality of the recorded X-ray profiles. For this reason, all other

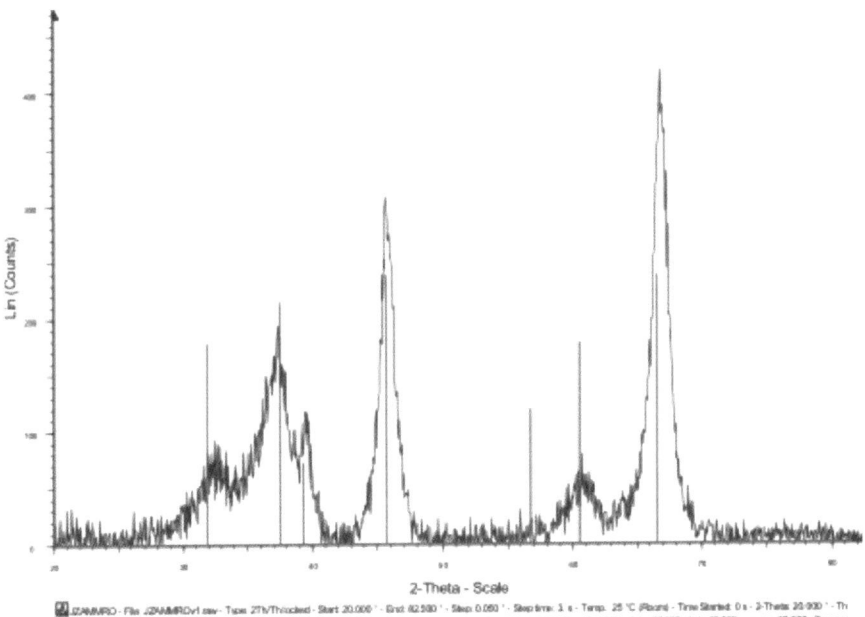

Figure 5.5 X-ray diffraction of γ-Al$_2$O$_3$ nanoparticles with broad peaks due to small crystallite size.

factors leading directly to the broadening of peaks should be eliminated. This can be achieved by using the Rietveld technique, which refers to theoretical diffraction profiles as modeled in the function of lattice parameters and the size of crystallites. These model profiles are adjusted to the real ones with a great benefit to the precision of measurements.

To a first approximation the crystallite size can be determined using the Scherrer equation of the following form:

$$D_{hkl} = \frac{K\lambda}{B \cos \theta_{hkl}} \tag{5.1}$$

where *hkl* are the Miller indices of lattice planes, which participate in the X-ray wave diffraction, D_{hkl} is the average crystallite size (measured in nanometres), K is Scherrer's constant (between 0.89 and 1.00), B is the full width of half maximum (in radians), λ is the length of the diffraction wave (in nanometres), and θ_{hkl} is Bragg's angle (in radians).

At first glance one may gain the impression from this formula that different values of the average grain size can be obtained for various combinations of *h*, *k*, and *l*. However, the differences are in the margin of error. It is worth mentioning that the full width of half maximum of an analyzed peak B should be computed taking into consideration the apparatus broadening, B_a, which can measured for a reference sample. In practice B is obtained by subtracting the width of half maximum of the reference sample from the measured width of half maximum. Further, it should be noted that the Scherrer equation disregards the strain broadening. The effect is described by Taylor's equation:

$$B_s = 4\varepsilon \, \mathrm{tg} \, \theta \tag{5.2}$$

where Bs is strain broadening, ε is the lattice strain, and θ is Bragg's angle.

Another approach to the determination of the crystallite size from X-ray profiles is based on the Williamson–Hall method, with the following formula:

$$\frac{B \cdot \cos \theta}{K \cdot \lambda} = \frac{1}{D} + \langle \frac{\Delta a}{a} \rangle \cdot \frac{4}{K \cdot \lambda} \sin \theta \tag{5.3}$$

where D is the average crystallite size (measured in nanometres), and $\Delta a/a$ is the lattice strain.

In the context of the present publication it should be underlined that the size of nanocrystals, which by their nature are smaller than 100 nm, with lattice distortions can be determined by analyzing the entire profile of the diffraction lines. In a simplified approach, the measurements are carried out on the width of the peaks on diffraction lines. The width of peaks can be calculated as the integral breadth or, as happens more frequently, as the full width at half maximum (FWHM).

The width of the diffraction peaks depends on the physical factors (the size of crystallites, the lattice distortion) as well as on the characteristics of the apparatus used. It is therefore necessary to define the "pure" diffraction profile by extracting the apparatus broadening. This can be achieved by using the reference samples, as mentioned earlier. The peaks of such samples should not show the physical broadening. In practice the silicon samples are used as this material is readily available in defect-free form. Alternatively, as a reference sample one can use coarse-sized variants of the analyzed nanoparticles.

5.3.2 Laser Diffraction

Laser diffraction is also known as low-angle laser light scattering (LALLS), Fraunhofer diffraction or Mie scattering. It takes place when a parallel, coherent, intense beam of laser light of fixed wavelength is passing through the dispersed set of particles. *Via* interactions with these particles the beam scatters at angles which depend on the particles' diameters. In fact, the scattering angle increases logarithmically with decreasing size of the particles. Also the intensity of scattered beams depends on the particle size. Larger particles scatter light at narrow angles with high intensity. Smaller ones do so at wider angles with low intensity.

A schematic diagram of the laser particle size analyzer and an example of diffraction fringes are shown in Figure 5.6. The Fourier lenses transform the scattered light to the diffraction rings recorded by the detector placed in the focal plane. Unscattered light remains focused on the optical axis.

The modern particle size analyzers are in fact much more sophisticated than the one shown schematically in Figure 5.6. They usually consist of a series of detectors (including back-scattered light detectors). Because of this, there is the possibility of performing measurements from around 0.02 through to 135°. The wavelength of the light is also important. Red and blue (smaller wavelengths) light sources make it possible to achieve the limit of resolution with sub-micron particles.

The laser diffraction patterns allow one to infer the particle size distribution based on the theoretical estimates of such patterns for a given population of the particles. Most estimates assume that the particles are spherical. If this is not the case, the results might be far from acceptable. For non-spherical

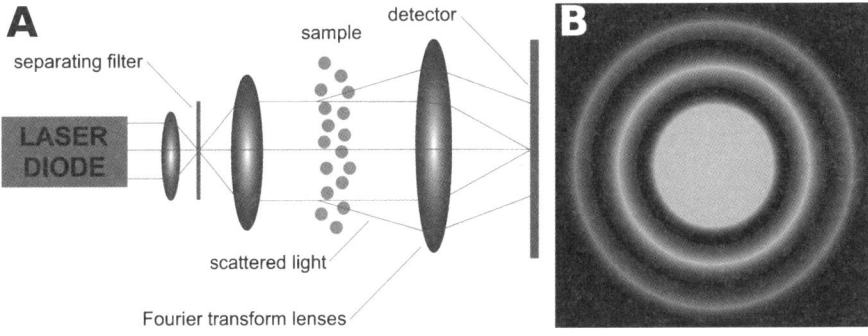

Figure 5.6 Diagram explaining the laser particle size analyzer (A) and an example of diffraction pattern (B).

particles special numerical codes can be developed providing theoretical estimates of the diffraction patterns.

The technique of laser diffraction offers an opportunity to measure the size of particles with diameters in the range 10 nm–3 mm. The samples can be in the form of aerosols, suspensions, emulsions, and sprays. The random dispersion of particles and good light transmission of the medium used in the measurements are essential. It should, however, be noted that this technique does not distinguish between free nanoparticles and their agglomerates and only the latter are measured.

5.3.3 Image Analysis

Computer-aided image analysis provides very efficient tools for quantitative descriptions of the populations of nanoparticles. Such analyses are usually based on 2D images obtained in microscopes and as a result their precision and reliability is strictly a function of the reliability/quality of these images. It should be noted in this context that no magic procedure may reduce the margin of the uncertainty due to the imaging errors (unlike in the action movies in which smart detectives correct blurred images – keep in mind they do only object recognition). Thus your analyses are only as good as your images! One should also keep in mind that the images must be taken of randomly selected objects and the general rule is to take more images and process them with less effort on the precision side.

The quantitative analyses provide descriptors of the elements imaged using a large spectrum of geometrical and non-geometrical parameters. However, before we start a brief introduction to their applications for nanoparticles, it is advisable to present the basics of computer image analysis, its capabilities and limitations.

5.3.3.1   First Step to Image Analysis – Acquisition

The first step is to obtain representative images of the nanoparticles. To this end we have a number of tools as described in section 5.2. However, although the equipment for materials structure imaging is more and more efficient and powerful, one needs to take into account image defects. As said before, the images to be subjected to image analysis need to be as good as possible. Thus, all possible measures should be taken to reduce any source of potential image distortions, such as, for example, noise at the stage of acquisition.

5.3.3.2   Image Processing – Corrections

Although the images in general are as good as obtained, there is a wide spectrum of image-processing algorithms which can "reduce" some of their imperfections. It should be noted, however, that these algorithms in one way or another depend on the assumptions made with regard to the nature of the images to be processed. Some of these assumptions are obvious – for example, insufficient or non-uniform illumination. Some are far from evident, for example, smoothing of the particle surface.

The most widely used image-processing procedures are known as filters, which can be divided into the following:

- Enhancement – equalizing histogram, shade correction,
- Blur – Box, Nagao, Gaussian,
- Rank value – minimal, maximal, median, mode,
- Edge detection – for example, Sobel, Roberts, Previtt, Canny–Deriche filters, and
- Frequency domain– Fourier transformation, low-pass/high-pass filters.

Enhancement filters are used when the image was made under poor lighting conditions. If the image is too dark or too light, and the objects are not clearly visible, the procedures of histogram equalizing can make them easier to apprehend by human beings (who distinguish only 40 to 60 levels of grey scale). Histogram equalizing transforms the initial, narrow grey level scale into the whole range from 0 to 255.

Shade correction is a procedure for correcting uneven background caused by a light source or improper exposition of the probe during the image acquisition. The shadow effect is usually unnoticed by the observer. On the other hand it may have a serious, negative effect on the results of detection. An efficient way to test the potential shadow effect on the image is based on profiling by plotting a pixel grey level along the test line through the image. The shade-correction algorithms extract the image of the shadow and correct the acquired image based on it. As a result it is possible to obtain an image with the same intensity of background in the entire area of the image analysis.

Blur filters like the Box filter, Nagao or Gaussian are usually used to correct "noise" in delineating the contours. The filtered images might be used as auxiliary ones. These image "distortions" are not recommended if the value of the pixel grey level is important as research data.

Rank value filters are very helpful for reducing a different kind of noise. These procedures are based on analyzing the grey level of a given pixel in conjunction with the grey levels of its neighborhood pixels (4 or 8) and ranking them from the lowest to the highest. For a minimal filter, the pixel will receive the minimum value of the rankings; for a maximal filter, the maximum; for a median, a new pixel grey level will be equal to the value in the middle of the rankings. In practice the minimal filter gives good results in reducing a dark noise, the maximum filter for light noise, and the median in cases when the noise is both dark and light (pepper and salt). The huge advantage of these filters, in contrast to the blur ones, is that they do not generate new values of the grey level and, despite the slight blur, the objects retain sharp edges.

Edge-detection filters are used to perform detection of edges of the features of interest in the image. Edges are lines delineating the objects against the background. Edge-detection filters are very common in image processing. They are based on calculating the image gradient and transform the initial image into

Figure 5.7   Example of an edge-detection result (before and after).

the one with edges being the lightest image element, corresponding to the highest gradients. This operation could be a first step to perform object detection. Depending on the character of the edge lines it is possible to choose the most appropriate filter or use a couple of them in combination (Figure 5.7).

Frequent-domain filters transform the initial image by detecting repeatable characteristic frequencies and amplitudes (for example, typical of noise) which can be reduced in this way. This function can also be used to texture analyses, advance smoothing or sharpening of the image or to extract from the image only low-/high-frequency parts. Fourier transformation is a reversible function, which means that after processing an image, it can be transformed back to the initial image.

5.3.3.3   Detection

When the image is well prepared for object detection, this task can be performed in a different way. The edge detection mentioned above is one of the frequently applied methods. Another one, based on analysis of the histogram, is called thresholding.

There are many methods of fully automatic thresholding, for example, based on the entropy, maximum contrast and moment methods. Semi-automatic methods require an operator intervention. Thresholding is in fact a very simple transformation of the grey level images into binary, with pixels being either black or white (0 or 1). The crucial part in the detection by thresholding is selecting the right value of the low and high thresholds, which define the range of the grey levels corresponding with the object on the image.

5.3.3.4   Correcting the Detection Results

Sometimes the binary images still contain some artifacts (black objects of grey level 1) which are of interest in terms of the measurements to be carried out. Those artifacts can be corrected using numerous operations of mathematical morphology. The most commonly used can include: erosion, dilation, opening, closing, border killing, hole filling, split clustering, watershedding, skeleton formation , SKIZ, *etc.* (see Figure 5.8).

Erosion is an operation that causes the removal of the most external layer of the pixels from the all objects. The feature of this transformation is that all objects change size and shape.

Dilation is the opposite operation to erosion as it causes the addition of a layer of the pixels to each of the objects on the image. It thickens each object of the image.

Opening is a combination of erosion and dilation executed in that order. After this operation all objects regain their shape and size. However, small isolated objects are removed by erosion and are not restored in the subsequent dilation.

Closing is the opposite operation to opening. It combines dilation followed by erosion. It smoothes edges and connects objects placed close to each other.

There are many morphological filters available and some are quite complicated. They can all be used to extract relevant information from the binary images, however, at the expense of distorting the initial image (Figure 5.8). The final binary image is lastly the subject of quantitative analyses, in essence based on counting black pixels. Binary images are quantified using dedicated software, which in standard versions allows for

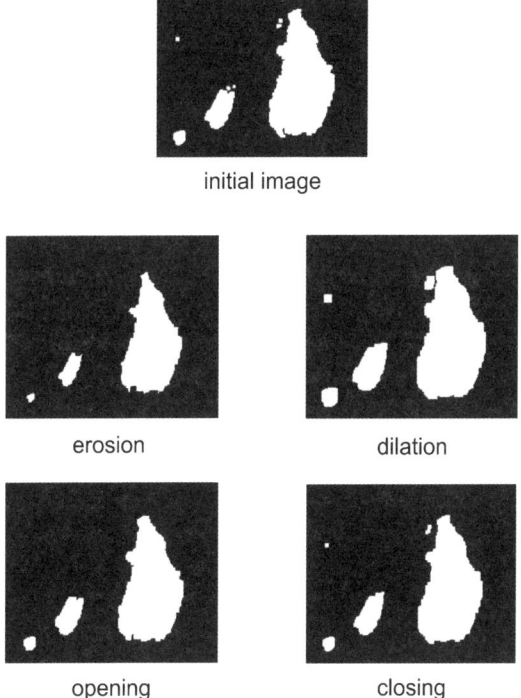

Figure 5.8 Morphological filters: erosion, dilation, opening and closing.

sizing particles, measuring their perimeter and computing their shape factors. One needs to remember, however, that the numbers obtained are only as good as the initial images. Also, it should be kept in mind that the precision of measuring geometrical features of the objects revealed in the images, here images of the nanoparticles, can be high, but only with regard to the ones pictured. If the purpose of our activity is to infer the features of an entire population, it is always better to analyze more randomly selected images than excel in the precision of analyzing a few of them.

5.3.4 Parameters Describing the Size, Size distribution and Shape of nanoparticles

According to a widely accepted, although not formalized, understanding nanoparticles are pieces of solids "with dimensions in the range from 1 to 100 nanometres". This definition is not very precise, as the size of 3-D objects can be described by one figure only in the case of either extremely elongated objects (length) or spherical (diameter). Nevertheless, it rightly drives the attention to the size of nanoparticles, which in more general terms can be defined more precisely by their volume, V. This in turn can be used to calculate the so-called equivalent diameter, D_e, which is the diameter of the spherical particle of the same volume, to reduce the dimensionality of the size parameter. Further, a collection of nanoparticles (powder) can be described in terms of the size distribution functions, $f(V_i)$ and $f(D_{e,i})$ and characterized by the mean values, $E(V)$ and $E(D_e)$. In a more precise approach, the standard deviation, $SD(V)$ and $SD(D_e)$ can be added or coefficients of variation $CV(V)$ and $CV(D_e)$ defined as E/SD.

The volume and in turn the equivalent diameter can be determined for individual nanoparticles from their images using techniques for 3-D reconstruction. Such a reconstruction usually requires at least three images taken at a different "illumination" angles. It should be remembered at this point that a single image is never sufficient to estimate the volume (size) of an object as it does not specify its shape. Estimates of the volume can be relatively easily obtained from obvious geometrical formula once the shape of nanoparticles can be approximated by one of the solids, such as spheres, rods, disks.

However, it must be remembered that the geometry, in particular the volume of the nanoparticles of interest, may vary and must be approached "statistically". It means that (a) some number of particles must be investigated to obtain meaningful estimates and (b) the investigated particles must be sampled randomly from their population. Methods of random sampling are described in a number of treatises on statistics. A very simple solution, not always applicable, which can be adopted when dealing with images is numbering them randomly and selecting every third for volume measurements.

In view of their distinct feature, which is a high fraction of surface atoms, nanoparticles should be characterized also in terms of the surface area, S.

Again, this parameter for individual nanoparticles can be determined from their images *via* 3-D reconstruction or using a formula based on the shape approximation. Once the surface is determined, a specific surface area, or in other words, surface-to-volume ratio, S_V, can be calculated which is of prime importance for analyses of the surface activity of nanoparticles.

Finally, it should be noted that in some applications the population of the nanoparticles is also described in terms of their density, N_V, *i.e.* their number in a given reference volume. Despite its apparent simplicity, the estimation of N_V requires quite a rigorous approach.

A description of the shape of nanoparticles is much more complicated than the description of their size. A standard approach is to classify them into some broad categories represented by conventional geometrical objects, such as spheres, cubes, disks, and needles. Spheres and cubes have their shape uniquely defined and do not need further specification. This is not true for the disks and needles, which can be more precisely described by the ratio of their thickness/length (t or l), to the radius of the maximum section, R.

Particle shape can also be described in terms of geometrical properties such as the degree of elongation, l/R, and of surface-area-to-volume ratio. Information about the shape is relatively easily retrieved from images. In performing such an exercise, one should remember the need for 3-D reconstruction.

5.4 Summary

The very definition of nanoparticles, referring to their size, underlines the importance of their geometry. In the context of the present book it should be noted that both the geometrical and the physical properties of nanoparticles may induce toxicity even in the case of otherwise inert materials. Thus experimental methods for investigating the structure and properties of nanoparticles have been presented in this chapter with the intention that more details on each method can be found elsewhere. The list of these methods is quite impressive and generally they allow for comprehensive characterization of any nanoparticles. However, the reader should be cautioned that none of them provides a complete set of information. For example, from X-ray diffraction one can obtain information on the average crystallite size. However, size distribution, shape and a degree of agglomeration cannot be determined just from X-ray diffractions. Similarly, laser diffraction allows one to characterize agglomerate size but not size of individual particles. In this context direct imaging using microscopic methods (TEM, SEM, AFM) seems to be one of the best options, providing information about the size, size distribution and shape of individual particles and agglomerates. However, the statistics in such studies is much worse (only a limited number of objects is measured) and frequently insufficient for statistically meaningful conclusions. Therefore, in order to have the full picture of the population of nanoparticles, it is recommended to use a set of techniques which enable one to characterize various aspects of particle size and properties.

Further Reading

1. L. V. Azaroff, *Elements of X-ray Crystallography*, McGraw-Hill Book Company, 1968.
2. D. Brandon and W. D. Kaplan, *Microstructural Characterization of Materials*, Wiley, 2nd edn, 2008.
3. B. D. Cullity, *Elements of X-ray Diffraction*, Addison-Wesley Publishing Company, Inc., 2nd edn, 1978.
4. S. Ghosh and D. Dimiduk, Computational Methods for Microstructure–Property Relationships, Springer, 2011.
5. J. Goldstein, D. E. Newbury, D. C. Joy, C. E. Lyman, P. Echlin, E. Lifshin, L. Sawyer and J. R. Michael, *Scanning Electron Microscopy and X-ray Microanalysis*, Springer, 2003.
6. K. J. Kurzydlowski and B. Ralph, *The Quantitative Description of the Microstructure of Materials*, CRC Press, Boca Raton, 1995.
7. E. Meyer, H. J. Hug and R. Bennewitz, *Scanning Probe Microscopy: The Lab on a Tip*, Springer, 2003.
8. S. Morita, Roadmap of Scanning Probe Microscopy, Springer, 2007.
9. J. Ohser and K. Shladitz, 3D Images of Material Structure. Processing and Analysis, Wiley-VCH, 2009.
10. V. K. Pecharsky and P. Y. Zavalij, Fundamentals of Powder Diffraction and Structural Characterization of Materials, Springer, 2005.
11. J. C. H. Spence, *High-Resolution Electron Microscopy*, Oxford University Press, 2003.
12. B. E. Warren, *X-ray Diffraction*, Dover Publications, Inc., New York, 1990.
13. R. Wiesendanger, Scanning Probe Microscopy: Analytical Methods, Springer, 1998.
14. D. B. Williams and C. B. Carter, Transmission Electron Microscopy – A Textbook for Materials Science, Springer, 2009.
15. L. Wojnar, Image Analysis: Application in Materials Engineering, CRC Press, 1998.
16. N. Yao and Z. L. Wang, *Handbook of Microscopy for Nanotechnology*, Springer, 2005.

CHAPTER 6

Nanoinformatics for Safe-by-Design Engineered Nanomaterials

CARLOS P. ROCA*[a], ROBERT RALLO[b], ALBERTO FERNÁNDEZ[a] AND FRANCESC GIRALT[a]

[a] Departament d'Enginyeria Quimica, Universitat Rovira i Virgili, Tarragona, Catalunya, Spain; [b] Departament d'Enginyeria Informatica i Matematiques, Universitat Rovira i Virgili, Tarragona, Catalunya, Spain
*E-mail: carlosp.roca@urv.cat

6.1 Introduction

Nanotechnology, often referred to as the next industrial revolution, has a considerable socioeconomic impact. Its benefits are expected to have significant effects on almost all industrial sectors and areas of society. In a recent communication, the European Commission defined the term nanomaterial as a natural, incidental or manufactured material containing particles, in an unbound state or as an aggregate or as an agglomerate and where, for 50% or more of the particles in the number size distribution, one or more external dimensions is in the size range 1–100 nm. According to this definition, a significant number of consumer products containing nanoscale components in their formulation (primarily silver, carbon, zinc, silica, titania and gold) can be currently found on the market.

The widespread use of nano-enabled products increases the risk of accidental releases and broadens potential nanomaterial exposure scenarios. Occupational exposure is also of concern due to the increased rate and volume of nanomaterial

RSC Nanoscience & Nanotechnology No. 25
Towards Efficient Designing of Safe Nanomaterials: Innovative Merge of Computational Approaches and Experimental Techniques
Edited by Tomasz Puzyn and Jerzy Leszczynski
© The Royal Society of Chemistry 2012
Published by the Royal Society of Chemistry, www.rsc.org

production. Therefore, the safe application of nanotechnology requires the careful consideration and understanding of the potential risks that may result from the introduction of engineered nanomaterials (ENMs) into the environment. Contrary to what occurs with natural nanoparticles, living organisms are not adapted to the presence of this new kind of stressor. Hence, a fundamental knowledge of the interactions of ENMs and biological systems has to be generated to ensure the correct and responsible use of ENMs and to minimize their potential environmental impact and adverse effects on human health. Compared with bulk chemicals, engineered nanomaterials have a very diverse landscape of possible structural configurations, including multiple combinations of core, surface and functional modifications. As a consequence of this heterogeneity, the study and understanding of nano–bio interactions is a challenging task, which requires a multidisciplinary approach involving environmental engineering and chemistry, material science, computer science, biology and medicine.

The understanding of the mechanisms that govern the adverse effects of ENMs on biological systems is of fundamental importance to implement new paradigms for the design of safe nanomaterials.[1,2] This chapter presents an integrated data analysis approach that combines ENM information with data describing the biological function of the target living organism.

6.2   Nanoinformatics for ENM Data Management

Data scarcity is a common situation that hampers the understanding of mechanisms which govern the impact of nanomaterials on biological systems. Data analysis, combined with the development of computational nanotoxicity, constitutes the fundamental paradigm for the discovery and understanding of nano–bio interaction mechanisms, and particularly nanoparticle-mediated toxicity. Typically, nanotoxicity models are developed from components identified by domain experts. However, this approach is inherently limited in the case of nanotoxicology where a complete mechanistic understanding has not been developed yet. In situations where domain knowledge is insufficient or lacking, data mining and knowledge extraction techniques (*e.g.*, signature discovery) constitute a promising approach. For instance, the identification of relevant ENM signatures from the analysis of high-throughput data sets requires a novel computational toxicology approach[3] which must face challenges such as data heterogeneity, lack of well-established experimental protocols and data variability/ uncertainty. As a response to these challenges, nanoinformatics emerges as the research field which integrates the use of information technologies for data management with computational algorithms for data analysis and modeling. Nanoinformatics can be defined as the science and practice of determining which information is relevant to the nanoscale science and engineering community, and then developing and implementing effective mechanisms for collecting, validating, storing, sharing, analyzing, modeling, and applying that information.†

---

† Nanoinformatics 2020 Roadmap; http://eprints.internano.org/607/

Nanoparticle information consists of unstructured data obtained from multiple experimental sources that are organized as a collection of highly dimensional heterogeneous data sets. Currently, most of the publicly available data are embedded in scientific papers in a format that is not suitable for automatic data retrieval and analysis. However, initiatives to build curated data repositories – containing high-quality data sets – are starting to provide machine-readable data suitable for computational studies. For example, the caBIG Nanotechnology Working Group funded by the National Cancer Institute (NCI) under the Integrative Cancer Research Initiative aims to develop a common vocabulary/ontology for nanoparticle data and to promote data sharing and data access protocols. Central to this effort is the development of a common ID schema for nanoparticles to facilitate interoperability between different ENM data repositories. Current efforts in this area are directed at the adoption of an approach similar to that used with Life Science Identifier[4] in order to identify bioinformatics data objects. The cancer Nanotechnology Laboratory portal is an implementation of the principles and technologies developed within the caBIG program. This portal is designed to facilitate data sharing activities by providing support for the annotation of ENM with physicochemical and *in vitro* characterizations, as well as the distribution of these characterizations and associated protocols in a secure fashion. Other public data repositories include the JRC NANOhub,‡ the NHECD§ knowledge repository, and the Nanomaterial–Biological Interactions Knowledgebase.¶

The lack of interoperability between the different sources of nanomaterial information constitutes a challenge that hinders efficient data-driven knowledge extraction strategies (*e.g.*, searches across multiple repositories). Interoperability among repositories can be implemented at different levels and with different purposes. For instance, semantic interoperability can be implemented *via* well-established and validated ontologies such as the Nanoparticle Ontology (NPO)[5] which defines the set of concepts (*i.e.*, standard vocabulary) and relationships that are required to represent nanoscale objects and their related information. Similarly, interoperability for data sharing can be achieved *via* the definition of standard file formats and data structures. For instance, the ISA-TAB Nano[6] standard defines a tabular text format for nanoparticle data exchange by extending the ISA-TAB (Investigation–Study–Assay)[7] file format with a description of the nanomaterial used in the study.

The characterization of ENM also plays a fundamental role in the development of computational models for ENM toxicity.[8,9] Nevertheless, often reported ENM characteristics (*e.g.*, size and surface charge) could vary depending on the experimental protocol and even the type of instrumentation used. To reduce the uncertainty associated with the utilization of different data

---

‡ http://www.nanpira.eu
§ http://nhecd.jrc.ec.europa.eu
¶ http://nbi.oregonstate.edu/

sets, the Minimum Information of Nanoparticle Characterization (MINChar) is a community effort to define standards that will ensure that ENMs are characterized in a way that facilitates the informed interpretation of data and cross-comparison of experimental results. The MINChar set includes the following nine parameters: agglomeration/aggregation, chemical composition, crystal structure, particle size, purity, shape, surface area, surface charge and surface chemistry.[10]

While computational chemistry and toxicology are fully developed for chemicals, they are still under development for nanoparticles. Initiatives like the U.S. Environmental Protection Agency (EPA) *ToxCast^{TM}* program and *OpenTox* have already identified the main issues (*i.e.*, ensuring unified access to distributed toxicological resources including data, computer models, validation and reporting) and provided the basis for the development of a predictive toxicology framework for chemicals. The study of these initiatives is therefore fundamental to defining the foundation of an integrated framework for predictive nanotoxicity.

6.3 Discovery of Nano–Bio Interaction Mechanisms for Safe-by-Design Strategies

The environmental and human health hazard potential associated with engineered nanomaterials has been poorly assessed. This is mainly due to a limited understanding of the toxicological mechanisms of action of ENMs. Recent efforts in the development of high-throughput screening (HTS) assays[11,12] open new opportunities for understanding how ENMs interact at cellular and sub-cellular levels, and to discover how specific physicochemical properties contribute to the identification of adverse effect pathways. However, preliminary studies on the environmental distribution of nanomaterials[13,14] delineate exposure scenarios involving much lower nanomaterial concentrations than the ones typically used for *in vitro* HTS nanotoxicity assays. Genomic and proteomic studies demonstrated that exposure to low concentrations of ENMs may disrupt basic biological functions at the cellular and sub-cellular levels[15–17] which does not always translate to observable cytotoxicity effects. These perturbations, however, may act as early indicators of much severe nanoparticle impacts. As a consequence, there is a need to relate ENM properties with the biological responses induced after exposure to low concentrations and with the adverse effects observed *in vitro* and *in vivo* at higher concentrations.

Understanding the mechanisms that govern the interactions of ENM with biological systems entails an integrated approach that has to combine intrinsic ENM information (*e.g.*, structure and physicochemical properties) with data describing the biological system. Figure 6.1 describes a conceptual framework to identify safety features that can be included within the design of a new generation of "safe" engineered nanomaterials. The first layer of the framework is related to data generation/gathering. Two separate entities,

ENMs and biological systems, are the main data sources for the discovery of nano–bio interaction mechanisms. ENM characterization data should include information regarding structure and physicochemical properties. These properties should be characterized as produced (*i.e.*, pristine ENMs) and also in relation to their *in situ* transformations (*e.g.*, ageing). Similarly, a variety of "-omics" techniques can be used to characterize diverse building blocks of biological systems (*e.g.*, genome, proteome, metabolome). Finally, integrated information regarding the effects observed when the biological systems are exposed to the ENMs (*e.g.*, toxicity end-points, alterations on cell signaling pathways, *etc.*) is also required.

The process of data collection involves the generation of experimental data and the gathering of information from existing sources (*e.g.*, data repositories and literature crawlers). The second layer is related to computational data analyses and modeling. These processes involve the application of nanoinformatics and bioinformatics techniques for ENMs and biological data, respectively (Figure 6.1). The outcome of this merging is the development and validation of computational nanotoxicity models which describe the observed effects when organisms are exposed to ENMs. It should be noted that models are developed by simultaneously considering the properties of ENMs and the "omics" alterations induced on the target organism. Mechanistic knowledge is extracted at the third level of this conceptual model by using advanced techniques such as complex network analysis. The outcome of this 3rd level analysis is the discovery of rule sets that represent/explain the interaction mechanisms between ENMs and biological systems. Finally, the understanding of these interaction mechanisms should guide the design of safe ENMs.

Two different aspects of the techniques required for the implementation of the conceptual framework depicted in Figure 6.1 are included below as

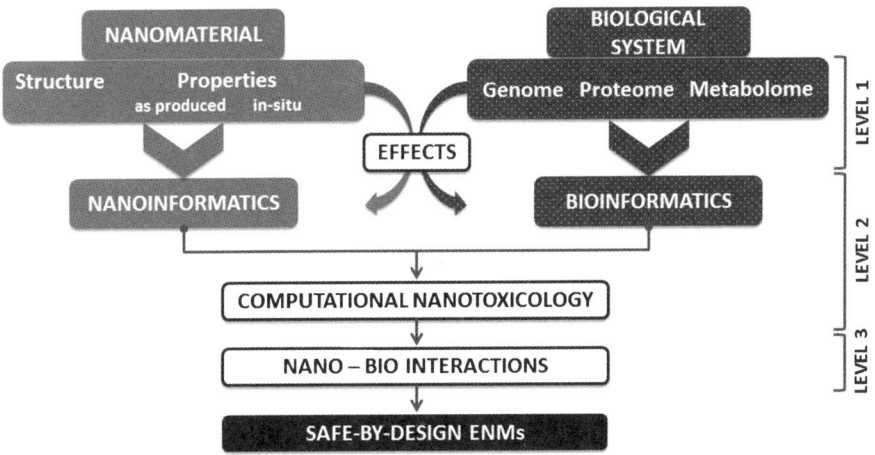

Figure 6.1 Components of a conceptual framework for safe-by-design ENMs.

two case studies. The purpose of these examples is to illustrate the potential of the proposed approach with data already published and analyzed in the literature.

6.3.1   Case Study 1: Self-organizing Maps (SOM) Analysis of ENM Data Sets

Data mining is a fundamental step for nanotoxicity model development since computational nanotoxicology relies heavily on available data. Therefore, it is essential to extract knowledge about the structure and relationships between the variables describing the analyzed system by exploring the information space. There is a wide array of data mining techniques[18] which can be used to explore ENM data sets and to convert information into useful mechanistic knowledge and models *via* the appropriate etiological hypothesis.

Data clustering techniques constitute the most commonly used tools for multidimensional data exploration and visualization. For instance, through row and/or column clustering heat-maps[19] provide ordered representations of data that facilitate the identification of similarity patterns. Alternative approaches, providing more straightforward visualization and interpretation, can be developed using diverse machine-learning techniques.

The SOM approach[20] is a clustering technique that can be used to visualize patterns embedded within ENM data while preserving their topological properties. SOM is based on an unsupervised competitive learning approach with the adaptation process being entirely data-driven and where map units compete to become specific detectors of certain data features. An n-dimensional weight vector (prototype) represents each map unit, where n is equal to the dimension of the input space. Each unit has a topological neighborhood determined by the shape of the SOM grid lattice, which can be either rectangular or hexagonal. The granularity (*i.e.* the size) of the map determines its accuracy and generalization capabilities. Once the adaptation completes, the SOM approaches the clustering structure of the data. This data structure can be visualized over the SOM grid by displaying the distances between the prototype vectors (distance matrices) or by projecting features onto the SOM space (component planes). Distance matrices provide a two-dimensional representation of the clustering structure of data. Labeling the map with auxiliary data enhances the visualization of these clusters. Details of the SOM algorithm and its application to the analysis of ENM HTS data sets can be found elsewhere.[21,22]

The use of SOM for mining ENM datasets is illustrated using the data presented by Puzyn *et al.*[23] that corresponds to the toxicity (half maximal effective concentration, EC_{50}) for a set of 17 metal oxide (MeO) nanoparticles with respect to bacteria *Escherichia coli.* La_2O_3 was excluded from the analysis since their nanodescriptors were incomplete. The final set of 16 MeO nanoparticles were characterized by 12 descriptors that describe the variability

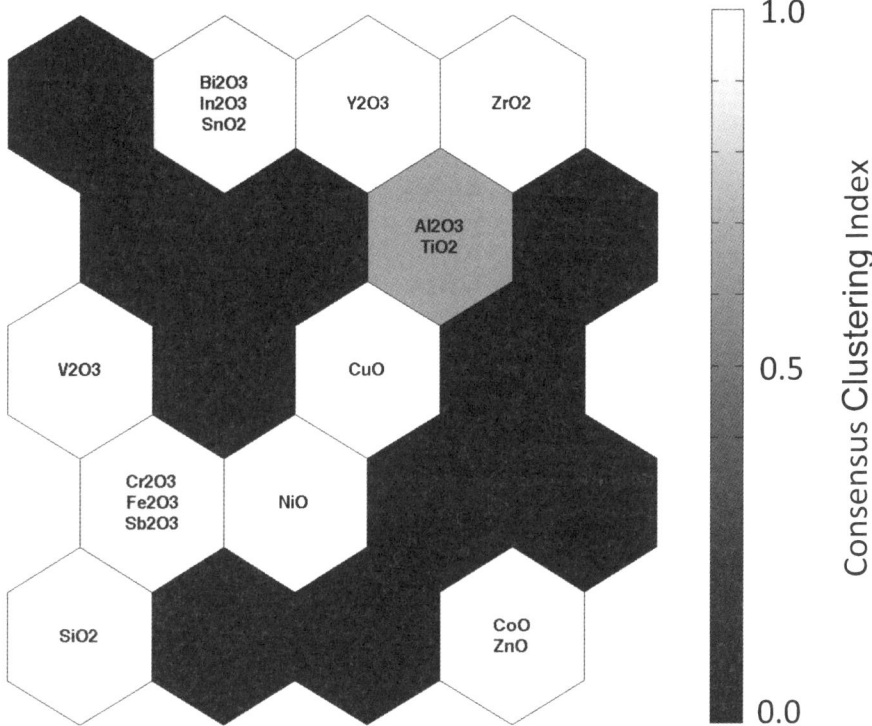

Figure 6.2 SOM clustering of the 12 descriptors corresponding to the set of 16 MeO nanoparticles.[23] The stability of each cluster (SOM unit) is indicated using a grayscale code according to the value of its consensus clustering index.

in their structure.[23] Nanoparticle descriptors were computed using quantum-chemical PM6 methods. Details of the experimental data generation and nanoparticle descriptor calculations can be found elsewhere.[23]

Remarkably, the MeO having the most significant impact on bacteria viability (*i.e.*, $1/EC_{50} > 3$) are located in a close neighborhood of the SOM, as shown by Figure 6.2. Although the toxicity information was not used during SOM development, the clustering of nanodescriptors is able to capture the toxicity pattern. Inspection of similarities in the projections of each descriptor onto the SOM lattice (component planes in Figure 6.3) reveals that some of the nanodescriptors contain redundant information. Examples of redundant descriptors that can be identified from the SOM projection are CA (area of the oxide cluster) with CV (volume of the oxide cluster); and TE (total energy of the oxide cluster) with EE (electronic energy of the oxide cluster). These results confirm that the SOM approach can be used to identify redundancy, *i.e.*, for dimensionality reduction.

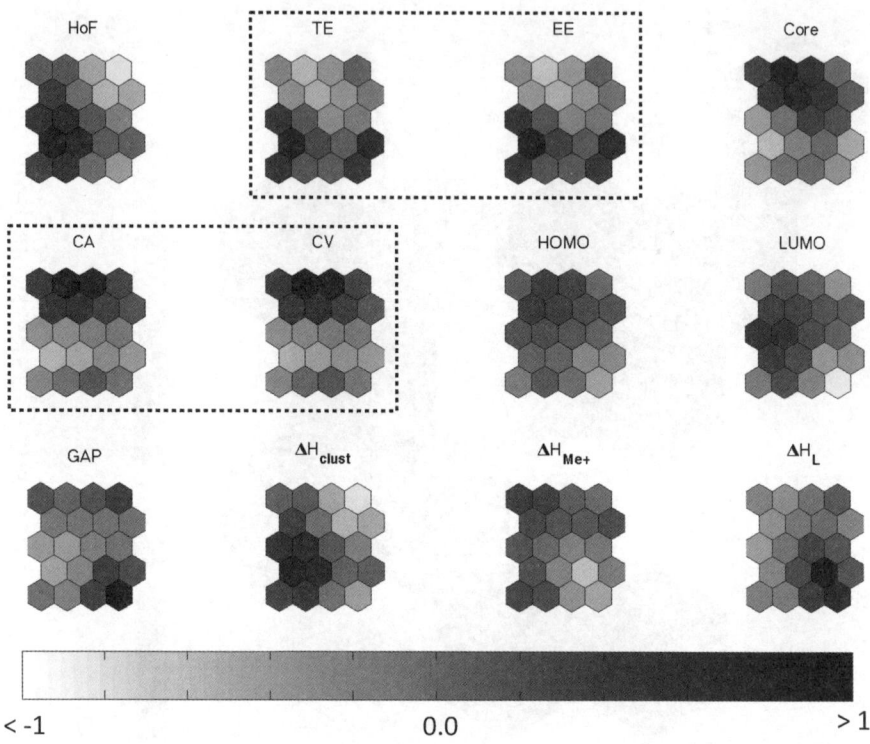

Normalized Descriptors

Figure 6.3   Projection of the normalized values of the set of 12 descriptors onto the SOM lattice (component planes). Boxes indicate descriptors that account for the same type of information (redundant).

The SOM approach can also be used for the integrated analysis of the nano–bio interactions (Figure 6.4) by adding information regarding the biological endpoint (*i.e.*, $1/EC_{50}$) to the initial set of nanodescriptors. A toroidal SOM with a hexagonal lattice of 5 × 4 units was developed from the normalized 12 descriptors and the EC_{50} endpoint. Figure 6.5 compares the projection of each nanodescriptor with the projection of the toxicity endpoint. Visual inspection of Figure 6.5 reveals that ΔH_L (lattice energy of the oxide) and ΔH_{Me+} (enthalpy of formation of a gaseous cation) are the descriptors with a distribution more similar to the endpoint projection. Thus, the SOM approach also facilitates the identification of candidate descriptors for developing nanotoxicity models. The above results are in agreement with the published model.[23]

Notwithstanding the limited information available in the above example, the SOM provides a suitable tool for the analysis and data mining of high-dimensional ENM data sets. The topology preserving properties of the SOM analysis facilitate the identification of similarities between ENMs (*i.e.*, formation of ENM categories) which can set the groundwork for implementing read-across methods.

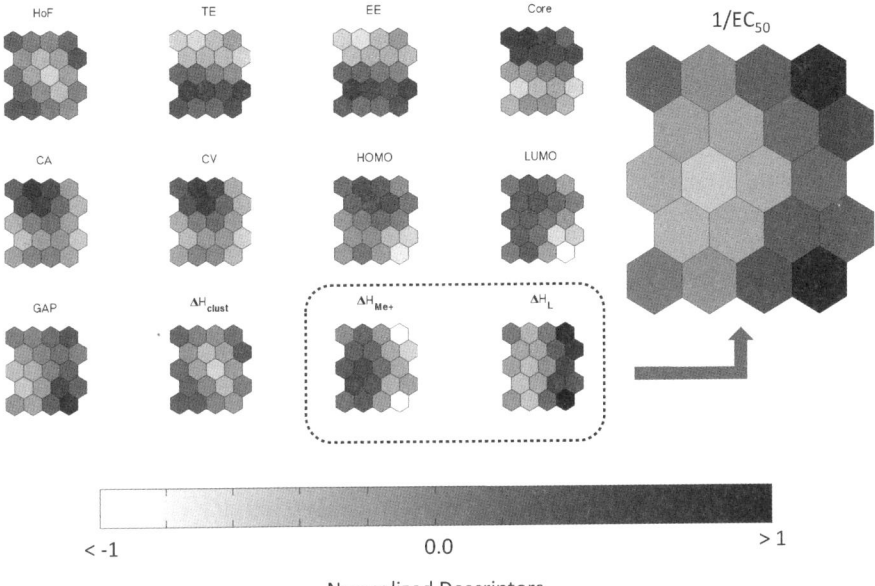

Figure 6.4 Component planes corresponding to the integrated SOM analysis of ENM descriptors and the biological endpoint. All variables (descriptors and endpoint) are normalized. The two descriptors that are most similar to the endpoint are enclosed in a dashed box.

6.3.2 Case Study 2: Systems Biology Approach for the Analysis of Nano–Bio Interactions

Basic and applied sciences, such as biology and chemistry, share many similarities from the point of view of systems organization. General principles that govern the structure and behavior of biological and chemical systems can be discovered with the help of systems engineering and computer science.[26,27] The growth in size and variety of biological and chemical databases has made the study and interpretation of the available information increasingly challenging. Many biological systems, ranging from cells to populations, can be described in terms of networks of interactions between their individual constituents. Additionally, modeling these systems as complex networks facilitates the understanding of their temporal behavior, with the help of complex systems tools such as non-linear dynamics and stochastic models. Despite the complexity of real networks, it has been found in many cases that their topology and evolution are governed by common organizational principles.[28,29]

An example of the application of complex network theory to biology which has attracted a great deal of attention is the study of food webs, where nodes represent species in an ecosystem and connecting links indicate that one species preys on the other. Although this kind of network of interactions between species is known to be well described by their topological properties, there is still much to be learned about their dynamical behavior.[30–34] The same applies to the chemical space, where complex networks can provide novel insights in a wide range of applications such as, for example, the relationships between chemicals linked by chemical reactions.[35]

Neuroscience is another field where the complex networks approach has found wide application. The best known example is the network of synaptic connections between neurons in the nematode *Caenorhabditis elegans*.[36,37] The network structure of the brain at the scales of functional areas and pathways has been also investigated.[38]

In the cell, all kinds of structural and functional processes are ruled by the intricate interaction of genes, proteins and other molecules. A classic example of a cellular network is the network of metabolic pathways,[39–42] which is a representation of chemical substrates, products and enzymes connected by metabolic reactions. Other interesting examples of networks in molecular biology are the networks of protein interactions[43–45] and the networks of gene regulation.[46,47] The expression of a gene is controlled, among other factors, by the presence of regulatory proteins, in such a manner that the genome itself forms a control network whose nodes are genes and links represent the regulatory dependence between genes.

The proposed systems biology approach with complex networks is illustrated by the analysis of the microarray data collected in a high-throughput nanotoxicity assay on the nematode *C. elegans* exposed to Ag nanoparticles.[48] The network representation of the genome is used to analyze the properties of the subset of most differentially expressed genes resulting from the experiment. To this end, a network representation[49] of the entire genome of *C. elegans* has been built, where the nodes are the gene transcripts present in the microarray, connected when they are annotated to the same pathway. The connecting links between nodes are weighted by the number of shared pathways between the two genes. Self-links are excluded from the analysis. In the context of this genetic network, the subset of the most differentially expressed genes identified in the microarray experiment constitutes a subnetwork. Two types of data are thus needed in this approach, namely gene expression profiles and gene-pathway annotations for *C. elegans*.

Genetic expression profiles were obtained from publicly available micro-array raw data[48] for the wild type and after 24 h of exposure to nano-Ag at a concentration of 0.1 mg L^{-1}. Hybridization was performed with the Affymetrix GeneChip *C. elegans* Genome Array (Affymetrix, Santa Clara, CA).[48] The raw data contains 22 625 probe sets against 22 150 transcripts, among which 20 783 have Entrez Gene annotation, resulting in a total number of 17 088 unique Entrez Gene IDs. These data were processed with R[50] and the Bioconductor package.[51] Preprocessing was performed following the robust multiarray average (RMA) methodology.[52]

The 100 most differentially (over-or under-) expressed genes were selected. The scatter plot of differential *vs.* average expression values of all gene transcripts is depicted in Figure 6.5. It shows that most of the 100 selected genes have more than 4-fold differential expression. Note that this dataset has the special feature of not having replicates, hence no measure of statistical significance can be used to discriminate between different transcripts corresponding to the same gene and, particularly, to select the most differentially expressed genes. Data were also preprocessed with the GeneChip RMA (GCRMA)[53] and the variance stabilization normalization (VSN)[54] methods, obtaining compatible results (more than 60% overlap between selected gene transcripts).

Gene-pathway annotations for *C. elegans* were obtained from the Kyoto Encyclopedia of Genes and Genomes (KEGG).[55,56] Pathway annotations for this species are only available for approximately 10% of genes. This is a limitation that will be encountered for most end-points and hence it is appropriate for the current illustration of the methodology. The pathway *cel01100*, which is a global map of metabolic pathways, has also been excluded. As a result, the genetic network that will be considered as representative of the genome of *C. elegans* comprises 2547 gene transcripts, which correspond to 1857 unique gene IDs connected by 123 pathways. The subset of the most differentially expressed genes is reduced, accordingly, from 100 to 19 gene transcripts, all of them associated with different gene IDs. The network has a

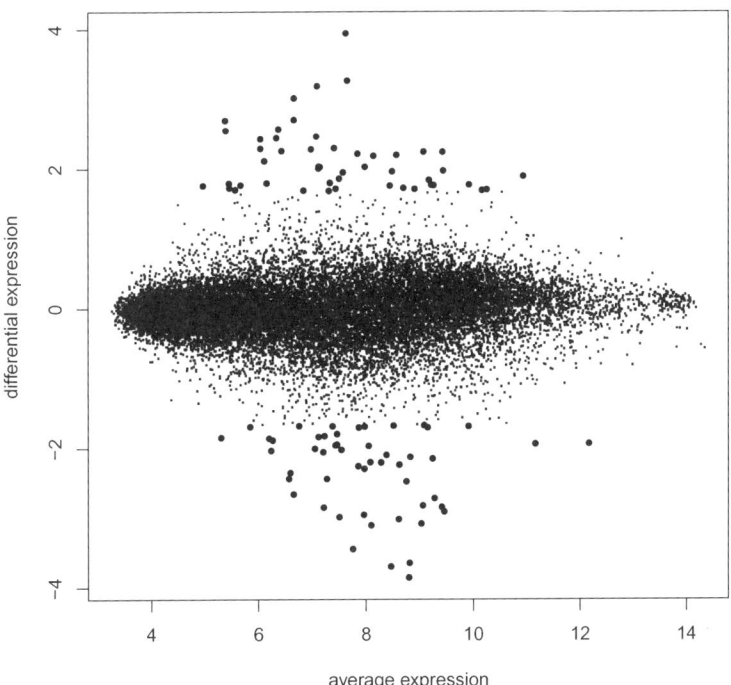

Figure 6.5   Scatter plot of $\log_2$ values of differential *vs.* average expression of *C. elegans* gene transcripts, when the nematode is exposed to Ag nanoparticles.[46] Larger points correspond to the 100 most differentially (over- or under-) expressed gene transcripts.

main connected component, *i.e.*, a large set to which the majority of nodes belong and in which there exists a connecting path between every pair of nodes. The main connected component includes 2502 gene transcripts and contains all the 19 most differentially expressed genes. The graphical representation of this main connected component obtained with the package *Cytoscape*[57] is depicted in Figure 6.6 with the 19 most differentially expressed genes highlighted. The network is densely connected, with both mean and median degree (average number of connections per node) larger than 100. The network topology is clearly not simple, with important differences in the connectivity patterns between nodes and also between regions of the network.

The properties of the subset of 19 most differentially expressed nodes are investigated within the context of the proposed network representation of the *C. elegans* genome by looking at biologically meaningful topological differences. The analysis of the network topology was performed with the R package *igraph*.[58]

One of the most important topological properties of any network is the degree distribution, *i.e.*, the distribution of the number of connections for each node. Figure 6.7 shows the degree distribution of the 2502 gene transcripts and also that of the subset of the 19 most differentially expressed genes. The large

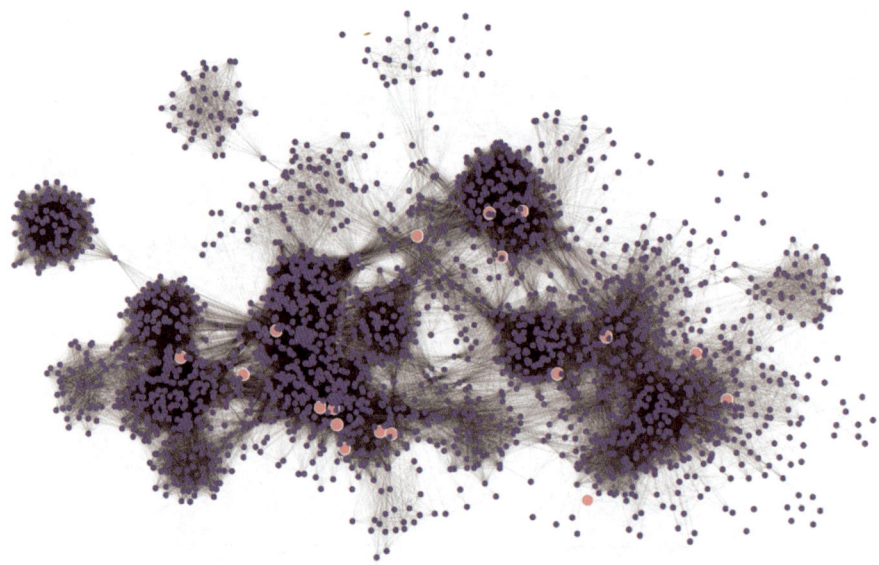

Figure 6.6 Graphical representation of the main connected component of the network formed by the annotated gene transcripts of *C. elegans*. Highlighted nodes (pink) show the 19 most differentially expressed genes, as a result of a nanotoxicity experiment with Ag nanoparticles.[46] In this network representation genes are connected when they are annotated to the same pathway, which gives rise to a large and non-trivial pattern of connectivity.

heterogeneity in the number of connections that gene transcripts have is remarkable. On the other hand, it is not clear that the subset under study features any relevant difference concerning the degree distribution. A statistical test yields the result that the difference between the main connected component and the subset is, in this respect, not significant (Kolmogorov–Smirnov statistic, two-sided, distribution obtained by bootstrap of 10 000 samples, $p = 0.267$).

Another important metric to characterize the network is the distance between nodes, both for the entire network and for the subset of most differentially expressed genes. The distance between nodes is defined as the minimum path length among all the paths connecting them. By definition, each

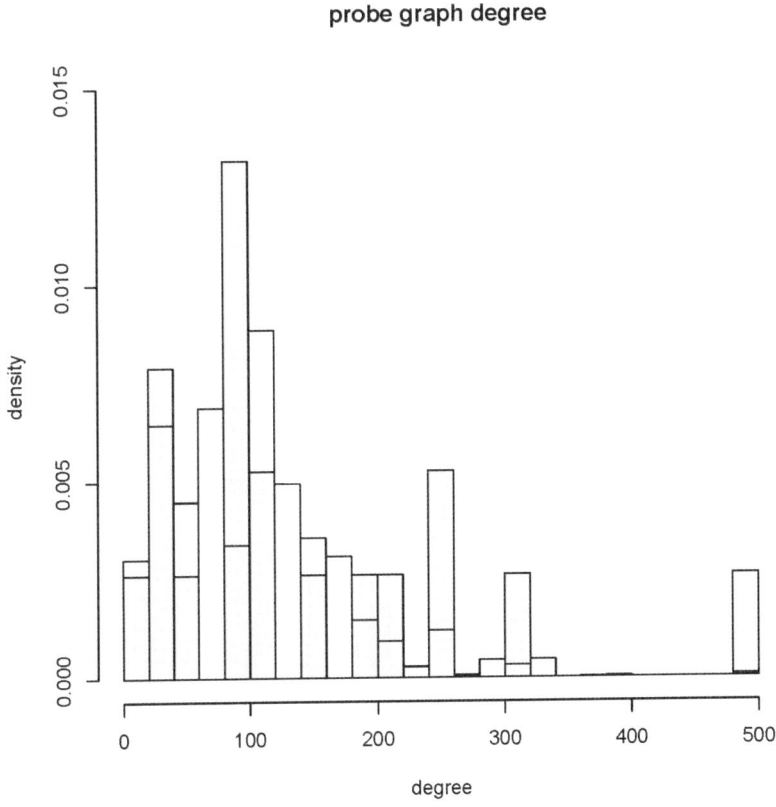

Figure 6.7   Distribution of degree (number of connections) for all nodes (gene transcripts) of the main connected component of *C. elegans* genetic network (in white), compared to the degree distribution corresponding to the 19 most differentially expressed genes when the nematode is exposed to Ag nanoparticles (in gray). Note the large connectivity and also the great heterogeneity in both cases. A statistical test yields the result that the difference between the distributions is not significant ($p = 0.267$, see main text).

link has a length equal to the inverse of its weight, *i.e.*, the length of a link connecting two gene transcripts is the inverse of the number of pathways shared by the two genes.

Figure 6.8 shows the histogram of distances among gene transcripts in the full network compared to the distances between the 19 most differentially expressed genes. In this case the most differentially expressed genes are nearer than genes in the full network. A statistical test confirms that this difference is significant with $p = 0.027$ (Kolmogorov–Smirnov statistic, two-sided, distribution obtained by bootstrap of 10 000 samples).

To interpret the meaning of this proximity among the 19 most differentially expressed genes, all the shortest paths connecting those 19 nodes are considered and the number of times each pathway occurs in a shortest path counted. Only 48 of the 123 *C. elegans* pathways appear at least once in a

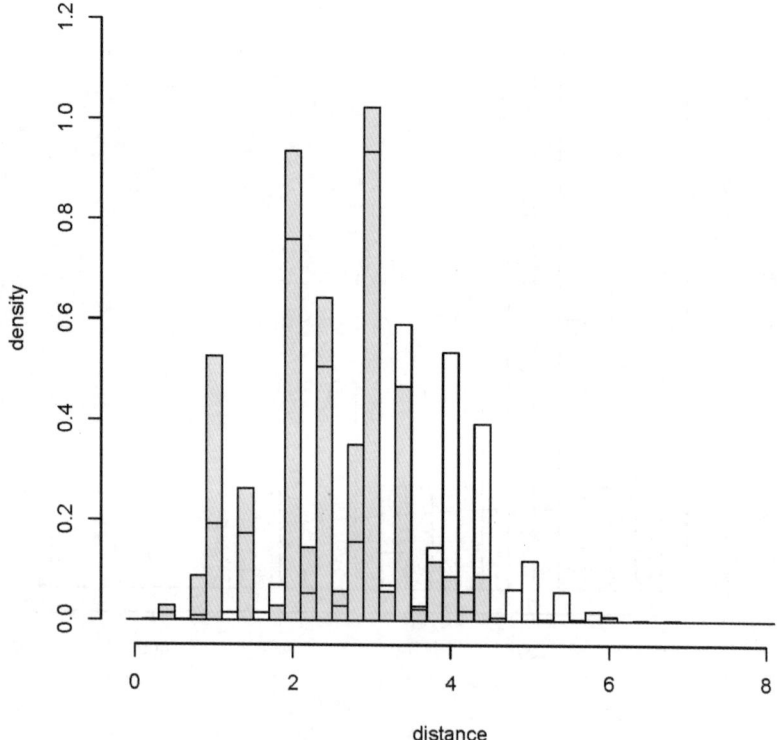

Figure 6.8   Distribution of distances between any two nodes (gene transcripts) of the main connected component of *C. elegans* genetic network (in white), compared to the distances among the 19 most differentially expressed genes when the nematode is exposed to Ag nanoparticles (in gray). The most differentially expressed genes appear to be closer among them. A statistical test confirms that the difference is significant ($p = 0.027$, see main text).

shortest path. Figure 6.9 shows the cumulative density distribution of the number of occurrences of those 48 pathways. Some of them (27%) appear more than 1000 times and half of them appear less than 300 times. Attention is focused on those occurring most frequently since they are the most likely candidates to be related to effects produced by the exposure of the nematode to Ag nanoparticles. Table 6.1 shows the top ten pathways emerging from this analysis. Four of them (highlighted in Table 6.1) are pathways associated with xenobiotic metabolism and related cellular processes. Very remarkably, these four pathways do not show up so prominently when the 100 most differentially expressed genes are considered separately. Table 6.1 shows that only four genes in the 19-subset are annotated to any of the four pathways, and none of them has been associated with toxicity effects, according to the WormBase database.[59] Their differential expression values are not particularly large either. The genes *cpr-1*, *cpr-2*, *fmo-2* and *vha-3* have differential expression values of -1.939, -1.706, -2.439 and -2.478, expressed as $\log_2$ of the ratio (exposed/wild type), respectively, and are ranked at positions 64th, 88th, 28th and 24th in the decreasing list of differential expression values.

The current analysis suggests that microarray data may contain biologically relevant information which remains hidden unless relationships between genes are considered on a broader scale than gene-by-gene. In network theory terms, this is equivalent to taking into account gene connections over the first neighborhood, *i.e.*, looking for phenomena which take place at mesoscopic network scales. From a biochemical perspective, this approach corresponds to focusing on effects or perturbations which involve groups of genes in related pathways, instead of targeting specific genes or particular pathways.

Table 6.1 Top ten pathways connecting the 19 most differentially expressed genes in the nanotoxicity experiment on *C. elegans*. Highlighted rows correspond to pathways which may be related to the exposure to nano-Ag. The rightmost column lists which of the 19 most differentially expressed genes are annotated to each of the highlighted pathways.

Kegg code	Occurrences	Name	Genes
cel00982	3036	Drug metabolism – cytochrome P450	*fmo-2*
cel03060	2161	Protein export	
cel04142	1826	Lysosome	*cpr-1, cpr-2, vha-3*
cel00040	1796	Pentose and glucuronate interconversions	
cel04145	1774	Phagosome	*vha-3*
cel00240	1565	Pyrimidine metabolism	
cel00983	1547	Drug metabolism – other enzymes	
cel00500	1456	Starch and sucrose metabolism	
cel04020	1429	Calcium signaling pathway	
cel00230	1337	Purine metabolism	

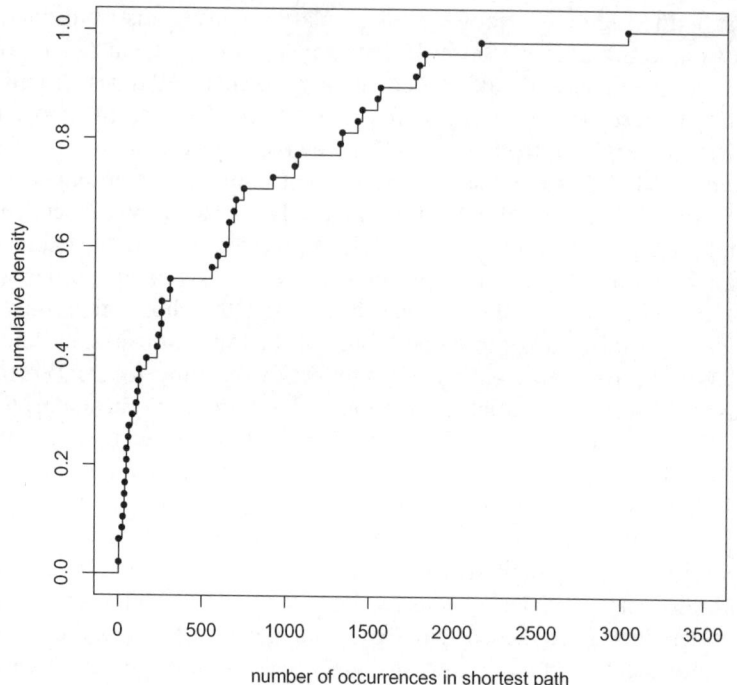

number of occurrences in shortest path

Figure 6.9 Cumulative density distribution of the number of occurrences of *C. elegans* pathways in shortest paths between the 19 most differentially expressed genes when the nematode is exposed to Ag nanoparticles.

6.4 Conclusions

The implementation of safe-by-design strategies requires an understanding of the biological processes triggered by the exposure to ENMs. The discovery of the mechanisms that govern these nano–bio interactions involves data on both nanomaterials and biological systems. Although specialized data repositories for ENMs are starting to emerge, there is still a need to develop and validate the computational basis for robust nanotoxicity models. The new field of nanoinformatics, together with well-established bioinformatics methods, provides the optimal framework for the integrated analysis of ENM and biological data sets. The mechanistic knowledge gained from this analysis will be fundamental for a new generation of safe-by-design ENMs.

Acknowledgements

This research was financially supported by the Generalitat de Catalunya (2009SGR-1529), AGAUR (InNaTox, 2010 CONES-00010) and the Spanish Ministry of Science and Innovation (MICINN, CTM2011-24303).

References

1. S. S. Tinkle, *Wiley Interdiscip. Rev.: Nanomed. Nanobiotechnol.*, 2010, **2**, 88–98.
2. S. George, S. Pokhrel, T. Xia, B. Gilbert, Z. Ji, M. Schowalter, A. Rosenauer, R. Damoiseaux, K. A. Bradley, L. Maedler and A. Nel, *ACS Nano*, 2010, **4**, 15–29.
3. R. J. Kavlock, G. Ankley, J. Blancato, M. Breen, R. Conolly, D. Dix, K. Houck, E. Hubal, R. Judson, J. Rabinowitz, A. Richard, R. W. Setzer, I. Shah, D. Villeneuve and E. Weber, *Toxicol. Sci.*, 2008, **103**, 14–27.
4. T. Clark, S. Martin and T. Liefeld, *Briefings Bioinf.*, 2004, **5**, 59–70.
5. D. G. Thomas, R. V. Pappu and N. A. Baker, *J. Biomed. Inf.*, 2011, **44**, 59–74.
6. D. G. Thomas, F. Klaessig, S. L. Harper, M. Fritts, M. D. Hoover, S. Gaheen, T. H. Stokes, R. Reznik-Zellen, E. T. Freund, J. D. Klemm, D. S. Paik and N. A. Baker, *Wiley Interdiscip. Rev.: Nanomed. Nanobiotechnol.*, 2011, **3**, 511–532.
7. P. Rocca-Serra, M. Brandizi, E. Maguire, N. Sklyar, C. Taylor, K. Begley, D. Field, S. Harris, W. Hide and O. Hofmann, *Bioinformatics*, 2010, **26**, 2354.
8. D. B. Warheit, *Toxicol. Sci.*, 2008, **101**, 183–185.
9. M. Auffan, J. Rose, J.-Y. Bottero, G. V. Lowry, J.-P. Jolivet and M. R. Wiesner, *Nat. Nanotechnol.*, 2009, **4**, 634–641.
10. J. W. Card and B. A. Magnuson, *J. Food Sci.*, 2009, **74**, VI–VII.
11. M. Andersen and D. Krewski, *Toxicol. Sci.*, 2009, **107**, 324.
12. S. George, T. Xia, R. Rallo, Y. Zhao, Z. Ji, S. Lin, X. Wang, H. Zhang, B. France, D. Schoenfeld, R. Damoiseaux, R. Liu, S. Lin, K. A. Bradley, Y. Cohen and A. E. Nel, *ACS Nano*, 2011, **5**, 1805–1817.
13. F. Gottschalk, T. Sonderer, R. Scholz and B. Nowack, *Environ. Sci. Technol.*, 2009, **43**, 9216–9222.
14. R. J. Miller, H. S. Lenihan, E. B. Muller, N. Tseng, S. K. Hanna and A. A. Keller, *Environ. Sci. Technol.*, 2010, **44**, 7329–7334.
15. Y.-Y. Tsai, Y.-H. Huang, Y.-L. Chao, K.-Y. Hu, L.-T. Chin, S.-H. Chou, A.-L. Hour, Y.-D. Yao, C.-S. Tu, Y.-J. Liang, C.-Y. Tsai, H.-Y. Wu, S.-W. Tan and H.-M. Chen, *ACS Nano*, 2011, **5**, 9354–9369.
16. N. Hanagata, F. Zhuang, S. Connolly, J. Li, N. Ogawa and M. Xu, *ACS Nano*, 2011.
17. L. Peng, A. J. Barczak, R. A. Barbeau, Y. Xiao, T. J. LaTempa, C. A. Grimes and T. A. Desai, *Nano Lett.*, 2010, **10**, 143–148.
18. F. Masulli and S. Mitra, *Information Fusion*, 2009, **10**, 211–216.
19. M. B. Eisen, P. T. Spellman, P. O. Brown and D. Botstein, *Proc. Natl. Acad. Sci. U. S. A.*, 1998, **95**, 14863–14868.
20. T. Kohonen, *Proc. IEEE*, 1990, **78**, 1464–1480.
21. R. Rallo, B. France, R. Liu, S. Nair, S. George, R. Damoiseaux, F. Giralt, A. Nel, K. Bradley and Y. Cohen, *Environ. Sci. Technol.*, 2011, **45**(4), 1695–1702.

22. R. Damoiseaux, S. George, M. Li, S. Pokhrel, Z. Ji, B. France, T. Xia, E. Suarez, R. Rallo, L. Mädler, Y. Cohen, E. M. V. Hoek and A. Nel, *Nanoscale*, 2011, **3**, 1345–1360.
23. T. Puzyn, B. Rasulev, A. Gajewicz, X. Hu, T. P. Dasari, A. Michalkova, H.-M. Hwang, A. Toropov, D. Leszczynska and J. Leszczynski, *Nat. Nanotechnol.*, 2011, **6**, 175–178.
24. S. Monti, P. Tamayo, J. Mesirov and T. Golub, *Mach. Learn.*, 2003, **52**, 91–118.
25. H. Shimodaira, *Ann. Stat.*, 2004, **32**, 2616–2641.
26. L. H. Hartwell, J. J. Hopfield, S. Leibler and A. W. Murray, *Nature*, 1999, **402**, C47–52.
27. M. E. Csete and J. C. Doyle, *Science*, 2002, **295**, 1664–1669.
28. R. Albert and A.-L. Barabási, *Rev. Mod. Phys.*, 2002, **74**, 47–97.
29. L. A. N. Amaral and J. M. Ottino, *Chem. Eng. Sci.*, 2004, **59**, 1653–1666.
30. R. V. Solé and J. M. Montoya, *Proc. R. Soc. London, Ser. B*, 2001, **268**, 2039–2045.
31. J. Camacho, R. Guimerà and L. A. N. Amaral, *Phys. Rev. Lett.*, 2002, **88**, 228102.
32. J. A. Dunne, R. J. Williams and N. D. Martinez, *Proc. Natl. Acad. Sci. U. S. A.*, 2002, **99**, 12917–12922.
33. S. L. Pimm, *Food Webs*, University of Chicago Press, Chicago, 2nd edn, 2002.
34. P. Jordano, J. Bascompte and J. M. Olesen, *Ecol. Lett.*, 2003, **6**, 69–81.
35. B. A. Grzybowski, K. J. M. Bishop, B. Kowalczyk and C. E. Wilmer, *Nat. Chem.*, 2009, **1**, 31–36.
36. J. G. White, E. Southgate, J. N. Thompson and S. Brenner, *Phil. Trans. R. Soc. London*, 1986, **314**, 1–340.
37. R. Milo, S. Shen-Orr, S. Itzkovitz, N. Kashtan, D. Chklovskii and U. Alon, *Science*, 2002, **298**, 824–827.
38. O. Sporns, *Complexity*, 2002, **8**, 56–60.
39. D. A. Fell and A. Wagner, *Nat. Biotechnol.*, 2000, **18**, 1121–1122.
40. H. Jeong, B. Tombor, R. Albert, Z. N. Oltvai and A.-L. Barabási, *Nature*, 2000, **407**, 651–654.
41. J. Stelling, S. Klamt, K. Bettenbrock, S. Schuster and E. D. Gilles, *Nature*, 2002, **420**, 190–193.
42. R. Guimerà and L. A. N. Amaral, *Nature*, 2005, **433**, 895–900.
43. H. Jeong, S. Mason, A.-L. Barabási and Z. N. Oltvai, *Nature*, 2001, **411**, 41–42.
44. S. Maslov and K. Sneppen, *Science*, 2002, **296**, 910–913.
45. G. Palla, I. Derényi, I. Farkas and T. Vicsek, *Nature*, 2005, **435**, 814–818.
46. N. Guelzim, S. Bottani, P. Bourgine and F. Kepes, *Nat. Genet.*, 2002, **31**, 60–63.
47. S. Shen-Orr, R. Milo, S. Mangan and U. Alon, *Nat. Genet.*, 2002, **31**, 64–68.
48. J.-y. Roh, S. J. Sim, J. Yi, K. Park, K. H. Chung, D.-y. Ryu and J. Choi, *Environ. Sci. Technol.*, 2009, **43**, 3933–3940.
49. M. E. J. Newman, *SIAM Rev.*, 2003, **45**, 167–256.

50. R Development Core Team, *R: A language and environment for statistical computing*, R Foundation for Statistical Computing, Vienna, Austria, 2008, http://www.R-project.org.
51. R. C. Gentleman, V. J. Carey, D. M. Bates, B. Bolstad, M. Dettling, S. Dudoit, B. Ellis, L. Gautier, Y. Ge, J. Gentry, K. Hornik, T. Hothorn, W. Huber, S. Iacus, R. Irizarry, F. Leisch, C. Li, M. Maechler, A. J. Rossini, G. Sawitzki, C. Smith, G. Smyth, L. Tierney, J. Y. H. Yang and J. Zhang, *Genome Biol.*, 2004, **5**, R80.
52. R. A. Irizarry, *Nucleic Acids Res.*, 2003, **31**, 15e–15.
53. Z. Wu, R. A. Irizarry, R. Gentleman, F. Martinez-Murillo and F. Spencer, *J. Am. Stat. Assoc.*, 2004, **99**, 909–917.
54. W. Huber, A. von Heydebreck, H. Sultmann, A. Poustka and M. Vingron, *Bioinformatics*, 2002, **18**, S96–S104.
55. M. Kanehisa, *Nucleic Acids Res.*, 2000, **28**, 27–30.
56. M. Kanehisa, S. Goto, Y. Sato, M. Furumichi and M. Tanabe, *Nucleic Acids Res.*, 2012, **40**, D109–114.
57. M. E. Smoot, K. Ono, J. Ruscheinski, P.-L. Wang and T. Ideker, *Bioinformatics*, 2011, **27**, 431–432.
58. G. Csardi and T. Nepusz, *InterJournal*, 2006, 1695.
59. T. W. Harris, I. Antoshechkin, T. Bieri, D. Blasiar, J. Chan, W. J. Chen, N. De La Cruz, P. Davis, M. Duesbury, R. Fang, J. Fernandes, M. Han, R. Kishore, R. Lee, H.-M. Müller, C. Nakamura, P. Ozersky, A. Petcherski, A. Rangarajan, A. Rogers, G. Schindelman, E. M. Schwarz, M. A. Tuli, K. Van Auken, D. Wang, X. Wang, G. Williams, K. Yook, R. Durbin, L. D. Stein, J. Spieth and P. W. Sternberg, *Nucleic Acids Res.*, 2010, **38**, D463–467.

CHAPTER 7

Interactions of Carbon Nanostructures and Small Gold Clusters with Nucleic Acid Bases and Watson–Crick Base Pairs and Nanocontacts Involving M_n–C_{60}–M_n (M = Au, Ag, and Pd; n = 2–8) System: Computational Elucidation of Structures and Characteristics

MANOJ K. SHUKLA[a], FRANCES HILL[a] AND
JERZY LESZCZYNSKI[b]

[a] US Army Engineer Research and Development Center, Environmental
Laboratory, Vicksburg, Mississippi, USA; [b] Interdisplinary Center for
Nanotoxicity, Department of Chemistry and Biochemistry, Jackson State
University, Jackson, Mississippi, USA

RSC Nanoscience & Nanotechnology No. 25
Towards Efficient Designing of Safe Nanomaterials: Innovative Merge of Computational
Approaches and Experimental Techniques
Edited by Tomasz Puzyn and Jerzy Leszczynski
© The Royal Society of Chemistry 2012
Published by the Royal Society of Chemistry, www.rsc.org

7.1 Introduction

Since the discovery and characterization of carbon fullerene (C_{60}) in 1985 by Kroto *et al.*,[1] carbon nanotubes (CNTs) in 1991 by Iijima[2] and graphene by Novoselov *et al.*[3] in 2004 the field of the nanosciences has grown enormously. The popularity of nano-research can be judged from the fact that generally all the scientific publishing companies or societies have at least one journal devoted solely to this area of research. For example, the American Chemical Society has three journals especially devoted to the publication of peer-reviewed articles in the nano-related field. Thus, it should not be an overstatement that it is difficult to envision any research and technological fields where the applications of nanomaterials have not been suggested or demonstrated.[4–10] Interestingly enough, even the suggestion that CNT ropes can be used for the construction of space elevators has been made.[11]

Single-walled carbon nanotubes (SWCNTs) have been shown to have antimicrobial activity and such a property originates from the metallic nature of the nanotube *i.e.* the bacterial cytotoxicity of the metallic SWCNT is much larger than the SWCNT of semiconducting nature.[12] Thus, not surprisingly, the application of metallic SWCNT instead of chemical disinfectants as an antimicrobial agent was suggested. Recently, the carbon-nanotube-based field-effect transistor has been used to monitor the dynamics of a single lysozyme molecule, an enzyme that hydrolyzes polysaccharides in bacterial cell walls.[13] Humic acid has been suggested to stabilize the dispersion of carbon nanotubes in aqueous solution.[14] Further, the role of humic acid in the dispersion–dissolution behavior of silver nanoparticles has also been investigated.[15] In addition, the ability to detect the presence of silver nanoparticles at very low concentrations has been demonstrated.[16]

Carbon nanotubes are hydrophobic, but covalent functionalization[17–21] can be used to make them soluble and thus functionalized nanotubes find applications in various fields such as sensors, medicinal chemistry, as a source of making polymer composites and many more. But, due to the functionalization, the sp^2 bonded networks of CNTs are disrupted. The resulting disruptions can be analyzed by resonance Raman spectroscopy. In such measurements the D-band around 1300 cm^{-1} shows an increase in the intensity and this intensity is decreased for the G-band which is located around the 1500 cm^{-1}.[17–21] Through accidental air leakage during the chemical vapor deposition of graphene on copper foil which led to the reaction of copper with the glass tube and the subsequent formation of a very thin layer of silica glass deposited on the graphene surface, the atomic structure of the amorphous material was able to be determined for the very first time.[22] Interestingly, the observed image of silica glass in this experiment obtained through atomic-resolution transmission electron microscopy, was found to be in agreement with the model proposed by Zachariasen long back in 1932.[23] In another investigation graphene was found to have anticorrosive properties when coated on nickel and copper surfaces.[24]

Although the properties of nanomaterials are mesmerizing, the novelties are also causing genuine concerns related to environmental and public health hazards.[25–31] Due to their small size it has been suggested that nanomaterials can penetrate through cellular membranes, travel to the heart through the blood stream and can be deposited in a targeted part of the body which may lead to tissue injury. In fact, the accumulation of nanoparticles in the food chain has already been detected by Judy *et al.*[32] who grew tobacco plants in a controlled environment with a high concentration of gold nanoparticles; the large accumulation of nanoparticles in the caterpillars who ate the tobacco plant was revealed. The presence of nanoparticles in the atmosphere can result in their deposition in lungs through inhalation. This can cause undesirable allergic reactions.[33–35] The health concerns related to nanoparticles, and particularly to carbon nanotubes or similar nanoparticles, originate from human experience with asbestos exposure which has been considered as the worst occupational health disaster in US history.[36] Actually, with small diameters and large aspect ratios the physiochemical properties of carbon nanotubes have been shown to be similar to asbestos fibers and International Agency of Research in Cancer has classified asbestos as a group I human carcinogen.[37] In fact, the longer multiwalled carbon nanotube (MWCNT) has been observed to have an asbestos-like toxic effect in mice.[38] Further, the functional groups, size and concentrations are also found to have significant influence on the toxicity of the MWCNTs.[39,40] On the other hand, Wang *et al.*[41] have shown that chronic exposure to single-walled carbon nanotubes causes the malignant transformation of epithelial cells of human lungs. Thus, proper information on the nature of nanomaterials and adequate safety procedures for their applications even in research laboratories are necessary to avoid human exposure. For example, the sonication of hydrophobic carbon-based nanomaterials in deionized water decreases while that of the hydrophilic carbon-based nanomaterials in moderately hard water containing natural surfactants significantly increases the airborne nanoparticulate matter of such materials, compared to those of the corresponding dry forms.[42]

Since the area of nanosciences currently covers a gigantic sea of research, it is almost impossible to review even a small fraction of research publications. Discussions of different aspects of nanomaterials, their structures, properties and interactions can be found in recent articles.[43–53] Therefore, in the current chapter we focus on a brief overview of results of computational research in the area of the interaction of C_{60} and SWCNT with nucleic acid bases and Watson–Crick base pairs, the interaction of small gold clusters with the DNA base guanine and the Watson–Crick guanine–cytosine base pair, and the nature of nanocontact between C_{60} and small clusters of gold, silver and palladium atoms.

7.2   Interaction of C_{60} with Nucleic Acid Bases and Watson–Crick Base Pairs

An important aspect of nanomaterials is the shape and size dependency of their optical and electronic properties.[54] The C_{50}, C_{60} and C_{70} molecules are all

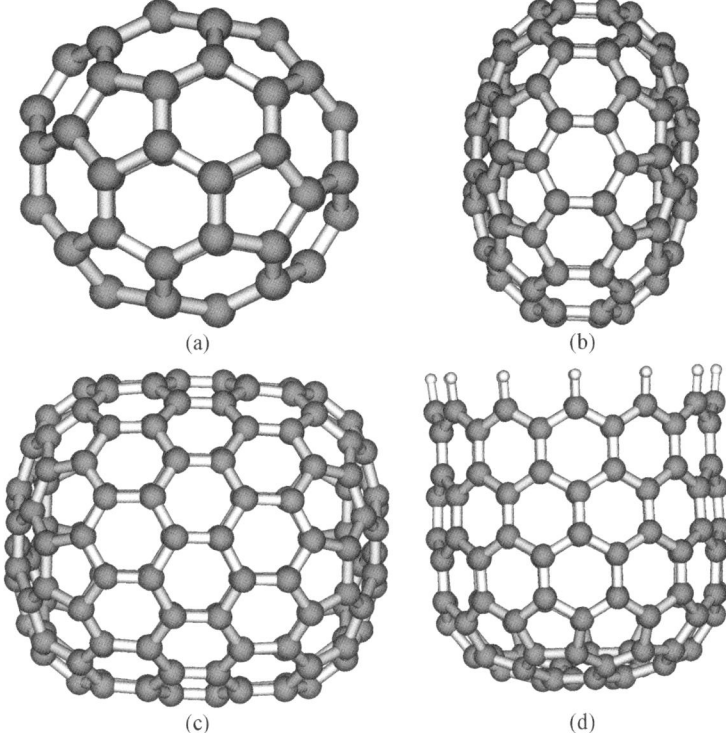

(a) (b)

(c) (d)

Figure 7.1 Structure of different carbon nanoclusters: (a) carbon fullerene (C_{60}), (b) carbon nano disk (C_{96}), (c) carbon nano capsule (C_{144}) and (d) carbon nano bowl ($C_{120}H_{12}$). Reprinted with permission from ref. 57. Copyright 2006 Elsevier.

called magic number fullerenes and their ionization potentials and dissociation energies are the largest among carbon clusters C_n (n = 40–70) as revealed by computational studies at the B3LYP/6-31G(d) level.[55] Hydrogenation has been found to decrease the first vertical ionization potential of fullerene.[56] We performed B3LYP/6-31G(d) level calculations on some carbon clusters namely C_{60}, C_{96} (disk), C_{144} (closed carbon nanotube or capsule), $C_{120}H_{12}$ (bowl or one end open CNT) (Figure 7.1).[57]

The cohesive energy (computed as the energy difference between total molecular energy and sum of isolated atomic energy) and the cohesive energy per atom was found to increase in magnitude with the size of the system. The energy gap between the highest occupied molecular orbital (HOMO) and lowest unoccupied molecular orbital (LUMO) was also predicted to decrease with the size of the cluster (Table 7.1). Based upon HOMO–LUMO energy gaps it was suggested that disk- and capsule-shaped carbon nanostructures can be promising alternatives for the silicon-based semiconductor material.[57] This suggestion was based on the computed agreement between the HOMO–LUMO energy gap of Si_{60} found to be 1.3 eV at the B3LYP/6-31G(d) level[57] to

Table 7.1   Computed first vertical ionization potential (IP, eV), vertical
electron affinity (EA, eV), HOMO–LUMO gap (H–L, eV), total
energy (TE, au), isolated atom energy (E_{iso}, a.u.), cohesive energy
(E_{coh}, a.u.) and the cohesive energy per atom (E_{coh}/A, a.u.) of
different carbon clusters at the B3LYP/6-31G(d) level.[57]

Structure	H–L	IP	EA	TE	E_{iso}[a]	E_{coh}	E_{coh}/A
C_{60}	2.77	7.24	1.75	−2286.17306	−37.7760	−19.6	−0.327
Disk (C_{96})	1.53	6.46	2.98	−3658.18759	−37.7760	−31.7	−0.330
Capsule (C_{144})	1.25	6.72	3.46	−5487.67305	−37.7760	−47.9	−0.333
Bowl ($C_{120}H_{12}$)	0.46	5.19	3.75	−4580.22882	−37.7760	−47.1	—[b]

[a]Hydrogen atom energy at the B3LYP/6-31G(d) level is −0.50027 hartree. [b]Due to the presence of
both carbon and hydrogen atoms the cohesive energy per atom was not computed.

the experimental band gap (about 1.2 eV) of silicon crystal at 0 K.[58] The first
vertical ionization potentials (electron affinities) were also revealed to decrease
(increase) with the size of the system (Table 7.1). Pogulay *et al.*[56] measured the
first ionization potential of C_{60} to be 7.65 ± 0.20 eV and Hertel *et al.*[59] have
measured it to be 7.54 ± 0.20 eV. On the other hand Brink *et al.*[60] have
determined the electron affinity of C_{60} to be 2.7 eV.

Since the increased utilization of nanomaterials raises concern about their
impact on biological systems, it is therefore logical to investigate their
interactions with genetic molecules. For example, it has been suggested that
nucleotide chain cleavage can be induced by fullerene derivatives and the
guanine site is the most susceptible for such reactions.[61–63] Several investigations have been performed to study the interaction of carbon nanomaterials
namely, C_{60}, CNT and graphene with nucleic acid bases and base pairs.[64–71] In
studying these types of systems, questions arise about the level of theory to be
suitably used. We performed detailed investigation on the interaction of C_{60}
with guanine at the B3LYP, ONIOM(MP2:B3LYP) and MP2 levels with 3-
21G(d), 6-31G(d) and 6-31G(d,p) basis sets.[72] It was concluded that stacking
interactions would dominate among these systems. The B3LYP method and
even MP2 approach with a small basis set such as 3-21G(d) would not be
suitable for such investigation. The problem with the MP2 method with a small
basis set is reflected by the fact that starting with a wrong geometrical
configuration in geometry optimization the final molecular geometry would
converge in the same input orientation. For example, in our investigation
where N7 and O6 sites of guanine interacted with the π-charge cloud of C_{60}
resulted in the same orientation as that predicted at the MP2/3-21G(d) level.
However, when the same initial geometrical orientation was considered at the
MP2/6-31G(d) level, the optimization yielded a stacking configuration with a
binding energy (computed as negative of interaction energy) of 8.1 kcal mol^{-1}
at the MP2/6-31G(d,p)//MP2/6-31G(d) level.[72] On the other hand, the
ONIOM(MP2/6-31G(d):B3LYP/6-31G(d)) level optimization where the most
important region was treated at the MP2 level and rest of the system at the B3LYP

level revealed the correct geometry (stacked complex) and the binding energy at the MP2/6-31G(d,p)//ONIOM level was computed to be 8.9 kcal mol^{-1}.[72]

Stacking interactions also play an important role in the interface of biological systems and nanosciences where it has been utilized for the electronic detection of enzyme dynamics[13] and has also been proposed to be applied to ultrafast DNA sequencing.[44] On the other hand, the importance of stacking interactions between carbon nanostructures such as CNTs, C_{60}, larger fullerenes, graphene and nucleic acid bases, base pairs and larger nucleic acid fragments is the main bottleneck for reliable theoretical investigation. These systems require electron-correlated methods such as MP2 for a reliable prediction of the geometry. However, this approach is not practical for systems of this size. Method such as the resolution of identity (RI) approximation of MP2 which provides results of similar accuracy and is less expensive than the regular MP2 is also not practical for such a large system.[73,74] As an example, we investigated the interactions of a zigzag (7,0) SWCNT of small diameter (about 5.5 Å, molecular formula $C_{112}H_{14}$) with nucleic acid bases and Watson–Crick base pairs (guanine (G), adenine (A), cytosine (C), thymine (T), uracil (U), GC and AT base pairs) (complex molecular formula: CNT-G ($C_{117}N_5OH_{19}$), CNT-A ($C_{117}N_5H_{19}$), CNT-C ($C_{116}N_3OH_{19}$), CNT-T ($C_{117}N_2O_2H_{20}$), CNT-U ($C_{116}N_2O_2H_{18}$), and CNT-GC ($C_{121}N_8O_2H_{24}$) and CNT-AT ($C_{122}N_7O_2H_{25}$)). These complexes have 1883, 1868, 1838, 1855, 1836, 2013 and 2015 basis functions respectively with the 6-31G(d) basis set.[75,76] Even for the ONIOM method, the application of the MP2 approach with a medium-sized basis set may not be practical for a large group of researchers if a reasonable size of the system in the interacting region is considered. However, recently developed meta-density functionals such as M05-2X and M06-2X from Truhlar's group[77] offer promising alternatives to study large systems with stacking interactions. We performed a theoretical investigation on interaction of C_{60} with nucleic acid bases and Watson–Crick base pairs at the M05-2X density functional theory (DFT) level with 6-311++G(d,p) basis set.[78] Interaction energies were corrected for the basis set superposition error (BSSE) using the Boys–Bernardi counterpoise correction scheme.[79] The deformation energy (computed as the energy difference between the monomer within the complex geometry and that of the separately optimized monomer) correction to interaction energy was also performed. It should be noted that significant deformation in C_{60} and nucleic acid base geometries was not revealed consequent to the complex formation. However, for the GC and AT base pairs the deformation energy was computed to be about 1.48 and 0.89 kcal mol$^-$ respectively.[78] Optimized geometries and intermolecular nearest heavy atom–heavy atom distance (in Å) of the complexes are shown in Figure 7.2.

It is clear that computed complexes are characterized by the stacking interaction between the π-charge clouds of both monomers. Geometries of the GC and AT base pairs are folded and such folding allows them to envelop the curvature of the C_{60}. The formation of such curvature is the result of stacking interactions between C_{60} and that of monomers of the base pairs which

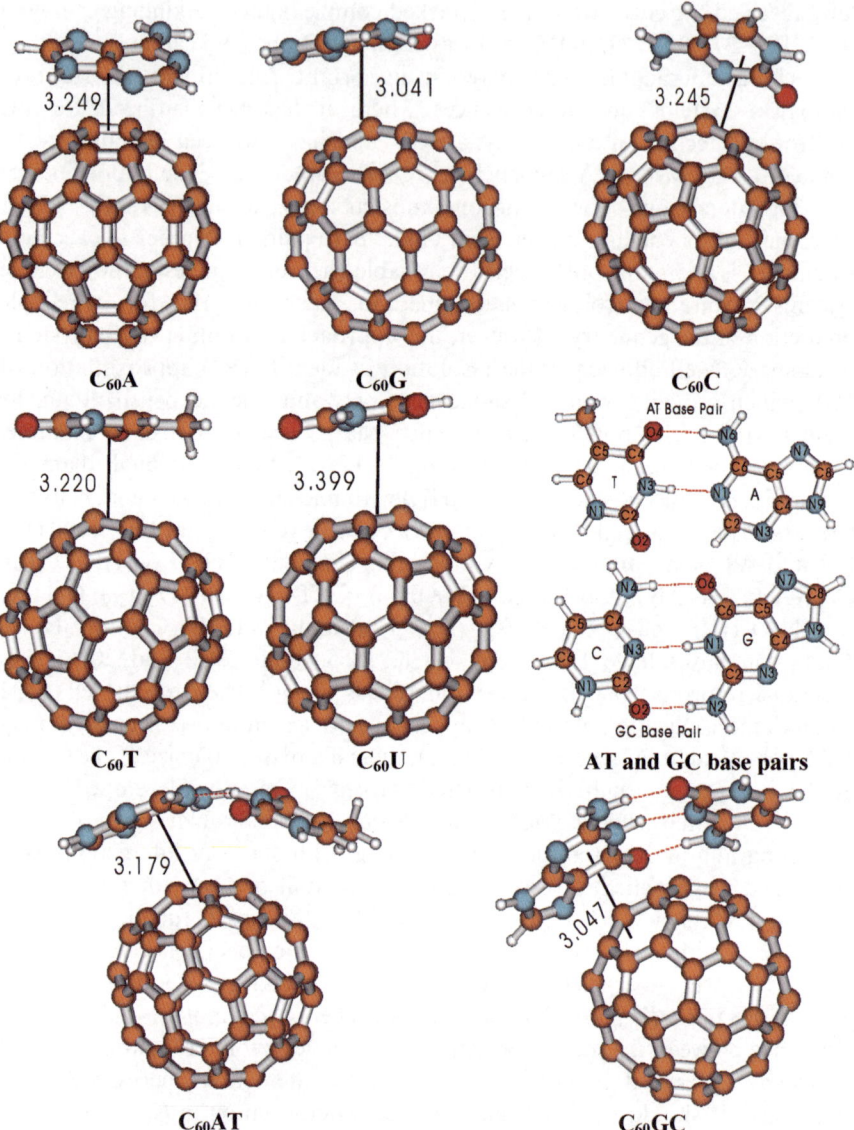

Figure 7.2 Optimized geometries of C$_{60}$-nucleic acid bases and Watson–Crick base pairs complexes and atomic numbering schemes of base pairs. The intermolecular nearest heavy atom–heavy atom distance is given in Å. Reprinted with permission from ref. 78. Copyright 2010 Elsevier.

produce an optimal interaction. Interestingly, a significant change of hydrogen bond distances in the base pairs before and after complex formation is not revealed (Table 7.2). Thus, C$_{60}$ may induce localized deformation in the structure of nucleic acids. In an earlier study using the molecular dynamics simulation, Zhao *et al.*[80] have shown the ability of C$_{60}$ to deform DNA

Table 7.2 Computed hydrogen bond distances (Å) of the AT and GC base pairs in the pristine condition and after complexation with the fullerene.[78]

Parameters	AT base pair[a]	$C_{60}AT$ complex[a]
NH2(A)...C4O(T)	1.960 (2.969)	1.962 (2.953)
N1(A)...N3H(T)	1.819 (2.859)	1.889 (2.889)
C2H(A)...O2(T)	2.803	2.802

Parameters	GC base pair[a]	$C_{60}GC$ complex[a]
C6O(G)...NH2(C)	1.792 (2.818)	1.814 (2.831)
N1H(G)...N3(C)	1.931 (2.955)	1.977 (2.972)
NH2(G)...O2O(C)	1.922 (2.937)	1.903 (2.915)

[a]Numbers in parentheses are the corresponding heavy atom–heavy atom distance.

structure. Our computed binding energies of the $C_{60}G$, $C_{60}A$, $C_{60}C$, $C_{60}T$ and $C_{60}U$ complexes were found to be about 4.99, 4.10, 4.26, 3.75 and 2.31 kcal mol^{-1} respectively. Thus, guanine forms the strongest and uracil forms the weakest stacking complex with fullerene. Further, the computed values of binding energies also indicated that the GC base pair should form a stronger complex than the AT base pair with C_{60}.[78] Deformation-corrected binding energies suggested that a GC base pair with a binding energy of 6.31 kcal mol^{-1} would form a stronger complex than the AT base pair with a binding energy of 6.00 kcal mol^{-1}. Here, we would like to point out that for the $C_{60}G$ complex the binding energy at the MP2/6-31G(d,p)//MP2/6-31G(d) level was found to be 8.1 kcal mol$^-$ which is about 3 kcal mol^{-1} larger than that obtained at the M05-2X/6-311++G(d,p) level. There appear to be two reasons for such a discrepancy: (i) stacking interactions are generally overestimated at the MP2 level and particularly for large basis sets,[81] and (ii) the M05-2X functional includes a dispersion correction empirically.

7.3 Interaction of CNTs with Nucleic Acid Bases and Watson–Crick Base Pairs

Detailed theoretical investigations were also performed on the interactions of nucleic acid bases with CNT by us[75] as well as by other researchers.[66–70] Further, we also extended such investigation to the Watson–Crick GC and AT base pairs.[76] Gowtham *et al.*[82] have studied the physisorption of nucleic acid bases on the surface of a zigzag (5,0) SWCNT using the local density approximation (LDA) with periodic boundary conditions. These authors reported that nucleic acid bases interact with CNTs in the following order: G > A > T > C > U. On the other hand Stepanian *et al.*[83,84] have reported the binding affinity of nucleic acid bases with CNT in the order of G > A > C ≈ T > U. And this binding preference was based upon the interaction of nucleic acid bases with SWCNT at the MP2 and various DFT functional levels where only a part of the CNT was considered. In other words, full geometry optimization of the complexes was not performed. In the case of interactions of adenine dinucleoside with zigzag and armchair CNTs Wang and Ceulemans[67] have found two types of interaction, namely (i) a stacking interaction between adenine and the CNT surface and (ii) a hydrogen bonding interaction between sugar and π-orbital of CNT, in the CNT–DNA complex formation.

We have investigated the interactions of nucleic acid bases and Watson–Crick base pairs with a small diameter zigzag (7,0) SWCNT which was saturated with hydrogen atoms to complete the valency of terminal carbon atoms at the M05-2X functional level.[75,76] The diameter of the CNT was determined using the formula:

$$D = \left(\sqrt{3}/\pi\right) a_{c-c} \left(n^2 + nm + m^2\right)^{1/2} \qquad (7.1)$$

Where, a_{c-c} is the average CC distance in CNT and m, n are integers. For the zigzag (7,0) SWCNT $n = 7$ and $m = 0$ and computed diameter was found to be about 5.5 Å. Molecular geometries of complexes were fully optimized using the 6-31G(d) basis set and are displayed in Figure 7.3.

The BSSE-corrected interaction energies were computed at the same theoretical level using various basis sets and are shown in Table 7.3. As expected, the characteristics of complexes are governed by the stacking interaction but their binding energies are small. Such small binding energies originate from the fact that a small diameter CNT was selected which does not provide enough flat surface for optimal stacking interaction with nucleic acid bases and base pairs. It is expected that larger diameter CNTs will interact more strongly with nucleic acid bases. In fact Umadevi and Sastry[65] have recently reported that nucleic acid bases interact with graphene more strongly than with the curved surface of CNTs. The shortest heavy atom–heavy atom stacking distance between the nucleic acid bases A, G, C, T and U and the CNT in the corresponding complexes was found to be about 3.204, 3.391, 3.118, 3.102 and 3.072 Å respectively.[75] The amino groups of G, A and C were also found to be non-planar in the complex and the amount of the non-planarity did not change significantly compared to that in the isolated system, although the amino group of adenine was revealed to be rotated.[75] Further, it is clear from the data shown in Table 7.3 that guanine interacts most strongly and uracil interacts least strongly with the zigzag (7,0) CNT. However, the binding energies of A and C show basis set dependency and their binding energies with CNT are very close to each other. Based upon results obtained using the 6-311G(d,p) and 6-31+G(d,p) basis sets it appears that adenine will form a slightly weaker complex at the outer surface of the zigzag (7,0) CNT than cytosine, and binding would follow the order of G > C > A > T > U. This prediction is in agreement with the experimental observation where a binding order of G > C > A > T nucleic acid bases with SWCNT was suggested.[85]

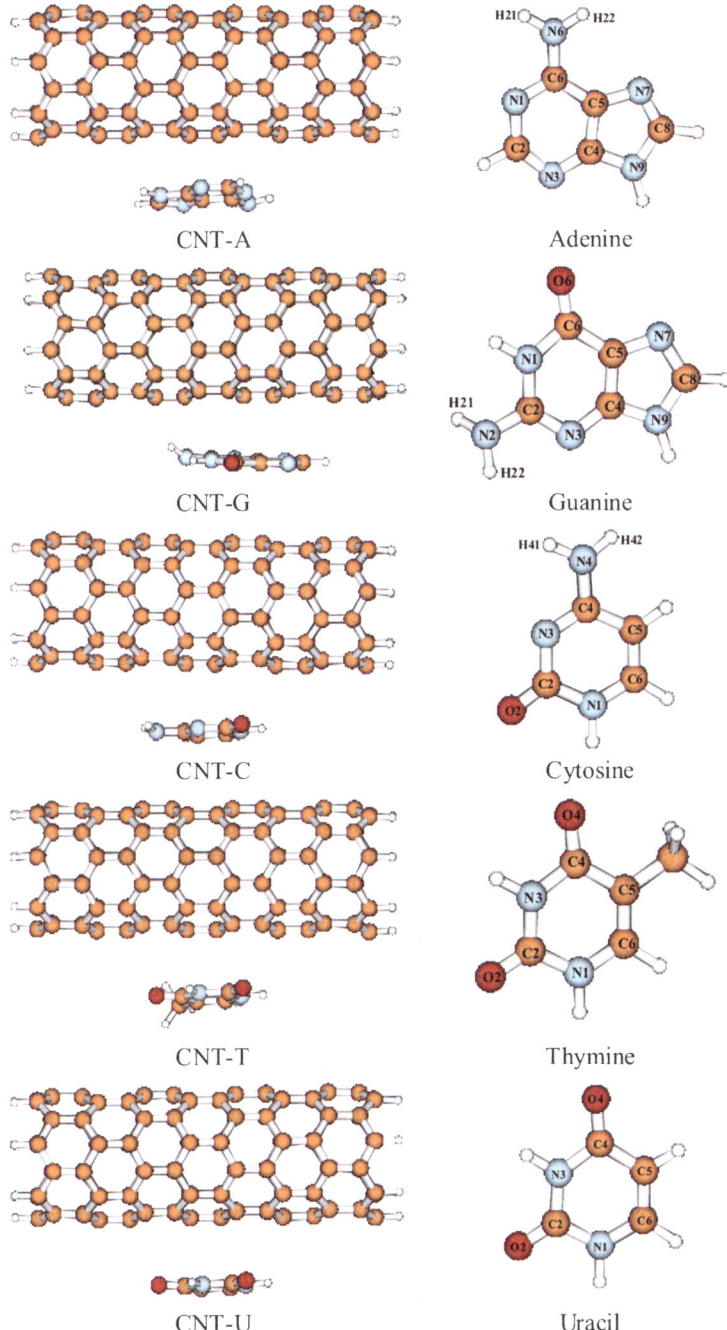

CNT-A   Adenine

CNT-G   Guanine

CNT-C   Cytosine

CNT-T   Thymine

CNT-U   Uracil

Figure 7.3 Optimized geometries of SWCNT–NAB complexes and atomic numbering schemes of nucleic acid bases (NABs). Reprinted with permission from ref. 75. Copyright 2009 Elsevier.

Table 7.3 BSSE-corrected binding energy (kJ mol^{-1}) of complexes at various basis sets, deformation energy of bases (kcal mol^{-1}) and total Mulliken charges (a.u.) on bases in the complex using the M05-2X/6-31G(d) optimized geometry.[75,76]

Complex	Binding energy				Deformation energy[a]	Mulliken charge[b]
	6-31G(d,p)	6-311G(d,p)	6-31+G(d,p)	cc-pVDZ		
CNT-G	26.5	35.05	32.9	31.3	0.9	0.042
CNT-A	18.2	26.06	24.0	23.4	0.4	0.033
CNT-C	18.1	26.14	25.1	22.6	0.8	0.031
CNT-T	17.6	24.80	23.7	21.4	0.3	0.021
CNT-U	17.1	23.82	23.0	20.8	0.5	0.020

[a]At M05-2X/6-31G(d,p)//M05-2X/6-31G(d). [b]At M05-2X/cc-pVDZ//M05-2X/6-31G(d).

Optimized geometries of complexes of zigzag (7,0) SWCNT with Watson–Crick GC and AT base pairs are shown in Figure 7.4. The BSSE-corrected interaction energies of these complexes were computed at the same level of theory and the 6-311G(dp), cc-pVDZ and cc-pVTZ basis sets using the two-body and three-body terms to shed light on the additive nature of the interaction energy. The nearest interatomic heavy atom–heavy atom stacking distance in the CNT-GC and CNT-AT complexes was revealed to be 3.059 and 3.174 Å respectively. A slight change among hydrogen bond distances of base pairs consequent to interaction with CNT was also revealed. It was predicted that both of the base pairs would form complexes of similar strength with the zigzag (7,0) SWCNT. Interesting results regarding the binding energies computed as two- and three-body systems were obtained. At the M05-2X/cc-pVDZ//M05-2X/6-31G(d) level the binding energies of about 174.9 and 110.1 kJ mol^{-1} of the CNT-GC and CNT-AT complexes respectively were obtained using the three-body interacting systems while the same using the two-body interacting systems were computed to be 50.4 and 50.3 kJ mol^{-1} respectively. On the other hand the BSSE-corrected binding energies of the GC and AT base pairs at the same theoretical level were found to be 124.8 and 60.5 kJ mol^{-1} respectively. By subtracting the binding energies of the GC and AT base pairs from the binding energies of the CNT-GC and CNT-AT complexes computed using the three-

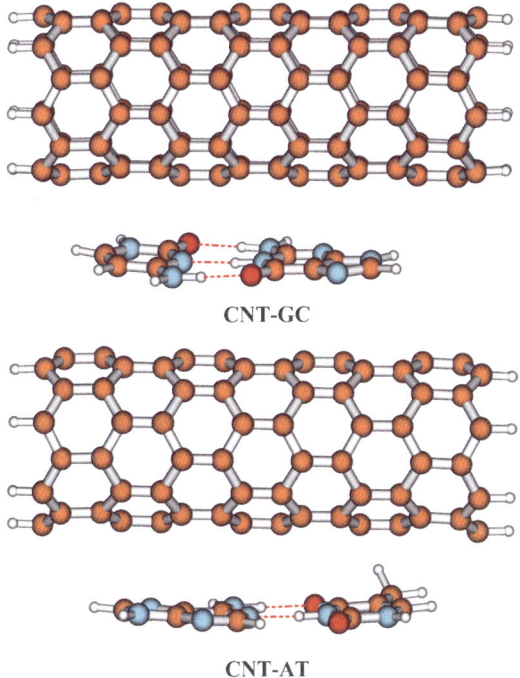

CNT-GC

CNT-AT

Figure 7.4   Optimized geometries of the CNT-GC and CNT-AT complexes obtained at the M05-2X/6-31G(d) level. Reprinted with permission from ref. 76. Copyright 2010 Elsevier.

body interacting systems one would get 50.1 and 49.6 kJ mol^{-1} which are similar to the binding energies of the CNT-GC and CNT-AT complexes computed using the two-body interacting systems. Thus, the major part of binding energy of the CNT-GC and CNT-AT complexes computed as three-body interacting systems originates from the hydrogen bond pairing of the nucleic acid bases guanine and cytosine in the GC base pair and adenine and thymine in AT. The HOMO and LUMO of the zigzag (7,0) CNT were located at the both ends of the CNT but in the complexes they are located at opposite ends of the CNT (Figure 7.5).

Such localization of HOMO and LUMO in complexes appears to be due to the interaction of base pairs with the CNT and the fact that the length of the considered CNT is only about 18 Å. It is expected that for very long CNTs the size of the base pair would be negligible and the predicted "end effect" of the HOMO and LUMO would not be observed.

In the interaction of nucleic acid bases and base pairs with C_{60} and CNTs, two factors have been suggested to play an important role for the stability of the complexes. The first one is the surface area involved in the interaction. It is

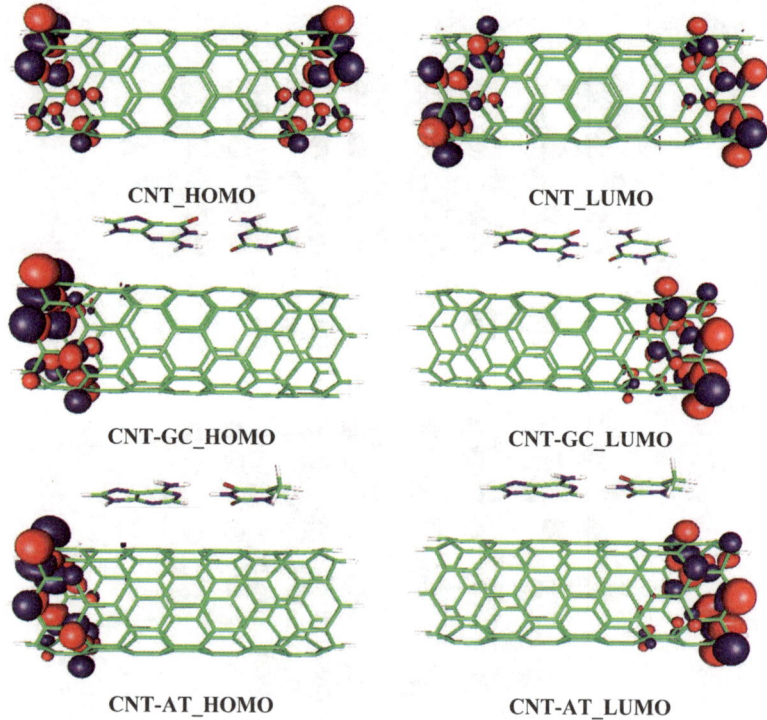

Figure 7.5 Highest occupied molecular orbital (HOMO) and lowest unoccupied molecular orbital (LUMO) of CNT, CNT-GC and CNT-AT complexes at the M05-2X/6-311G(d,p)//M05-2X/6-31G(d) level. Reprinted with permission from ref. 76. Copyright 2010 Elsevier.

expected that a curved surface would not provide a suitable coverage area for stacking interaction, contrary to the case of the flat surface. And therefore, it is not surprising that Umadevi and Sastry[65] have obtained maximum interaction between nucleic acid bases and a graphene sheet rather than with CNT surface. The values of the ionization potentials of nucleic acid bases and base pairs have been suggested as a second factor for the stability of such complexes. It is known that guanine has the lowest and uracil the highest first vertical ionization potential.[86,87] In fact, experimental values for the first vertical ionization potential for nucleic acid bases follow the order U (9.50) > T (9.14) > C (8.80) > A (8.48) > G (8.24) (where values in parentheses represent the corresponding first vertical ionization potential in eV).[88] Further, base pairing lowers the ionization potential and the Watson–Crick GC base pair has a lower ionization potential than the AT base pair.[89] Thus, in general it appears that the base with lower ionization potential forms a stronger complex compared to the base with higher ionization potential. Umadevi and Sastry have also correlated the aromaticity of nucleobases rings with their binding affinity to CNTs.[65] On the other hand, Gowtham *et al.* have found a good correlation between the polarizability of nucleic acid bases and their binding energies with CNTs.[82]

7.4 Interaction of Small Gold Clusters with the Nucleic Acid Base Guanine and the Watson–Crick Guanine–Cytosine Base Pair

Gold nanoparticles occur in a variety of shapes and sizes such as spheres, rods and prisms, which results in unique optical properties that have been utilized in the areas of biosensors, photocatalysts, diagnostics and medicinal applications.[90–95] Gold nanorods absorb in the near-infrared region which provides the basis for various biomedical applications, since tissues generally do not absorb in this region.[96] For interesting work on numerous aspects of gold, readers are referred to a series of articles by Pyykko.[97–99] Gold has very interesting properties. For example, it is highly stable and resistant to oxidation[100] but smaller gold clusters show a strong shape-dependence for electronic properties.[101,102] The tetrahedral form of the Au_{20} gold cluster has a HOMO–LUMO energy gap of about 1.82 eV which is significantly large compared to the energy gap observed for other forms of the same cluster, generally ranging between 0–0.7 eV.[101,102] On the other hand, photoelectron spectroscopy investigations have shown that with few exceptions (Au_2 and Au_6 clusters) the Au_{20} cluster has the largest HOMO–LUMO energy gap among the coinage metal clusters.[103] Interestingly, the physical dimension has been shown to have significant influence on the kinetics and intracellular uptake of gold nanoparticles.[104]

There have been some recent investigations exploring the interactions of small gold clusters with nucleic acid bases and base pairs.[105–108] Such investigations are important due to the fact that the formation of gold aggregates in cells from gold nanoparticles or nanorods have been suggested.[95,96,109] Further, the accumulation of gold nanoparticles in the food chain has also been reported.[32] The DFT level of theory has been used to study interactions of small gold clusters involving up to 6 atoms with nucleic acid bases and base pairs.[110,111] In these investigations Kryachko and Remacle[110,111] have found that gold clusters interact with the oxygen and nitrogen centers and the presence of Au...HN hydrogen bonding was also revealed. On the other hand, Kumar *et al.*[112] have studied clusters of 4 and 8 gold atoms interacting with the guanine–cytosine (GC) and adenine–thymine (AT) base pairs at the DFT level. Although different binding sites of bases were considered, the maximum number of gold atoms in each cluster was limited to only four. However, significant modification on the binding of gold clusters in the anionic complexes was revealed. In our investigation into the interactions of small gold clusters with an even number of gold atoms Au_n (n = 2, 4, 6, 8, 10, 12) with the N7 site of the guanine (G) and the Watson–Crick GC base pair, the DFT level of theory was employed using the B3LYP functional and the standard 6-31G(d) basis set for all but Au element, for which the LANL2DZ effective core potential was used. The combined basis set hereafter in discussion of this section will be denoted as 6-31G(d)∪LANL2DZ.[105] Kumar *et al.*[112] have found that the singlet state of the GC–Au_4 complex is about 2.4 eV more stable than the triplet state. Therefore, we performed our investigation only for the ground singlet state.

Ground state optimized geometries of the Au_n-G and Au_n-GC complexes are shown in Figure 7.6. In these complexes gold clusters are coordinated to guanine and the GC base pair through the N7 site of the former molecule. Table 7.4 shows the Au1–N7 coordination distance (or anchor distance) and the hydrogen bond distance of base pair with and without coordination with gold clusters. Here, Au1 represents the gold atom involved in direct interaction with the N7 site of guanine. In addition, the BSSE-corrected interaction energies, vertical ionization potentials and vertical electron affinities are shown in Table 7.5 and those of natural bond orbital (NBO) charges are depicted in Table 7.6.

It is evident that gold clusters form stable complexes with the guanine and the GC base pair. The Au1–N7 anchor distance was found to be in the range of 2.1 to 2.3 Å. It was generally revealed that with the increased size of the gold cluster, the O6(G)...NH2(C) hydrogen bond distance is also increased and N1H(G)...N3'(C) and NH2(G)...O2'(C) hydrogen bond distances are decreased. A significant amount of charge transfer, through the NBO charge analysis, from guanine or the GC base pair to the gold cluster was revealed (Table 7.6) and such charge transfer was larger for the GC–Au_n than the G–Au_n complexes. Although a larger amount of charge transfer was revealed using the Mulliken analysis by Kumar *et al.,*[112] the mode of interaction of the gold cluster was quite different than those studied by us.[105] An important result that was revealed is that the Au1 atom in direct interaction with the N7 site of the guanine and the GC base pair is generally electron deficient in these

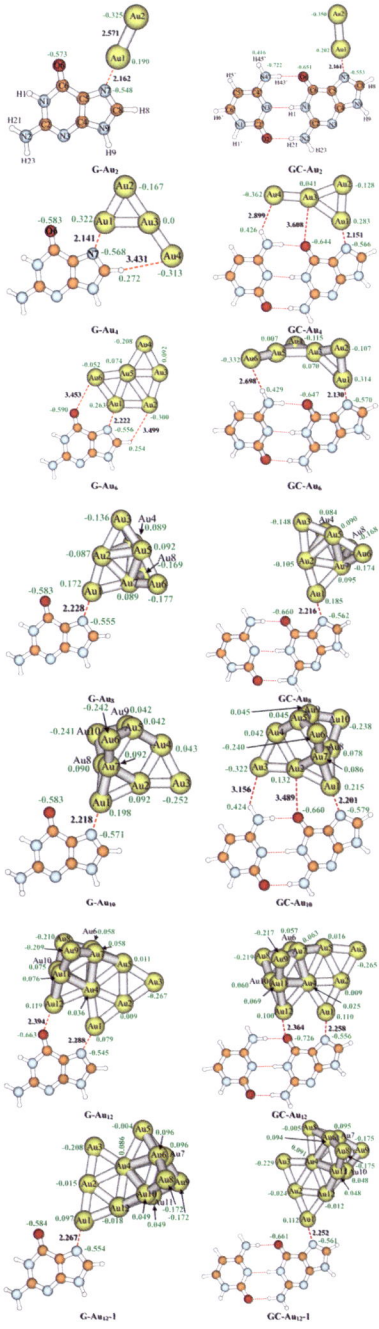

Figure 7.6 Optimized geometries and atomic numbering schemes of G–Au$_n$ and GC–Au$_n$ complexes. NBO charges for selected atoms are given in green while selected gold atoms and base distances are in Å. Reprinted with permission from ref. 105. Copyright 2009 American Chemical Society.

Table 7.4   Hydrogen bond and corresponding heavy–heavy atom distances (in Å) in the GC and GC–Au$_n$ complexes and the N7–Au1 distance in the G–Au$_n$ and GC–Au$_n$ complexes obtained at the B3LYP level using the 6-31G(d)∪LanL2DZ basis set. The N7...Au1 parameter for G–Au$_n$ is shown in parentheses.[105]

| | *Hydrogen bond distance* | | | |
Complex	*O6(G)...NH2(C)/ O6(G)...N4'(C)*	*N1H(G)...N3'(C)/ N1(G)...N3'(C)*	*NH2(G)...O2'(C)/ N2(G)...O2'(C)*	*N7...Au1*
GC	1.782 / 2.818	1.918 / 2.950	1.913 / 2.937	
GC–Au$_2$	1.818 / 2.849	1.888 / 2.925	1.861 / 2.887	2.161 (2.162)
GC–Au$_4$	1.816 / 2.842	1.858 / 2.900	1.842 / 2.871	2.151 (2.141)
GC–Au$_6$	1.864 / 2.887	1.857 / 2.899	1.816 / 2.846	2.130 (2.222)
GC–Au$_8$	1.826 / 2.856	1.886 / 2.924	1.858 / 2.885	2.217 (2.228)
GC–Au$_{10}$	1.847 / 2.863	1.873 / 2.913	1.842 / 2.870	2.201 (2.218)
GC–Au$_{12}$	1.903 / 2.922	1.902 / 2.943	1.810 / 2.839	2.258 (2.288)
GC–Au$_{12}$-1	1.824 / 2.854	1.887 / 2.924	1.859 / 2.885	2.252 (2.267)

complexes. In the interface region, the Au1 atom was found to be involved in donation of electronic charge to the N7 site and that of the other gold atoms. Charge transfer to the N7 site was confirmed by the comparison of charge at the N7 site of complexes with that in the pristine guanine and the GC base pair. In the radical anionic form (vertical) of complexes, generally gold clusters are found to have more than one unit of electronic charge. However, the Au1 atom was generally found to have positive charge, though amount is smaller than in the neutral complexes. On the other hand, in the radical cationic form (vertical) generally gold clusters have more positive charge than the complementary monomer (or base pair) in the complex. Additionally, the

Table 7.5   BSSE-corrected and uncorrected (in parentheses) interaction energy (E_{int}, kcal mol^{-1}), dipole moment (μ, debye), vertical first ionization potential (IP$_v$, eV) and vertical electron affinity (EA$_v$, eV) of the G–Au$_n$ and GC–Au$_n$ complexes.[105]

| | *X = Guanine* | | | | *X = GC base pair* | | | |
Complex	E_{int}	μ	*IP$_v$*	*EA$_v$*	E_{int}	μ	*IP$_v$*	*EA$_v$*
X		6.5	7.7	−1.8		6.1	6.9	−0.8
X–Au$_2$	−24.2 (−28.9)	13.7	7.5	−0.3	−27.3 (−32.3)	12.0	7.1	0.1
X–Au$_4$	−33.3 (−38.5)	13.7	7.4	0.9	−35.9 (−42.1)	9.5	7.1	1.0
X–Au$_6$	−19.3 (−25.1)	15.7	6.9	1.4	−43.9 (−51.2)	9.5	7.1	1.5
X–Au$_8$	−18.7 (−24.0)	15.4	6.9	1.5	−21.8 (−27.4)	15.2	6.8	1.4
X–Au$_{10}$	−24.3 (−29.7)	18.9	6.6	1.4	−28.6 (−35.4)	14.4	6.6	1.5
X–Au$_{12}$	−25.4 (−32.8)	21.8	6.6	1.5	−30.1 (−38.5)	19.4	6.5	1.6
X–Au$_{12}$-1	−15.7 (−20.9)	15.7	6.6	2.1	−18.8 (−24.3)	15.3	6.5	2.0

Table 7.6 Total NBO charges on each monomers of the neutral (N), vertical radical cation (RC) and vertical radical anion (RA) of G–Au$_n$ and GC–Au$_n$ complexes. The Mulliken spin density on gold clusters for radical cations is shown in parentheses.[105]

	X = Guanine					X = GC base pair								
	N	RC		RA		N			RC			RA		
Complex	Au	Au	G	Au	G	Au	G	C	Au	G	C	Au	G	C
X							-0.029	0.029		0.899	0.101		-0.194	-0.806
X-Au$_2$	-0.135	0.530 (0.689)	0.470	-0.761	-0.239	-0.148	0.097	0.051	0.363 (0.511)	0.542	0.095	-0.583	0.008	-0.424
X-Au$_4$	-0.158	0.693 (0.875)	0.307	-1.102	0.102	-0.166	0.110	0.056	0.493 (0.662)	0.424	0.083	-1.086	0.048	0.038
X-Au$_6$	-0.131	0.802 (0.975)	0.198	-1.077	0.077	-0.163	0.111	0.052	0.519 (0.677)	0.401	0.080	-1.098	0.057	0.041
X-Au$_8$	-0.127	0.787 (0.941)	0.213	-1.089	0.089	-0.141	0.088	0.053	0.646 (0.799)	0.285	0.069	-1.027	0.055	-0.028
X-Au$_{10}$	-0.137	0.825 (1.002)	0.175	-1.051	0.051	-0.156	0.094	0.062	0.794 (0.994)	0.137	0.069	-1.086	0.037	0.049
X-Au$_{12}$	-0.164	0.777 (0.975)	0.223	-1.013	0.013	-0.191	0.115	0.076	0.731 (0.962)	0.183	0.086	-1.060	0.033	0.027
X-Au$_{12}$-1	-0.117	0.802 (0.939)	0.198	-1.085	0.085	-0.131	0.078	0.053	0.761 (0.911)	0.176	0.063	-1.097	0.050	0.047

Au1 atom generally has more positive charge than that in the corresponding neutral complexes. It should be noted that the analysis discussed above is based upon NBO charges. The electronic affinity and ionization phenomena in the studied complexes were explained in terms of the localization of the HOMO and LUMO of neutral and highest singly occupied molecular orbital (SOMO) and lowest singly unoccupied molecular orbital (SUMO) of radical species and that of the ionization properties of monomers. It was found that for the GC base pair, the HOMO was localized on guanine and the LUMO on cytosine and this is in agreement with the well-known fact that guanine has a lower ionization potential and electron affinity than cytosine.[89,113–116] In the G–Au$_n$ and GC–Au$_n$ complexes, the HOMO and LUMO are mainly localized on the gold clusters, except for GC–Au$_2$ where the LUMO was localized on the cytosine. The localization of orbitals is consistent with the computed results that electron attachment in the studied complexes will be at the gold clusters and will be also accompanied by some electronic charge transfer from the base or base pair to the gold clusters. One-electron ionization will also take place at the gold clusters in the studied complexes and following oxidation some electronic charge transfer will also take place, mainly from guanine to the gold cluster. This charge transfer will be larger for the smaller complexes.

It is evident from the geometries of the complexes shown in Figure 7.6 that the mode of interaction of gold clusters is generally different in guanine than in the GC base pair. In some cases, in the G–Au$_n$ complexes, the C8H site of guanine is also involved in direct interaction with a gold cluster. On the other hand, in the corresponding GC–Au$_n$ complexes, the orientation of the Au$_n$ cluster is changed resulting in direct interaction with an amino hydrogen of cytosine. Such interactions (either through the C8H site of guanine or amino hydrogen of cytosine) originate from the charge distributions among gold atoms in the Au$_n$ clusters as depicted in Figure 7.6. Obviously, a negatively charged gold atom would try to coordinate with suitable sites. The BSSE-corrected interaction energies of the G–Au$_n$ and GC–Au$_n$ complexes were found to be in the range of -16 to -33 and -19 to -44 kcal mol^{-1} respectively. Evidently, the GC base pair forms a stronger bond than the isolated guanine with gold clusters. The computed interaction energy can be regarded as in good agreement with the experimental data where the heat of desorption of guanine on a gold surface was found to be around 35 kcal mol^{-1} and that of 2′-deoxyguanosine was 29 kcal mol^{-1}.[117]

7.5 Nanocontacts Involving C$_{60}$ and Small Au, Ag and Pd Atomic Clusters

7.5.1 Au$_n$–C$_{60}$–Au$_n$ System

An important aspect of molecular/nano-electronics is the device minimization. However, the existing knowledge of bulk structure cannot be translated to nanodevices unless the structures and properties of nanosystems are also understood extensively and reliably. Thus, understanding the performance and transport properties of nanodevices require a reliable and thorough knowledge of structures and properties of metallic contact in such devices. The nature of nanocontacts has a significant influence on the electrical transport properties and the contact resistance presents a serious problem in understanding the intrinsic electrical properties of CNTs.[118–120] As discussed in the introduction, C_{60} may have various applications including in nanoelectronics. C_{60}-based single molecular transistors with gold as electrodes were fabricated by Park *et al.*[121] where the coupling of motion of center of mass of C_{60} with the frequency of oscillation between gold electrodes was about 1.2 THz, and that of a single electron hopping was indicated through the transport measurement. On the other hand, current-induced oscillation in Au–C_{60}–Au heterojunctions was investigated theoretically by Seideman and coworkers.[122–124] In these investigations C_{60} was placed between gold electrodes oriented along the (100) plane. The coordination of metals with the C_{60} surface is a complicated phenomenon. Such interactions significantly depend on the nature of metals used in the investigation.[125–128] Lyon and Andrews[126] have suggested that Au may possibly bind to the pentagonal ring of C_{60} and such suggestion was based upon the IR spectra of the Au–C_{60} complex. However, computational studies at the B3LYP level of calculation using the STO-3G* basis set for carbon and SDD for gold atom revealed a different type of coordination which agreed with the computed results for the silver complex. Very comprehensive investigations on the bonding of a palladium atom and the silver(I) ion with C_{60} surface were performed by Lichtenberger *et al.*[127,128] In these investigations five bonding sites of C_{60} with metals were considered and these sites include (i) single carbon atom (η^1-coordination), (ii) at the top of center of fused six-membered rings (η^2-coordination), (iii) at the top of center of fused five and six-membered rings, (iv) above the center of pentagonal ring (η^5-coordination), and (v) above the center of hexagonal ring (η^6-coordination); the η^2-coordination was favored among all bonding sites.

Several studies have been performed to theoretically investigate the electron transport in molecular and or nanoclusters devices placed between electrodes.[129–131] In another work, the performance of groups 10 and 11 metals as an electrode in the charge transport calculation was also investigated and the performance of Pd was predicted to be the best while that of Au and Ag were found to be significantly inferior among considered metals.[132] Theoretically, electron transport is computed by putting the molecule between idealized electrodes. However, it is well known that the nature of the contact has an important role in the performance of electronic devices. The main problem is that the actual atomic structure at the point of contact is not known. It is expected that the surface at the point of contact should be rough. Further, the atomic structure at the point of contact would be different in each sample even though they might have been fabricated under the same conditions and even at the same time. The roughness of the electrode surface at the point of contact in the case of C_{60}-electrodes can be modeled by placing a fullerene in between metallic clusters of various sizes. The monolayer of C_{60} adsorbed on the Ag(111) surface has been found to cause the reconstruction of the substrate surface where the resulting vacancies are occupied by the fullerene.[133] We would like to emphasize that the number of points of contact would also depend upon the size of atomic clusters on the electrode surface in contact with the material which is particularly important for nanodevices. Therefore, we performed a detailed theoretical investigation of the structures and properties of complexes involving C_{60} sandwiched between different sizes of gold (Au), silver (Ag), and palladium (Pd) atomic clusters that we used to model the nanocontacts.[134–136]

Two types of bonding configurations were considered. In the investigated systems Au clusters interacted with C_{60} through the center of the top of the fused six-membered ($\eta^{2(6)}$-coordination) and fused six and five-membered rings ($\eta^{2(5)}$-coordination), as depicted in Figure 7.7. In these complexes C_{60} was sandwiched between gold clusters forming Au_n–C_{60}–Au_n (n = 2–8) symmetric structures. Geometries were optimized at the B3LYP DFT level using the standard 6-31G(d) basis set for carbon atoms and LANL2DZ effective core potential (ECP) for the gold atoms. The C_{2h} symmetry was used for all but the Au_5–C_{60}–Au_5 complex for which C_2 symmetry was revealed. Interaction energies of complexes were computed from the three-body terms given by eqn (7.2):

$$E_{int} = E_{ABC} - E_{A(ABC)} - E_{B(ABC)} - E_{C(ABC)} \tag{7.2}$$

where E_{int} represents the interaction energy, E_{ABC} is the total energy of the complex, $E_{B(ABC)}$ represents the total energy of the C_{60} with ghost atoms in place of the rest of the system, $E_{A(ABC)}$ or $E_{C(ABC)}$ represents the total energy of either side of the gold clusters of the complex with ghost atoms for the rest of the system. Computed interaction energies were corrected for the BSSE using the Boys–Bernardi counterpoise correction scheme.[79] Binding energies were computed as the negative of the interaction energy.

Optimized geometries of the Au_n–C_{60}–Au_n complexes obtained under both of the $\eta^{2(5)}$- and $\eta^{2(6)}$-coordination schemes are shown in the Figures 7.8 and 7.9 respectively. The computed Au–C coordination distance, binding energy, Mulliken charge on each cluster, and HOMO–LUMO energy gap are shown in

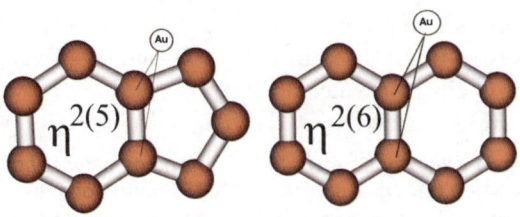

Figure 7.7   Bonding configurations of Au clusters with C_{60}. Reprinted with permission from ref. 134. Copyright 2008 American Chemical Society.

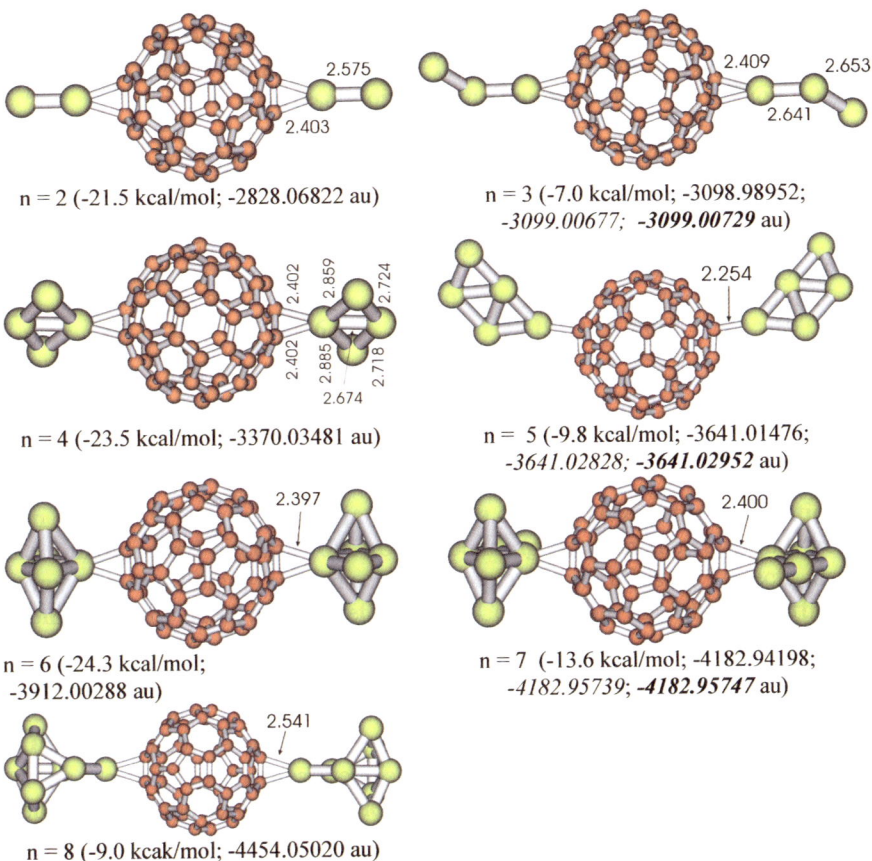

n = 2 (-21.5 kcal/mol; -2828.06822 au)

n = 3 (-7.0 kcal/mol; -3098.98952; -3099.00677; *-3099.00729* au)

n = 4 (-23.5 kcal/mol; -3370.03481 au)

n = 5 (-9.8 kcal/mol; -3641.01476; -3641.02828; *-3641.02952* au)

n = 6 (-24.3 kcal/mol; -3912.00288 au)

n = 7 (-13.6 kcal/mol; -4182.94198; -4182.95739; *-4182.95747* au)

n = 8 (-9.0 kcak/mol; -4454.05020 au)

Figure 7.8 Optimized geometries (singlet state) of Au_n–C_{60}–Au_n complexes in the $\eta^{2(5)}$-coordination. For each complex, the first value in parenthesis represents interaction energy in kcal mol^{-1}, the second is the optimized singlet ground state total energy, the third (in italics) is the single point ground state triplet energy and the last one (in bold italics) is the optimized triplet ground state total energy. Reprinted with permission from ref. 134. Copyright 2008 American Chemical Society.

Table 7.7. From this Table, it is clear that the Au–C coordination distance for the complexes with $\eta^{2(5)}$-coordination is larger than for the corresponding $\eta^{2(6)}$-coordinated complexes. Further, for the $\eta^{2(6)}$-coordinated complexes the coordination distance for larger complexes was found to be smaller than that in smaller complexes. Further, the binding energy of the $\eta^{2(6)}$-coordinated complexes is significantly larger than for the corresponding complexes of $\eta^{2(5)}$-coordination. It should be noted that Au_5–C_{60}–Au_5 complex is characterized by the η^1-coordination, *i.e.* coordinated with only one carbon atom.

Obviously, the $\eta^{2(6)}$ type of coordination appears to be more important than the $\eta^{2(5)}$ type of coordination. It is expected that such coordination may exist

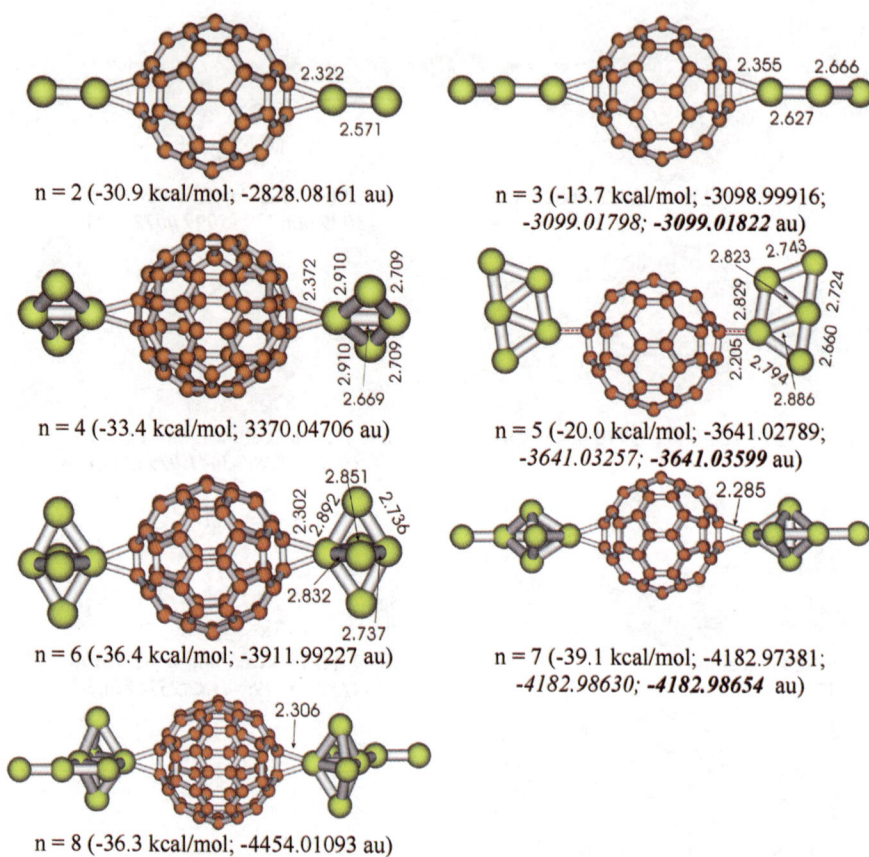

Figure 7.9 Optimized geometries (singlet state) of C_{60}-gold complexes in the $\eta^{2(6)}$-coordination. For each complex, the first value in parenthesis represents interaction energy in kcal mol^{-1}, the second is the optimized singlet ground state total energy, the third (in italics) is the single point ground state triplet energy and the last one (in bold italics) is the optimized triplet ground state total energy. Reprinted with permission from ref. 134. Copyright 2008 American Chemical Society.

in the real system and therefore in the further discussion the results related to $\eta^{2(6)}$ type of coordination will be provided. The growth of a fullerene thin film and monolayers on the gold surface has been studied by several researchers and it was predicted that the adsorption energy would be around 40–60 kcal mol^{-1}.[137–139] From the data shown in Table 7.7, we find that for the larger clusters in $\eta^{2(6)}$-coordination the binding energy is in the range of 35–39 kcal mol^{-1}. Thus, our computed binding energy can be regarded as being in good agreement with the experimentally estimated adsorption energy. Since consideration of a larger gold cluster in the Au_n–C_{60}–Au_n system would result in formation of more coordination sites with C_{60} and thus will increase the

Table 7.7 Computed Au–C and interacting C–C distances (Å), Mulliken charges (in a.u.) on each Au clusters, BSSE-corrected interaction energies (ΔE_{int}, kcal mol^{-1}), HOMO–LUMO energy gap (H–L, eV) and Fermi level energy (E_{FL}, eV) of $\eta^{2(5)}$- and $\eta^{2(6)}$- types of Au$_n$–C$_{60}$–Au$_n$ complexes.[134]

Complex	$\eta^{2(5)}$-coordination				$\eta^{2(6)}$-coordination					
n	Au–C	Charge	ΔE_{int}[a]	H–L[b]	Au–C	C–C	Charge	ΔE_{int}[a]	H–L[b]	E_{FL}[c]
2	2.403	−0.155	−21.5	2.2	2.322	1.433	−0.139	−30.9	2.37	−5.06
3	2.409	−0.128	−7.0	0.23 (1.53)	2.355	1.434	−0.200	−13.7	0.23 (1.60)	−5.03
4	2.402	−0.194	−23.5	1.60	2.372	1.434	−0.197	−33.4	1.80	−4.91
5	2.205	−0.095	−9.8	0.28 (1.22)	2.205	–	−0.048	−20.0	0.50 (1.25)	−4.75
6	2.397	−0.205	−24.3	1.50	2.302	1.444	−0.143	−36.4	2.12	−4.86
7	2.400	−0.228	−13.6	0.21 (1.41)	2.285	1.450	−0.088	−39.1	0.31 (1.35)	−4.96
8	2.541	−0.174	−9.0	1.89	2.306	1.442	−0.181	−36.3	2.01	−5.07

[a]Interaction energy calculation as a three-body term; Mulliken charges are on each gold cluster of the complex. [b]Values in parentheses correspond to those obtained from the single point energy calculation of triplet state using the reference singlet ground state optimized geometry. [c]The computed value for C$_{60}$ is −4.61 eV.

binding energy, it is expected that such a consideration will improve the quality of the agreement between the calculated binding energy and the estimated adsorption energy.

The results of Mulliken charge analysis shown in Table 7.7 suggest that in general there is a significant amount of charge transfer from C_{60} to the gold clusters in the studied complexes. On the other hand the distribution of Mulliken charges in the form of a color plot along with the charges at important sites shown in Figure 7.10 suggest that generally the most negative Mulliken charges are localized on gold clusters, except for the Au_6–C_{60}–Au_6 complex where the interacting C–C atoms have the largest negative charge. The Au atom involved in direct coordination with C_{60} generally has the largest amount of positive charge while the complementary interacting C–C atoms generally have a smaller amount of negative charge. Thus, at the interface

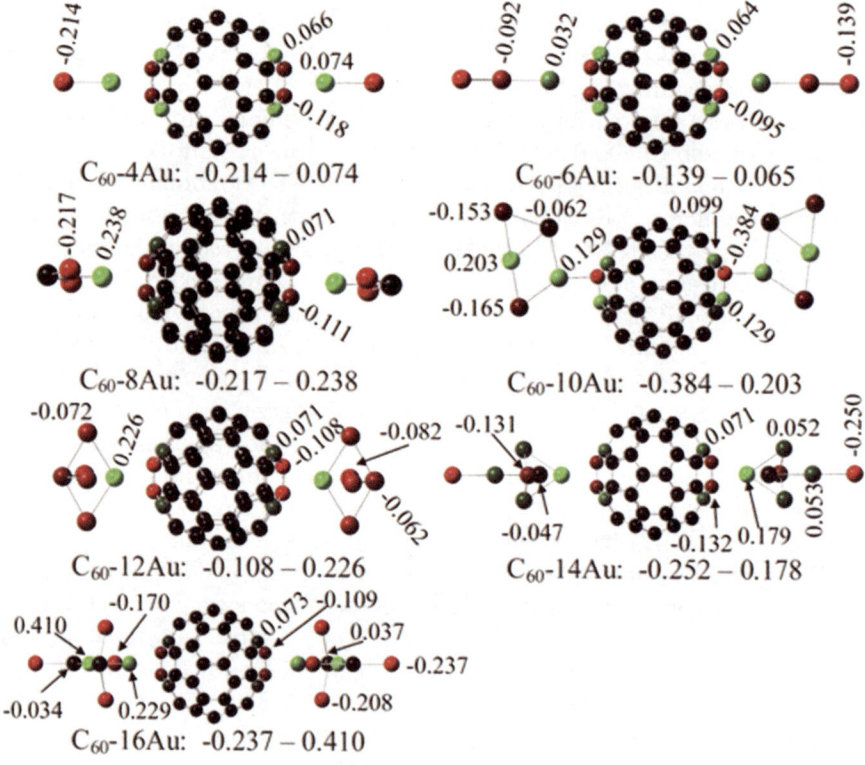

Figure 7.10 Distribution of Mulliken charges in the $\eta^{2(6)}$-coordinated C_{60}-Au complexes. The red color represents the location of the largest negative while the green color represents the largest positive charge. The range of charge (▬▬▬▬▬) in each plot is given next to the name of the corresponding complex. The Mulliken charges (in a.u.) for major contributing atoms are also given. Reprinted with permission from ref. 134. Copyright 2008 American Chemical Society.

there is a charge transfer from the directly interacting Au atom to other gold atoms of the cluster as well as to the interacting C–C atoms. This analysis is also in agreement with experimental results where charge transfer from the gold surface to C_{60} was suggested in the C_{60}–metal contact region.[126,140] Further, there is only a localized effect of charge transfer on the C_{60} in these complex formations and most of the carbon atoms (except those in the coordination region) have approximately zero Mulliken charge.

Ab initio calculations play an important role in the prediction of the charge transport of molecular systems. The delocalized molecular orbitals which are located near the Fermi energy level play an essential role in the charge transport due to their ability to provide a good conduction pathway through the molecular junction. The π-charge cloud associated with C–C bond provides a pathway for the charge transport which is essentially true for the C_{60} also since the diagonal path inside the cage cannot be used as a conduction pathway.[141] The importance of these orbitals from the point of view of charge transport in molecular devices also emerges from the frontier orbital analysis of extended π-conjugated molecular system.[142] It was revealed that the direction as well as the amount of charge transport can be generally accounted from the nature and amplitude of frontier orbitals. This is particularly applicable for systems where electrodes form weak contact with the molecular system. Table 7.7 also shows the computed HOMO–LUMO energy gap of the Au_n–C_{60}–Au_n complexes. The HOMO–LUMO energy gap of C_{60} at the B3LYP/6-31G(d) level was found to be 2.77 eV while the experimentally determined band gap of solid C_{60} is reported to be 2.3 ± 0.1 eV.[143] Thus it would be a good approximation to model the HOMO–LUMO energy gap of C_{60} as the band gap of C_{60} in the solid phase. It is evident that the band gap of C_{60}-Au complexes shown in the Table 7.7 is smaller than that of the pristine C_{60}. Further, the band gap is significantly smaller for complexes involving an odd number of atoms on each side of C_{60} than for those involving an even number of gold atoms on each side of the C_{60}. Generally, the HOMO–LUMO energy gap was found to decrease with an increase in the size of the gold cluster. We would like to mention that for those complexes where odd numbers of gold atoms on each side of C_{60} are involved in the complex formation the singlet ground state was found to be less stable than the corresponding triplet state. For example, for the Au_3–C_{60}–Au_3 complex the singlet state was found to be about 0.5 eV less stable than the triplet state. However, significant changes in the geometries of singlet and triplet states were not observed. Further, in the triplet state the energy difference between the highest occupied and lowest unoccupied α-orbitals would correspond to the band gap and these data are also shown in the Table 7.7.

By taking the average of HOMO and LUMO orbital energies Witek *et al.*[144] computed the Fermi level energy (E_{FL}) of various carbon clusters (C_{28}, C_{60} and C_{70}). The E_{FL} value of −4.7 eV was revealed for C_{60} and C_{70} at the BLYP/cc-pVTZ and BLYP/3-21G levels and about −5.0 eV at the SCC-DFTB level, but depending upon the theoretical level the corresponding value for C_{28} was predicted in the range of -5.1 to -5.2 eV. Our computed E_{FL} of C_{60} at the B3LYP/6-31G(d) level at -4.6 eV is in good agreement with that reported by Witek *et al.*[144] The data presented in Table 7.7 show that in the $Au_n–C_{60}–Au_n$ system the E_{FL} values are significantly affected by the size of the cluster. Further, since the Fermi energy of the Au metal is about -5.3 eV,[132] the computed E_{FL} values of studied complexes (in the range of -5.1 to -4.8 eV) are near to the average value of the E_{FL} of bulk Au and that of the C_{60}.[144] Additionally, it is well known that the HOMO of C_{60} is five-fold degenerate with h_u symmetry while that of LUMO is triply degenerate with t_{1u} symmetry.[145] We also performed a detailed analysis of frontier orbitals of studied systems and compared results with those obtained for the isolated C_{60} system. Several degenerate types of occupied and unoccupied orbitals similar to the degenerate HOMO and LUMO of C_{60} were obtained. Few orbitals which are largely delocalized throughout the system and located near the corresponding Fermi energy level were also identified and were suggested to actively participate in the charge transport in the investigated system.

7.5.2 $Ag_n–C_{60}–Ag_n$ System

The structure and properties of the $Ag_n–C_{60}–Ag_n$ ($n = 2–4, 6–8$) system within the $\eta^{2(6)}$ type of coordination scheme were also investigated to understand the nanocontact involving the silver cluster–C_{60} interface.[135] Geometries of the complexes were optimized at the B3LYP level under C_{2h} symmetry and with a 6-31G(d) basis set for the carbon atoms and LANL2DZ ECP for silver atoms; the combined basis set will hereafter be denoted as 6-31G(d)∪LANL2DZ. Interaction energies were corrected for BSSE and were computed using eqn (7.2). Further, the BSSE-corrected interaction energies were also computed at the same theoretical level using the 6-311G(d) basis set for carbon atoms and the Stuttgart RSC 1997 ECP for the Ag atoms from the basis set exchange library;[146,147] the combined basis set, hereafter, will be denoted as 6-311G(d)∪RSC. Optimized geometries of the complexes are shown in Figure 7.11.

Further, in the case of these systems the complexes with odd number of silver atoms on each side of C_{60} were also revealed to have the ground state triplet as a more stable form than the corresponding singlet state. For example, at the B3LYP/6-31G(d)∪LANL2DZ level, the triplet states of the $Ag_3–C_{60}–Ag_3$ and $Ag_7–C_{60}–Ag_7$ complexes are predicted to be about 0.12 (0.59) and 3.91 (4.49) kcal mol^{-1} respectively more stable than the corresponding singlet state; the values in parentheses correspond to the optimized triplet geometries. Thus, it is also evident from the vertical (singlet state reference) and relaxed state similarity in the relative stability that geometries of the triplet state do not differ significantly, compared to the respective singlet ground state geometry.

Important computed parameters of the studied $Ag_n–C_{60}–Ag_n$ complexes are shown in Table 7.8. It is evident that the Ag1–C coordination distance is in the range of 2.409–2.730 Å, but a well-defined trend cannot be established. The

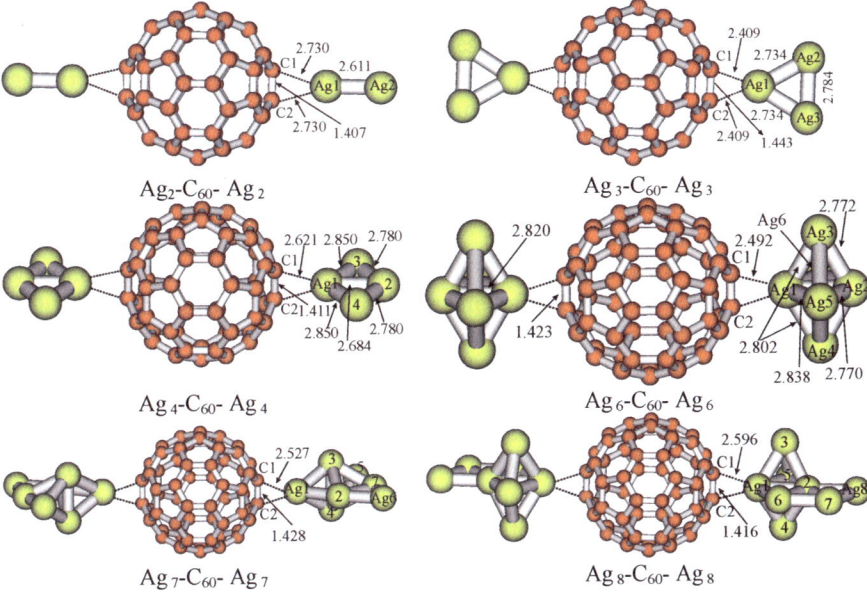

Figure 7.11 Optimized geometries of Ag_n–C_{60}–Ag_n complexes at the B3LYP/6-31G(d)∪LANL2DZ level. Geometrical parameters are in Å. Reprinted with permission from ref. 135. Copyright 2012 American Chemical Society.

largest coordination distance corresponds to Ag_2–C_{60}–Ag_2 and the smallest one for the Ag_3–C_{60}–Ag_3 complex. On the other hand, the coordinating C–C distance shows dependency on the Ag–C coordination distance; the larger Ag–C distance triggers the smaller C–C distance and *vice versa*. In the pristine C_{60} the C–C distance corresponding to the fused six-membered rings is found to be 1.395 Å at the B3LYP/6-31G(d) level which is in good agreement with the value of 1.401 Å obtained using an electron diffraction technique.[148] The increase in the coordinating C–C bond distance due to interaction with metal clusters is also revealed for the Au_n–C_{60}–Au_n and Pd_n–C_{60}–Pd_n complexes.[134,136] With respect to the small size cluster formation both the Ag and Au elements are shown to have different behavior.[149] DFT calculations have shown that Au tends to favor a two-dimensional structure for clusters involving up to 13 atoms while, starting with the heptamer, three-dimensional structures dominate for Ag clusters. Similar behavior has been predicted between metal clusters in the Au_n–C_{60}–Au_n and Ag_n–C_{60}–Ag_n complexes.[134,135] For example, the Ag clusters involving three atoms in the corresponding system form a triangular structure which was not observed for a similar cluster involving gold atoms (Figures 7.9 and 7.11). Similarly, the structures of Ag clusters in the Ag_7–C_{60}–Ag_7 and Ag_8–C_{60}–Ag_8 complexes were found to be slightly different than those of the gold clusters in the corresponding Au–C_{60} complexes. Further, based upon the binding energy data shown in Table 7.8

Table 7.8 Computed Ag–C and interacting C–C distances (Å), charges (Mulliken and NBO in a.u.) on each side of Ag cluster, BSSE-corrected binding energies (ΔE_{BE}, kcal mol^{-1}), HOMO–LUMO energy gap (H–L, eV) and Fermi energy (E_{FL}, eV) of Ag$_n$-C$_{60}$-Ag$_n$ complexes at the B3LYP/6-31G(d)ULANL2DZ level.[135]

n	Ag–C	C–C	Mulliken	NBO	ΔE_{BE}	H–L[b]	E_{FL}
2	2.730	1.407	−0.131 (0.113)	−0.008 (−0.004)	4.3 (5.5)	1.33 (1.22)	−4.42 (−4.71)
3	2.409	1.443	0.170 (0.604)	0.457 (0.474)	29.6 (37.3)	0.49 [0.58] (0.52)	−3.65 (−3.91)
4	2.621	1.411	−0.154 (0.208)	0.016 (0.002)	10.1 (11.6)	0.69 (0.62)	−4.14 (−4.40)
6	2.492	1.423	−0.086 (0.467)	0.157 (0.080)	16.9 (20.7)	1.24 (1.18)	−4.15 (−4.37)
7	2.527	1.428	0.083 (0.565)	0.335 (0.355)	4.1 (9.7)	0.35 [0.55] (0.38)	−3.77 (−3.97)
8	2.596	1.416	−0.094 (0.424)	0.124 (0.066)	8.1 (11.3)	1.16 (1.09)	−4.11 (−4.33)

[a]Interaction energy calculation as a three-body term; charges are on each side of the Ag cluster of the complex. The corresponding C–C distance in isolated C$_{60}$ is 1.395 Å. Values in parentheses for Mulliken, NBO, ΔE_{BE} and H–L, E_{FL} are obtained at the B3LYP/6-311G(d)USTG//B3LYP/6-31G(d)ULANL2DZ level.
[b]Values in square brackets correspond to those obtained from the single point energy calculation of the triplet state using the reference singlet ground state optimized geometry.

and comparison with the corresponding results for the $Au_n–C_{60}–Au_n$ system (Table 7.7) it is evident that in general Ag clusters form weaker complexes with C_{60} than do the Au clusters. From the charge analysis it was revealed that at the interface the Ag atom involved in direct interaction with C_{60} loses electronic charge while the interacting C–C atoms gain electronic charge. Further, the amount of charge on these complementary atomic centers at the interface depends upon the nature of the silver clusters.

The HOMO–LUMO energy gaps of $Ag_n–C_{60}–Ag_n$ complexes are shown in Table 7.8. Since for $Ag_3–C_{60}–Ag_3$ and $Ag_7–C_{60}–Ag_7$ complexes the triplet ground state was predicted to be more stable than the singlet ground state, the HOMO–LUMO energy gaps of these complexes were determined as the energy difference between the SOMO and SUMO of the α-spin. The HOMO–LUMO energy gap of the $Ag_n–C_{60}–Ag_n$ complexes are revealed to be significantly reduced when compared to that for isolated C_{60} and such reduction is comparatively larger than those found for the corresponding $Au_n–C_{60}–Au_n$ complexes. From the Fermi level energy of $Ag_n–C_{60}–Ag_n$ complexes shown in Table 7.8 and comparison of such data with the corresponding $Au_n–C_{60}–Au_n$ complexes depicted in Table 7.7, it is predicted that the Fermi level of the former complexes will move upward while those of the corresponding latter complexes will go down with respect to the E_{FL} of isolated C_{60}. Further, analysis of frontier orbitals of the studied complexes suggested the location of important orbitals which are generally delocalized throughout the molecules and located near the corresponding Fermi energy level.

7.5.3 $Pd_n–C_{60}–Pd_n$ System

Palladium forms a variety of complexes which can be used in several applications such as a hydrogenation catalyst,[150] gas sensor and gas filter.[151,152] The Pd–C_{60}-based sensors can detect organic molecules whose concentration is comparable to that present in the atmosphere.[151] For example, toluene has been reported to be adsorbed on $C_{60}–Pd_n$ polymer-type materials and this adsorption is largest on $C_{60}Pd_2$ for concentration as small as 1000 ppb of the organic molecule.[151] Further, based upon an experimental investigation Pd has been suggested to form better contact than Au.[153] As discussed earlier, in theoretical investigations Pd has been suggested to be a superior electrode material compared to other metals.[132] Therefore, we also performed a theoretical investigation to understand the characteristics of nanocontacts in the $Pd_n–C_{60}–Pd_n$ ($n = 2–4, 6–8$) system, where geometries of the complexes were optimized at the B3LYP level with the 6-31G(d) basis set for carbon and LANL2DZ ECP for Pd under C_{2h} symmetry.[136]

In these complexes, palladium clusters form a complex with C_{60} through an $\eta^{2(6)}$- type of coordination, as discussed earlier. Interaction energies were computed using eqn (7.2) and BSSE correction was performed using the Boys–Bernardi counterpoise correction scheme.[79] The binding energy was defined as the negative of the interaction energy.

The computed binding energy, Pd–C and interacting C–C distances, Mulliken charges, HOMO–LUMO energy gap and the Fermi level energy of the investigated complexes are presented in Table 7.9 and the optimized geometries are shown in Figure 7.12. It is clear that the Pd_2–C_{60}–Pd_2 complex has the largest and Pd_6–C_{60}–Pd_6 complex has the smallest binding energy. Further, binding energies of all of the studied complexes are in the range of 44–64 kcal mol^{-1} (Table 7.9).

By the comparison of binding energies of the Pd_n–C_{60}–Pd_n with the corresponding Au_n–C_{60}–Au_n and Ag_n–C_{60}–Ag_n complexes one finds that Pd forms a stronger complex with C_{60} than do Au and Ag (Tables 7.7 and 7.8). However, in general, smaller Pd clusters have larger binding energies while the opposite is true for Au where smaller clusters are found to have smaller binding energies with C_{60}. The larger binding ability of smaller Pd clusters with C_{60} might be related to the reactivity of the palladium where a $C_{60}Pd_2$ complex is found to have the best adsorptivity towards toluene.[151] The tracer diffusion of the C_{60} adsorbed on the Pd(110) surface involves a barrier of about 1.4 $\pm$ 0.2 eV (32.3 $\pm$ 4.6 kcal mol^{-1}).[154] This tracer diffusion barrier height is in the lower range of the computed binding energy of C_{60}–Pd complexes in our work. With regard to the Pd–C coordination distance among the investigated complexes, the Pd_2–C_{60}–Pd_2 complex has the smallest and the Pd_6–C_{60}–Pd_6 has the largest value and it correlates well with the computed binding energy

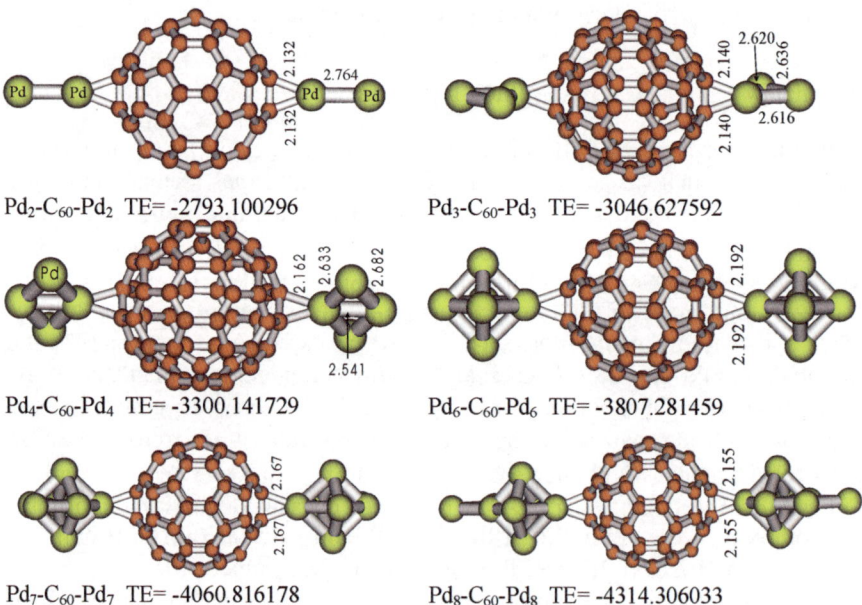

Pd_2-C_{60}-Pd_2 TE= -2793.100296 Pd_3-C_{60}-Pd_3 TE= -3046.627592

Pd_4-C_{60}-Pd_4 TE= -3300.141729 Pd_6-C_{60}-Pd_6 TE= -3807.281459

Pd_7-C_{60}-Pd_7 TE= -4060.816178 Pd_8-C_{60}-Pd_8 TE= -4314.306033

Figure 7.12   Optimized geometries of Pd_n–C_{60}–Pd_n complexes. Geometrical parameters are in Å. Total energy (TE) of the respective complexes is given in hartree. Reprinted with permission from ref. 136. Copyright 2009 American Chemical Society.

Table 7.9 Computed Pd–C and interacting C–C distances (Å), Mulliken charges (in a.u.) on each side of Pd cluster, BSSE-corrected binding energies (ΔE_{BE}, kcal mol^{-1}), HOMO–LUMO energy gap (H–L, eV) and Fermi energy (E_{FL}, eV)^a of Pd$_n$–C$_{60}$–Pd$_n$ complexes.[136]

n	Pd–C	C–C	Charge	ΔE_{BE}	H–L	E$_{FL}$
2	2.132	1.461	0.116	63.5	1.53	−4.53
3	2.140	1.455	−0.003	60.1	1.61	−4.62
4	2.162	1.450	−0.004	45.4	1.59	−4.65
6	2.192	1.449	0.049	43.7	1.20	−4.75
7	2.167	1.453	0.043	51.5	1.28	−4.53
8	2.155	1.455	0.032	55.6	1.34	−4.54

^{a}Interaction energy calculation as a three-body term; Mulliken charges are on each side of the Pd cluster of the complex. The corresponding C–C distance in isolated C$_{60}$ is 1.395 Å.

that indicates the former complex forms the strongest and latter one forms the weakest complex. The interaction of palladium also causes the elongation of the interacting C–C bond (in the range of 1.449–1.461 Å) from its value of 1.395 Å in the isolated C$_{60}$ (at the B3LYP/6-31G(d) level). Further, elongation of the Pd–C coordination distance and that of the interacting C–C bond in the Pd$_n$–C$_{60}$–Pd$_n$ complexes is comparatively smaller than the corresponding parameters in the corresponding Au$_n$–C$_{60}$–Au$_n$ complexes. We would also like to mention that for Pd$_n$–C$_{60}$–Pd$_n$ complexes the ground state was found to have an electronic singlet configuration.

Mulliken charge analysis suggested that generally there is charge transfer from the Pd clusters to C$_{60}$ and such transfer is significant for the Pd$_2$–C$_{60}$–Pd$_2$ complex (Table 7.9). This result correlates very well with the reactivity of the C$_{60}$Pd$_2$ complex toward the toluene where palladium was suggested to be partially positively charged.[151] Further, in the contact region the interacting C–C atoms have negative charges and other neighboring carbon atoms directly connected to the C–C atoms have a positive charge while the rest of the C$_{60}$ atoms have charges close to zero (Figure 7.13).

However, unlike the Au$_n$–C$_{60}$–Au$_n$ and Ag$_n$–C$_{60}$–Ag$_n$ systems, a clear trend about charge distribution on the coordinating Pd atom is not revealed. The HOMO–LUMO energy gap of the studied complexes was found to be significantly smaller than that of the isolated C$_{60}$ (Table 7.9). The HOMO–LUMO energy gaps of the corresponding Au$_n$–C$_{60}$–Au$_n$ complexes were predicted to be considerably larger (Table 7.7). Such difference between the Pd$_n$–C$_{60}$–Pd$_n$ and Au$_n$–C$_{60}$–Au$_n$ systems should be related to the fact that Pd clusters form more stable complexes with C$_{60}$ than the gold clusters. On the other hand the computed Fermi level energies of the C$_{60}$–Pd complexes were found to be in the range of −4.53 to −4.75 eV (Table 7.9). A few important orbitals which are delocalized throughout the system and located near the Fermi energy level and may play an important role in the charge transport were also identified through the frontier molecular orbital analysis.

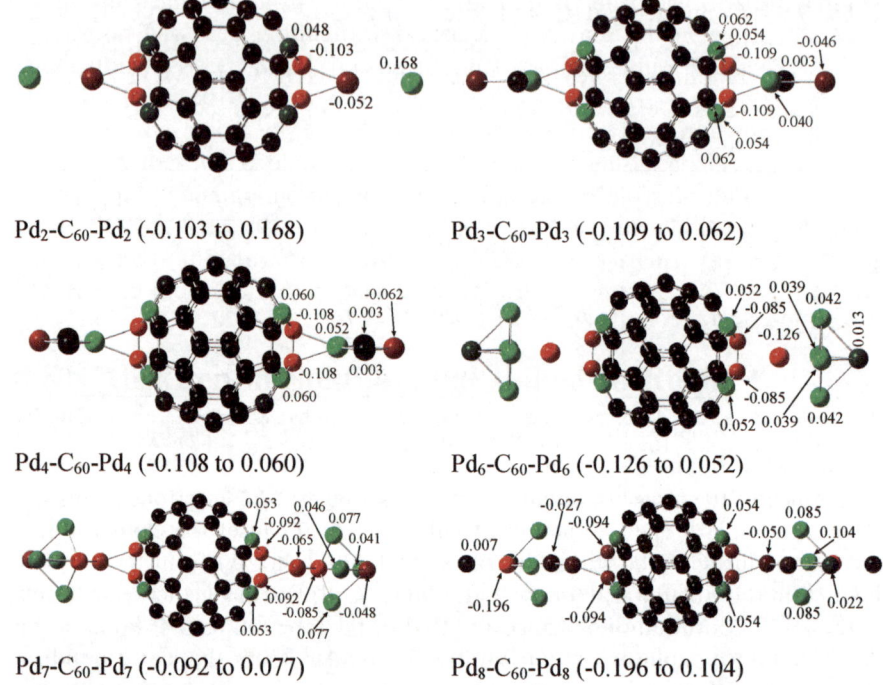

Pd$_2$-C$_{60}$-Pd$_2$ (-0.103 to 0.168)   Pd$_3$-C$_{60}$-Pd$_3$ (-0.109 to 0.062)

Pd$_4$-C$_{60}$-Pd$_4$ (-0.108 to 0.060)   Pd$_6$-C$_{60}$-Pd$_6$ (-0.126 to 0.052)

Pd$_7$-C$_{60}$-Pd$_7$ (-0.092 to 0.077)   Pd$_8$-C$_{60}$-Pd$_8$ (-0.196 to 0.104)

Figure 7.13   Mulliken charge distributions in the Pd$_n$–C$_{60}$–Pd$_n$ complexes. The range of charge (in a.u.) distribution for each complex is also given where with red being the most negative and green being the most positive charge. Reprinted with permission from ref. 136. Copyright 2009 American Chemical Society.

7.6 Conclusions

Materials at the nanoscale have properties which are significantly different from the corresponding bulk structure. The novelty of nanomaterials also brings genuine concerns related to their potential hazardous effects on the environment and on human health. Thus, a thorough understanding of the properties of nanostructures is needed for their safe use with minimal environmental impact for future generations. There are tremendous needs for rigorous evaluation of nanoscience using theoretical approaches. Accurate calculations are hampered by the size of the systems. For example, the interactions of carbon nanostructures such as C$_{60}$, CNT and graphene with nucleic acids and base pairs are governed by stacking interactions and correct predictions using electron-correlated levels of theory (which do not have empirical parameters) are required. Unfortunately, this approach is still not very feasible for extensive calculations. On the other hand, reliable information about such systems is important in view of the suggested ultrafast DNA sequencing as proposed by Kim and coworkers[44] and that of the first electronic evaluation of dynamics of enzyme.[13] In both cases stacking interactions have been shown to play a dominant role. It has been recently reported that gold nanoparticles can also accumulate in the food chain.[32] Our investigation of interactions of small gold clusters with guanine and the guanine–cytosine base pair has suggested that gold clusters can oxidize the base and base pair while electron attachment will also take place at the gold cluster of the complex. Experimentally, it is quite difficult to determine the atomic structure of contact regions and theoretical methods provide an excellent avenue for such investigations. They are also considered an efficient tool for studying various aspects of nanomaterials and, with constantly increasing speed and power of computers, computational methods will certainly become a method of choice for future investigations.

Acknowledgements

The use of trade, product, or firm names in this report is for descriptive purposes only and does not imply endorsement by the U.S. Government. The tests described and the resulting data presented herein, unless otherwise noted, were obtained from research conducted under the Environmental Quality Technology Program of the United States Army Corps of Engineers and the Environmental Security Technology Certification Program of the Department of Defense by the USAERDC. Permission was granted by the Chief of Engineers to publish this information. The findings of this report are not to be construed as an official Department of the Army position unless so designated by other authorized documents. The authors thank Dr Andrea Scott of USACE and Dr Leonid Gorb of SpecPro, Inc. for their editorial comments.

References

1. H. W. Kroto, J. R. Heath, S. C. O'Brien, R. F. Curl and R. E. Smalley, *Nature*, 1985, **318**, 162.
2. S. Iijima, *Nature*, 1991, **354**, 56.
3. K. S. Novoselov, A. K. Geim, S. V. Morozov, D. Jiang, Y. Zhang, S. V. Dubonos, I. V. Grigorieva and A. A. Firsov, *Science*, 2004, **306**, 666.
4. Y. Choi, I. S. Moody, P. C. Sims, S. R. Hunt, B. L. Corso, I. Perez, G. A. Weiss and P. G. Collins, *Science*, 2012, **335**, 319.
5. B. M. Venkatesan, D. Estrada, S. Banerjee, X. Jin, V. E. Dorgan, M.-H. Bae, N. R. Aluru, E. Pop and R. Bashir, *ACS Nano*, 2012, **6**, 441.
6. D. F. Moyano and V. M. Rotello, *Langmuir*, 2011, **27**, 10376.
7. C. D. Vecitis, K. R. Zodrow, S. Kang and M. Elimelech, *ACS Nano*, 2010, **4**, 5471.
8. A. N. Khlobystov, *ACS Nano*, 2011, **5**, 9306.
9. T. Puzyn, B. Rasulev, A. Gajewicz, X. Hu, T. P. Dasari, A. Michalkova, H.-M. Hwang, A. Toropov, D. Leszczynska and J. Leszczynski, *Nat. Nanotechnol.*, 2011, **6**, 175.

10. J. Leszczynski, *Nat. Nanotechnol.*, 2010, **5**, 633.
11. A. G. Mamalis, L. O. G. Vogtländer and A. Markopoulos, *Precis. Eng.*, 2004, **28**, 16.
12. C. D. Vecitis, K. R. Zodrow, S. Kang and M. Elimelech, *ACS Nano*, 2010, **4**, 5471.
13. Y. Choi, I. S. Moody, P. C. Sims, S. R. Hunt, B. L. Corso, I. Perez, G. A. Weiss and P. G. Collins, *Science*, 2012, **335**, 319.
14. M. A. Chappell, A. J. George, K. M. Dontsova, B. E. Porter, C. L. Price, P. Zhou, E. Morikawa, A. J. Kennedy and J. A. Steevens, *Environ. Pollut.*, 2009, **157**, 1081.
15. M. A. Chappell, L. F. Miller, A. J. George, B. A. Pettway, C. L. Price, B. E. Porter, A. J. Bednar, J. M. Seiter, A. J. Kennedy and J. A. Steevens, *Chemosphere*, 2011, **84**, 1108.
16. A. R. Poda, A. J. Bednar, A. J. Kennedy, A. Harmon, M. Hull, D. M. Mitrano, J. F. Ranville and J. Steevens, *J. Chromatogr., A*, 2011, **218**, 4219.
17. R. C. Haddon, *Acc. Chem. Res.*, 2002, **35**, 997.
18. C. A. Dyke and J. M. Tour, *J. Phys. Chem. A*, 2004, **108**, 11151.
19. D. Tasis, N. Tagmatarchis, A. Bianco and M. Prato, *Chem. Rev.*, 2006, **106**, 1105.
20. E. Vazquez and M. Prato, *ACS Nano*, 2009, **3**, 3819.
21. N. Karousis, N. Tagmatarchis and D. Tasis, *Chem. Rev.*, 2010, **110**, 5366.
22. P. Y. Huang, S. Kurasch, A. Srivastava, V. Skakalova, J. Kotakoski, A. V. Krasheninnikov, R. Hovden, Q. Mao, J. C. Meyer, J. Smet, D. A. Muller and U. Kaiser, *Nano Lett.*, 2012, **12**, 1081.
23. W. H. Zachariasen, *J. Am. Chem. Soc.*, 1932, **54**, 3841.
24. D. Prasai, J. C. Tuberquia, R. R. Harl, G. K. Jennings and K. I. Bolotin, *ACS Nano*, 2012, **6**, 1102.
25. A. J. Kennedy, M. S. Hull, A. J. Bednar, J. D. Goss, J. C. Gunter, J. L. Bouldin, P. J. Vikesland and J. A. Steevens, *Environ. Sci. Technol.*, 2010, **44**, 9571.
26. M. S. Hull, A. J. Kennedy, J. A. Steevens, A. J. Bednar, C. A. Weiss Jr. and P. J. Vikesland, *Environ. Sci. Technol.*, 2009, **43**, 4169.
27. M. M. Dahm, D. E. Evans, M. K. Schubauer-Berigan, M. E. Birch and J. E. Fernback, *Ann. Occup. Hyg.*, 2011, DOI: 10.1093/annhyg/mer110.
28. D. R. Johnson, M. M. Methner, A. J. Kennedy and J. A. Steevens, *Environ. Health Perspect.*, 2010, **118**, 49.
29. B. Nowack, *Environ. Pollut.*, 2009, **157**, 1063.
30. J. K. Stanley, J. G. Coleman, C. A. Weiss Jr. and J. A. Steevens, *Environ. Toxicol. Chem.*, 2010, **29**, 422.
31. J. G. Coleman, D. R. Johnson, J. K. Stanley, A. J. Bednar, C. A. Weiss Jr., R. E. Boyd and J. A. Steevens, *Environ. Toxicol. Chem.*, 2010, **29**, 1575.
32. J. D. Judy, J. M. Unrine and P. M. Bertsch, *Environ. Sci. Technol.*, 2011, **45**, 776.
33. A. Nel, T. Xia, L. Madler and N. Li, *Science*, 2006, **311**, 622.

34. R. R. Mercer, J. Scabilloni, L. Wang, E. Kisin, A. R. Murray, D. Schwegler-Berry, A. A. Shvedova and V. Castranova, *Am. J. Physiol.: Lung Cell. Mol. Physiol.*, 2008, **294**, L87.
35. R. R. Mercer, A. F. Hubbs, J. F. Scabilloni, L. Wang, L. A. Battelli, D. Schwegler-Berry and V. Castranova, *Part. Fibre Toxicol.*, 2010, **7**, 1.
36. R. F. Service, Nanotechnology's Public Health Hazard, *Science NOW*, May 2008, http://news.sciencemag.org/sciencenow/2008/05/20-01.html.
37. K. Donaldson, R. Aitken, L. Tran, V. Stone, R. Duffin, G. Forrest and A. Alexander, *Toxicol. Sci.*, 2006, **92**, 5.
38. A. Takagi, A. Hirose, T. Nishimura, N. Fukumori, A. Ogata, N. Ohashi, S. Kitajima and J. Kanno, *J. Toxicol. Sci.*, 2008, **33**, 105.
39. A. J. Kennedy, M. S. Hull, J. A. Steevens, K. M. Dontsova, M. A. Chappell, J. C. Gunter and C. A. Weiss Jr., *Environ. Toxicol. Chem.*, 2008, **27**, 1932.
40. A. J. Kennedy, J. C. Gunter, M. A. Chappell, J. D. Goss, M. S. Hull, R. A. Kirgan and J. A. Steevens, *Environ. Toxicol. Chem.*, 2009, **28**, 1930.
41. L. Wang, S. Luanpitpong, V. Castranova, W. Tse, Y. Lu, V. Pongrakhananon and Y. Rojanasakul, *Nano Lett.*, 2011, **11**, 2796.
42. D. R. Johnson, M. M. Methner, A. J. Kennedy and J. A. Steevens, *Environ. Health Perspect.*, 2010, **118**, 49.
43. Z. Sun, D. K. James and J. M. Tour, *J. Phys. Chem. Lett.*, 2011, **2**, 2425.
44. Y. Cho, S. K. Min, J. Y. Lee, W. Y. Kim and K. S. Kim, in *Practical Aspects of Computational Chemistry I*, eds. J. Leszczynski and M. K. Shukla, Springer, New York, 2012, p. 319.
45. N. L. Rangel, P. A. Leon-Plata and J. M. Seminario, in *Practical Aspects of Computational Chemistry I*, eds. J. Leszczynski and M. K. Shukla, Springer, New York, 2012, p. 347.
46. S. Irle, A. J. Page, B. Saha, Y. Wang, K. R. S. Chandrakumar, Y. Nishimoto, H.-J. Qian and K. Morokuma, in *Practical Aspects of Computational Chemistry II*, eds. J. Leszczynski and M. K. Shukla, Springer, New York, 2012.
47. T. Yumura and M. Kertesz, in *Practical Aspects of Computational Chemistry II*, eds. J. Leszczynski and M. K. Shukla, Springer, New York, 2012.
48. J. Huang, A. Beste, J. Younker, A. Vazquez-Mayagoitia, E. Cruz-Silva, M. Fuentes-Cabrera, J. Jakowski, A. Lopez-Bezanilla, V. Meunier and B. G. Sumpter, in *Practical Aspects of Computational Chemistry II*, eds. J. Leszczynski and M. K. Shukla, Springer, New York, 2012.
49. A. K. Geim and K. S. Novoselov, *Nat. Mater.*, 2007, **6**, 183.
50. A. Rodriguez-Fortea, S. Irle and J. M. Poblet, *Wiley Interdiscip. Rev.: Comput. Mol. Sci.*, 2011, **1**, 350.
51. L. Dai, in *Self Organized Organic Semiconductors: From Materials to Device Applications*, ed. Q. Li, John Wiley & Sons, Inc., New Jersey, 2011, p. 165.
52. L. Hu, D. S. Hecht and G. Gruner, *Chem. Rev.*, 2010, **110**, 5790.
53. A. C. Dillon, *Chem. Rev.*, 2010, **110**, 6856.

54. R. Saito, G. Dresselhaus and M. S. Dresselhaus, *Physical Properties of Carbon Nanotubes*, Imperial College Press, London, 1999.
55. G. Sanchez, S. Diaz-Tendero, M. Alcami and F. Martin, *Chem. Phys. Lett.*, 2005, **416**, 14.
56. A. V. Pogulay, R. R. Abzalimov, S. K. Nasibullaev, A. S. Lobach, T. Drewello and Y. V. Vasil'ev, *Int. J. Mass Spectrom.*, 2004, **233**, 165.
57. M. K. Shukla and J. Leszczynski, *Chem. Phys. Lett.*, 2006, **428**, 317.
58. J. Singleton, *Band Theory and Electronic Properties of Solids*, Oxford University Press Inc., New York, USA, 2001.
59. I. V. Hertel, H. Steger, J. de Vries, B. Weisser, C. Menzel, B. Kamke and W. Kamke, *Phys. Rev. Lett.*, 1992, **68**, 784.
60. C. Brink, L. H. Andersen, P. Hvelplund, D. Mathur and J. D. Voldstad, *Chem. Phys. Lett.*, 1995, **233**, 52.
61. H. Tokuyama, S. Yamago, E. Nakamura, T. Shiraki and Y. Sugiura, *J. Am. Chem. Soc.*, 1993, **115**, 7918.
62. F. Prat, C.-C. Hou and C. S. Foote, *J. Am. Chem. Soc.*, 1997, **119**, 5051.
63. R. Bernstein, F. Prat and C. S. Foote, *J. Am. Chem. Soc.*, 1999, **121**, 464.
64. S. Gowtham, R. H. Scheicher, R. Ahuja, R. Pandey and S. P. Karna, *Phys. Rev. B*, 2007, **76**, 33401.
65. D. Umadevi and G. N. Sastry, *J. Phys. Chem. Lett.*, 2011, **2**, 1572.
66. W. Lv, *Chem. Phys. Lett.*, 2011, **514**, 311.
67. H. Wang and A. Ceulemans, *Phys. Rev. B*, 2009, **79**, 195419.
68. A. Das, A. K. Sood, P. K. Maiti, M. Das, R. Varadarajan and C. N. R. Rao, *Chem. Phys. Lett.*, 2008, **453**, 266.
69. Y. Wang and Y. Bu, *J. Phys. Chem. B*, 2007, **111**, 6520.
70. Y. Wang, *J. Phys. Chem. C*, 2008, **112**, 14297.
71. J. Zou, W. Ling and S. Zhang, *Int. J. Numer. Meth. Eng.*, 2010, **83**, 968.
72. M. K. Shukla and J. Leszczynski, *Chem. Phys. Lett.*, 2009, **469**, 207.
73. F. Weigend, M. Haser, H. Patzelt and R. Ahlrichs, *Chem. Phys. Lett.*, 1998, **294**, 143.
74. F. Weigend, A. Kohn and C. Hattig, *J. Chem. Phys.*, 2001, **116**, 3175.
75. M. K. Shukla, M. Dubey, E. Zakar, R. Namburu, Z. Czyznikowska and J. Leszczynski, *Chem. Phys. Lett.*, 2009, **480**, 269.
76. M. K. Shukla, M. Dubey, E. Zakar, R. Namburu and J. Leszczynski, *Chem. Phys. Lett.*, 2010, **496**, 128.
77. Y. Zhao, N. E. Schultz and D. G. Truhlar, *J. Chem. Theory Comput.*, 2006, **2**, 364.
78. M. K. Shukla, M. Dubey, E. Zakar, R. Namburu and J. Leszczynski, *Chem. Phys. Lett.*, 2010, **493**, 130.
79. S. F. Boys and F. Bernardi, *Mol. Phys.*, 1970, **19**, 553.
80. X. Zhao, A. Striolo and P. T. Cummings, *Biophys. J.*, 2005, **89**, 3856.
81. J. Sponer, K. E. Riley and P. Hobza, *Phys. Chem. Chem. Phys.*, 2008, **10**, 2595.
82. S. Gowtham, R. H. Scheicher, R. Pandey, S. P. Karna and R. Ahuja, *Nanotechnology*, 2008, **19**, 125701.

83. S. G. Stepanian, M. V. Karachevtsev, A. Yu. Glamazda, V. A. Karachevtsev and L. Adamowicz, *Chem. Phys. Lett.*, 2008, **459**, 153.
84. S. G. Stepanian, M. V. Karachevtsev, A. Yu. Glamazda, V. A. Karachevtsev and L. Adamowicz, *J. Phys. Chem. A*, 2009, **113**, 3621.
85. F. Albertorio, M. E. Hughes, J. A. Golovchenko and D. Branton, *Nanotechnology*, 2009, **20**, 395101.
86. V. M. Orlov, A. N. Smirnov and Y. M. Varshavsky, *Tetrahedron Lett.*, 1976, **48**, 4377.
87. C. E. Crespo-Hernandez, R. Arce, Y. Ishikawa, L. Gorb, J. Leszczynski and D. M. Close, *J. Phys. Chem. A*, 2004, **108**, 6373.
88. V. M. Orlov, A. N. Smirnov and Y. M. Varshavsky, *Tetrahedron Lett.*, 1976, **48**, 4377.
89. C. E. Crespo-Hernández, D. M. Close, L. Gorb and J. Leszczynski, *J. Phys. Chem. B*, 2007, **111**, 5386.
90. C. J. Murphy, L. B. Thompson, A. M. Alkilany, P. N. Sisco, S. P. Boulos, S. T. Sivapalan, J. A. Yang, D. J. Chernak and J. Huang, *J. Phys. Chem. Lett.*, 2010, **1**, 2867.
91. M.-C. Daniel and D. Astruc, *Chem. Rev.*, 2004, **104**, 293.
92. Y. Nagasaki, *Chem. Lett.*, 2008, **37**, 564.
93. B. D. Chithrani, A. A. Ghazani and W. C. W. Chan, *Nano Lett.*, 2006, **6**, 662.
94. X. Huang, I. H. El-Sayed, W. Qian and M. A. El-Sayed, *Nano Lett.*, 2007, **7**, 1591.
95. S. Wang, W. Lu, O. Tovmachenko, U. S. Rai, H. Yu and P. C. Ray, *Chem. Phys. Lett.*, 2008, **463**, 145.
96. H. Takahashi, Y. Niidome, T. Niidome, K. Kaneko, H. Kawasaki and S. Yamada, *Langmuir*, 2006, **22**, 2.
97. P. Pyykko, *Chem. Soc. Rev.*, 2008, **37**, 1967.
98. P. Pyykko, *Inorg. Chim. Acta*, 2005, **358**, 4113.
99. P. Pyykko, *Angew. Chem., Int. Ed.*, 2004, **43**, 4412.
100. H.-G. Boyen, G. Kästle, F. Weigl, B. Koslowski, C. Dietrich, P. Ziemann, J. P. Spatz, S. Riethmüller, C. Hartmann, M. Möller, G. Schmid, M. G. Garnier and P. Oelhafen, *Science*, 2002, **297**, 1533.
101. J. Li, X. Li, H.-J. Zhai and L.-S. Wang, *Science*, 2003, **299**, 864.
102. E. S. Kryachko and F. Remacle, *Int. J. Quantum Chem.*, 2007, **107**, 2922.
103. K. J. Taylor, C. L. Pettiette-Hall, O. Cheshnovsky and R. E. Smalley, *J. Chem. Phys.*, 1992, **96**, 3319.
104. B. D. Chithrani, A. A. Ghazani and W. C. W. Chan, *Nano Lett.*, 2006, **6**, 662.
105. M. K. Shukla, M. Dubey, E. Zakar and J. Leszczynski, *J. Phys. Chem. C*, 2009, **113**, 3960.
106. P. J. Mohan, A. Datta, S. S. Mallajosyula and S. K. Pati, *J. Phys. Chem. B*, 2006, **110**, 18661.
107. P. Sharma, H. Singh, S. Sharma and H. Singh, *J. Chem. Theory Comput.*, 2007, **3**, 2301.
108. A. Martinez, *J. Phys. Chem. A*, 2009, **113**, 1134.

109. E. E. Connor, J. Mwamuka, A. Gole, C. J. Murphy and M. D. Wyatt, *Small*, 2005, **1**, 325.

110. E. S. Kryachko and F. Remacle, *Nano Lett.*, 2005, **5**, 735.

111. E. S. Kryachko and F. Remacle, *J. Phys. Chem. B*, 2005, **109**, 22746.

112. A. Kumar, P. C. Mishra and S. Suhai, *J. Phys. Chem. A*, 2006, **110**, 7719.

113. M. K. Shukla and J. Leszczynski, *J. Biomol. Struct. Dyn.*, 2007, **25**, 93.

114. N. A. Richardson, S. S. Wesolowski and H. F. Schaefer, III, *J. Am. Chem. Soc.*, 2002, **124**, 10163 and references cited therein.

115. D. M. Close, *J. Phys. Chem. A*, 2004, **108**, 10376.

116. T. Caruso, M. Carotenuto, E. Vasca and A. Peluso, *J. Am. Chem. Soc.*, 2005, **127**, 15040.

117. L. M. Demers, M. Ostblom, H. Zhang, N.-H. Jang, B. Liedberg and C. A. Mirkin, *J. Am. Chem. Soc.*, 2002, **124**, 11248.

118. S. Frank, P. Poncharal, Z. L. Wang and W. A. de Heer, *Science*, 1998, **280**, 1744.

119. A. Bachtold, M. S. Fuhrer, S. Plyasunov, M. Forero, E. H. Anderson, A. Zettl and P. L. McEuen, *Phys. Rev. Lett.*, 2000, **84**, 6082.

120. J. Nygard, D. H. Cobden and P. E. Lindelof, *Nature*, 2000, **408**, 342.

121. H. Park, J. Park, A. K. L. Lim, E. H. Anderson, A. P. Alivisatos and P. L. McEuen, *Nature*, 2000, **407**, 57.

122. R. Jorn and T. Seideman, *J. Chem. Phys.*, 2006, **124**, 84703.

123. C.-C. Kaun and T. Seideman, *Phys. Rev. Lett.*, 2005, **94**, 226801.

124. S. Alavi, B. Larade, J. Taylor, H. Guo and T. Seideman, *Chem. Phys.*, 2002, **281**, 293.

125. A. N. Andriotis and M. Menon, *Phys. Rev. B*, 1999, **60**, 4521.

126. J. T. Lyon and L. Andrews, *ChemPhysChem*, 2005, **6**, 229.

127. D. L. Lichtenberger, L. L. Wright, N. E. Gruhn and M. E. Rempe, *Synth. Met.*, 1993, **59**, 353.

128. D. L. Lichtenberger, L. L. Wright, N. E. Gruhn and M. E. Rempe, *J. Organomet. Chem.*, 1994, **478**, 213.

129. J. J. Palacios, *Phys. Rev. B*, 2005, **72**, 125424.

130. H. Nakamura and K. Yamashita, *J. Chem. Phys.*, 2006, **125**, 194106.

131. N. Sergueev, D. Roubtsov and H. Guo, *Phys. Rev. Lett.*, 2005, **95**, 146803.

132. J. M. Seminario, C. E. De La Cruz and P. A. Derosa, *J. Am. Chem. Soc.*, 2001, **123**, 5616.

133. H. I. Li, K. Pussi, K. J. Hanna, L.-L. Wang, D. D. Johnson, H.-P. Cheng, H. Shin, S. Curtarolo, W. Moritz, J. A. Smerdon, R. McGrath and R. D. Diehl, *Phys. Rev. Lett.*, 2009, **103**, 056101.

134. M. K. Shukla, M. Dubey and J. Leszczynski, *ACS Nano*, 2008, **2**, 277.

135. M. K. Shukla, M. Dubey, E. Zakar and J. Leszczynski, *J. Phys. Chem. C*, 2012, **116**, 1966.

136. M. K. Shukla, M. Dubey, E. Zakar and J. Leszczynski, *J. Phys. Chem. C*, 2009, **113**, 11351.

137. E. I. Altman and R. J. Colton, *Surf. Sci.*, 1992, **279**, 49.

138. E. I. Altman and R. J. Colton, *Phys. Rev. B*, 1993, **48**, 18244.
139. C.-T. Tzeng, W.-S. Lo, J.-Y. Yuh, R.-Y. Chu and K.-D. Tsuei, *Phys. Rev. B*, 2000, **61**, 2263.
140. R. J. Baxter, P. Rudolf, G. Teobaldi and F. Zerbetto, *ChemPhysChem*, 2004, **5**, 245.
141. T. Ono and K. Hirose, *Phys. Rev. Lett.*, 2007, **98**, 26804.
142. K. Yoshizawa, T. Tada and A. Staykov, *J. Am. Chem. Soc.*, 2008, **130**, 9406.
143. R. W. Lof, M. A. van Veenendaal, B. Koopmans, H. T. Jonkman and G. A. Sawatzky, *Phys. Rev. Lett.*, 1992, **68**, 3924.
144. H. A. Witek, S. Irle, G. Zheng, W. A. de Jong and K. Morokuma, *J. Chem. Phys.*, 2006, **125**, 214706.
145. A. L. Balch and M. M. Olmstead, *Chem. Rev.*, 1998, **98**, 2123.
146. D. Feller, *J. Comput. Chem.*, 1996, **17**, 1571.
147. K. L. Schuchardt, B. T. Didier, T. Elsethagen, L. Sun, V. Gurumoorthi, J. Chase, J. Li and T. L. Windus, *J. Chem. Inf. Model.*, 2007, **47**, 1045.
148. K. Hedberg, L. Hedberg, D. S. Bethunde, C. A. Brown, H. C. Dorn, R. D. Johnson and M. de Vries, *Science*, 1991, **254**, 410.
149. H. M. Lee, M. Ge, B. R. Sahu, P. Tarakeshwar and K. S. Kim, *J. Phys. Chem. B*, 2003, **107**, 9994.
150. R. Yu, Q. Liu, K.-L. Tan, G.-Q. Xu, S. C. Ng, H. S. O. Chan and T. S.A. Hor, *J. Chem. Soc., Faraday Trans.*, 1997, **93**, 2207.
151. A. Hayashi, S. Yamamoto, K. Suzuki and T. Matsuoka, *J. Mater. Chem.*, 2004, **14**, 2633.
152. M. Kozlowski, R. Diduszko, K. Olszewska, H. Wronka and E. Czerwosz, *Vacuum*, 2008, **82**, 956.
153. L. T. Cai, H. Skulason, J. G. Kushmerick, S. K. Pollack, J. Naciri, R. Shashidhar, D. L. Allara, T. E. Mallouk and T. S. Mayer, *J. Phys. Chem. B*, 2004, **208**, 2827.
154. J. Weckesser, J. V. Barth and K. Kern, *Phys. Rev. B*, 2001, **64**, 161403.

CHAPTER 8

Theoretical Studies of Interactions in Nanomaterials and Biological Systems

HARALAMBOS TZOUPIS[a,b],
AGGELOS AVRAMOPOULOS*[a], HERIBERT REIS[a],
GEORGIOS LEONIS*[a], SERDAR DURDAGI[c],
THOMAS MAVROMOUSTAKOS[b],
GRIGORIOS MEGARIOTIS[a,d] AND
MANTHOS G. PAPADOPOULOS*[a]

[a] National Hellenic Research Foundation, Institute of Biology, Medicinal Chemistry and Biotechnolodgy, 48 Vas. Konstantinou, 11635, Athens, Greece; [b] National and Kapodistrian University of Athens, Department of Chemistry, Panepistimioupolis Zographou, Athens 15771, Greece; [c] University of Calgary, Department of Biological Sciences, Institute of Biocomplexity and Informatics, 2500 University Drive, AB T2N 1N4, Calgary, Canada; [d] National Technical University of Athens, Department of Chemical Engineering, 9 Heroon Polytechniou Street, 15780, Athens, Greece
*E-mail: aavram@eie.gr or gleonis@eie.gr or mpapad@eie.gr

8.1 Introduction

In this article, we review some of our recent work and related articles by other authors, which deal with various aspects of nanomaterials and biological complexes and in particular, the effects of interactions on selected properties.

RSC Nanoscience & Nanotechnology No. 25
Towards Efficient Designing of Safe Nanomaterials: Innovative Merge of Computational Approaches and Experimental Techniques
Edited by Tomasz Puzyn and Jerzy Leszczynski
© The Royal Society of Chemistry 2012
Published by the Royal Society of Chemistry, www.rsc.org

The second section deals with the electronic contribution to the (hyper)polarizabilities of Li@C_{60} and [Li@C_{60}]$^+$. These properties have been computed by employing a series of methods (*e.g.* UB3LYP, unrestricted Møller–Plesset second-order perturbation theory (UMP2)) and basis sets (*e.g.* 6-31+G*). The linear and non-linear optical (L&NLO) properties of Li@C_{60} and [Li@C_{60}]$^+$ are compared with those of C_{60} and [C_{60}]$^+$. The effect of the interaction of Li with the cage, on the properties of interest is discussed. We have also studied the L&NLO properties of Li@C_{60} as a function of the distance between the Li atom and the center of the cage. The nuclear relaxation (nr) contribution to the L&NLO properties of Li@C_{60} and [Li@C_{60}]$^+$ has also been computed. In the third section, we have considered the effect of encapsulation of Sc_2 in C_{72} on the L&NLO properties of Sc_2@C_{72}.

The fourth section deals with the L&NLO properties of Ti@C_{28}. Some spectra of Ti@C_{28} and $C_{28}H_4$ have been computed: (i) the UV-Vis spectra have been calculated by employing the time-dependent density functional theory (TDDFT) method with various functionals (*e.g.* BLYP, B3LYP), and (ii) the IR and Raman spectra have been determined in harmonic approximation using the B3LYP and CAM-B3LYP functionals. The static diagonal components of the electronic (hyper)polarizabilities of Ti@C_{28}, C_{28} and $C_{28}H_4$ have been calculated by various methods (*e.g.* restricted open-shell Møller–Plesset second-order perturbation theory (ROMP2)). The effect of the interaction of Ti with C_{28} on the L&NLO properties is discussed. The vibrational contributions (double harmonic approximation) to the L&NLO properties of Ti@C_{28} have been also computed and analyzed.

The fifth section reviews some recent works, which deal with the interaction between receptors and ligands. These interactions are important because (i) they promote our understanding of the biophysical properties of the ligand–receptor system and (ii) they are an essential tool for the design of novel inhibitors. We also discuss the interactions of two fullerene derivatives with human immunodeficiency virus type 1 aspartic protease (HIV-1 PR). The binding free energy is resolved into various contributions (*e.g.* electrostatic, van der Waals) by using the molecular mechanics Poisson–Boltzmann surface area (MM–PBSA) method.

The sixth section reviews an *in silico* approach to the design of novel fullerene inhibitors for HIV-1 PR. Experimental results confirmed the theoretical predictions based on docking calculations. One of the proposed derivatives has about three times better inhibitory binding than the most active, currently available, fullerene-based inhibitor.

The seventh section considers the interactions of the enzymes cyclooxygenase-2 (COX-2) and lipoxygenase-5 (LOX-5) with 4-[(2*S*)-2-(1*H*-imidazol-1-ylmethyl)-5-oxotetrahydro-1*H*-pyrrol-1-yl]methylbenzenecarboxylic acid (MMK16) by employing molecular dynamics and molecular docking techniques. MM–PBSA calculations were also used to study the interaction of aliskiren with renin in the eighth section.

The ninth section reviews the drug–membrane interactions. Two sartans, losartan and candesartan CV-11974, have been considered. Various experimental techniques have been used to study the above interactions (*e.g.* Raman and solid-state ^{31}P NMR spectroscopies).

8.2   Li@C$_{60}$

Since the discovery of C$_{60}$ in 1985,[1] fullerene-based materials have been extensively studied due to possible applications in nanotechnology.[2] Among these are the endohedral fullerenes in which an atom or a molecular unit is encapsulated inside the carbon cage. A possible use of an endohedral C$_{60}$ compound is that it can act as a memory device; if the trapped atom could be moved inside the cage between two stable positions, the position could encode the state of the device.[3]

A large number of studies has been performed to explore several effects associated with the stability, structure, spectroscopy, electronic and optical properties of several endohedral fullerene derivatives. These endohedral fullerenes include many atomic as well as molecular species trapped inside carbon cages, such as metal atoms, rare gas atoms and small molecules.[4–33] All of these studies mainly deal with the effect on the electronic structure and stability of the resulting endohedral system associated with the charge transfer and covalent interactions between the fullerene cage and the entrapped unit.

An interesting topic is the determination of the linear and non-linear optical properties of endohedral fullerenes. For this purpose, three systems have been selected: the endohedral fullerenes of Li@C$_{60}$,[Li@C$_{60}$]$^{+}$ and Sc$_2$@C$_{72}$. There is experimental evidence for Li@C$_{60}$.[34] Of interest is the motion of Li inside the C$_{60}$ cage which is associated with large amplitude vibrational motions and could induce large electronic and vibrational (hyper)polarizabilities. Whitehouse and Buckingham computed the vibrational contribution to the linear polarizability and second hyperpolarizability of the [Li@C$_{60}$]$^{+}$ cation by employing a simplified potential followed by a classical analysis to obtain temperature-dependent expressions.[35] Their results indicated that the vibrational contribution could be larger than the electronic one. However, a study on the vibrational polarizability of Kr@C$_{60}$ by Pederson *et al.*,[36] employing the double harmonic approximation, showed that this contribution was found to be quite small compared with the electronic one.

Regarding the electronic contribution to the first (β) and second (γ) hyperpolarizabilities of the neutral Li@C$_{60}$, one may note the studies by Campbell *et al.*[37,38] and Yaghobi *et al.*[39,40] Campbell *et al.* computed the diagonal components of the first hyperpolarizability employing an uncoupled Hartree–Fock scheme with the molecular orbitals obtained from the restricted open-shell Hartree–Fock procedure. In this study, an experimental measurement of the second harmonic generation response of Li@C$_{60}$ thin films is also reported.[37] In ref. 38 the diagonal components of the second hyperpolarizability are computed with the same approximation as in ref. 37. Moreover an experimental measurement of the second hyperpolarizability has also been obtained by employing the z-scan and the degenerate four-wave-mixing

(DFWM) procedures.[38] The authors concluded that the neutral Li@C_{60} has a response 3–5 times larger than that of pure C_{60}.

Recently, Reis *et al.* performed a detailed study of the (hyper)polarizabilities of neutral Li@C_{60} and its cation.[41] Details on the computational procedure, the methods and the software used to compute the electronic and vibrational contributions to the (hyper) polarizabilities are given in ref. 41. The first step aimed to determine the most stable geometric arrangement. For this purpose, geometry optimizations were carried out for near C_{3v} and C_{5v} symmetries employing the same level of approximation, allowing the cage to fully relax. In the case of I_{h-} symmetry, the cage was slightly distorted giving a C_s optimized structure with the Li atom slightly shifted (0.015 Å) from the center of the cage. However, it is known that this does not correspond to a minimum; it is associated with four imaginary frequencies. For the two minima of near-C_{5v} and near-C_{3v} symmetry the Li atom is located at 1.5 Å from the center of cage and the symmetry reduces again to C_s. The previous results are in a semi-quantitative agreement with the findings of Zhang *et al.*[42] The eccentric position of the trapped Li has been interpreted in terms of dispersion and repulsion interactions.[43] It was found that the approximate C_{3v} structure is stabilized by 3.4 kJ mol^{-1} (1.6 kJ mol^{-1}) than the C_{5v} one, employing the UB3LYP/6-311G*(UB3LYP/6-31G) method. The energy difference between the near-I_h and near-C_{3v} symmetry structures is 56.5 kJ mol^{-1} at the UB3LYP/6-31G level of theory. The property values we report were computed by employing the most stable near-C_{3v} structure.

Regarding the structure of the cation [Li@C_{60}]$^+$, at the near-C_{3v} symmetry optimized structure, the position of the Li atom is located at 1.4 Å from the center of the cage. However, it is noted that the cation structure is more spherical than the neutral one. For this we used the following quantity $\Delta I = ((I_x - I_y)^2 + (I_x - I_z)^2 + (I_y - I_z)^2)^{1/2}$, where I_i, $i = x,y,z$, corresponds to the principal component of the inertia tensor in the i-direction, with respect to the center of mass. The values of ΔI were computed as 1.4 and 31.5 g Å^2 mol^{-1} for the cation and the neutral structures, respectively.

The computed static electronic properties of Li@C_{60} and [Li@C_{60}]$^+$, along with the dipole moment direction (z), are presented in Table 8.1.

It is noted that augmentation of the basis set with diffuse functions has a large effect on the computed properties, especially on the second hyperpolarizability. Further addition of polarization functions is less important for α_{zz} and γ_{zzzz}. However, for β_{zzz} the polarization functions seem to offset the effect of the diffuse functions. A large change of the computed hyperpolarizabilities is observed on going from the cation to the neutral. This is attributed to the additional electron in a (formerly) unoccupied orbital, which is localized on the C_{60}.[41]

In Table 8.2 some components of the properties of Li@C_{60}, [Li@C_{60}]$^+$, C_{60}^- and C_{60} are shown, computed at the (U)B3LYP/6-31+G* level of theory. Our scope is to investigate the effect of the interaction on the (hyper)polarizabilities between the hypothetical non-interacting species, Li$^+$+C_{60} with [Li@C_{60}]$^+$ and

Table 8.1 The electronic contribution to the polarizability, first and second hyperpolarizability of Li@C$_{60}$ and [Li@C$_{60}$]$^+$, employing a series of methods and basis sets. All the computations were performed at the (U)B3LYP/6-31G optimized geometry. The property values are in a.u.

| Method | Li@C$_{60}$ | | | | | [Li@C$_{60}$]$^+$ | | | |
| | UB3LYP | UB3LYP | UB3LYP | UMP2 | | UMP2 | B3LYP | B3LYP | MP2 |
Property	6-31+G	6-31+G*	6-31G	6-31G		6-31+G	6-31+G*	6-31G	6-31G
α_{zz}	578.6	589.0	508	527		520	534	469	464
β_{zzz}	1839	1532	1540	1362		−53	−118	−237	−192
$\gamma_{zzzz} \times 10^{-3}$	64	66	−39			99	99	28	

Table 8.2 The electronic contribution to the components of the polarizability, first and second hyperpolarizability of Li@C_{60}, [Li@C_{60}]$^+$, [C_{60}]$^-$ and C_{60}, calculated at the (U)B3LYP/6-31+G* level of theory. All property values are in a.u.

	Li@C_{60}	[Li@C_{60}]$^+$	C_{60}[a]	[C_{60}]$^-$[b]
α_{xx}	560	534	550	576
α_{yy}	590	534	550	618
α_{zz}	589	534	551	620
β_{xxy}	98			
β_{xzy}		−58		
β_{yyy}	−290	0	0	−88
β_{yxz}	−1104			
β_{zzz}	1532	−118	−11	441
β_{zxz}		−52		
β_{zyz}	514	−52		
$\gamma_{xxxx} \times 10^{-3}$	293	102	136	211
$\gamma_{yyyy} \times 10^{-3}$	−20	102	135	−46
$\gamma_{zzzz} \times 10^{-3}$	66	99	136	−86

[a]Structure computed at the UB3LYP/6-31G optimized geometry of [Li@C_{60}]$^+$. [b]Structure computed at the UB3LYP/6-31G optimized geometry of Li@C_{60}.

Li$^+$+$C_{60}$$^-$ with [Li@C_{60}]. The properties of Li$^+$ are negligible[44] and are not presented.

It can be seen that the interaction of Li$^+$ with $C_{60}$$^-$ or C_{60} leads to a small reduction of the polarizability. However, the effect of the above interaction on the second hyperpolarizability of [Li@C_{60}]$^+$ is larger.

The reduction of the polarizability components of [Li@C_{60}]$^+$ may be attributed to the contraction of the electron density caused by the attraction of the cation. It is noted that the first hyperpolarizability arises from the asymmetry of the charge distribution and is enhanced in the resulting endohedral systems.

In order to study how the lithium position in C_{60} affects the electric properties of Li@C_{60}, a series of computations was performed, where we considered the property of Li@C_{60} as a function of the distance of Li from the center of mass of the cage (Figure 8.1). All computations were performed by using the B3LYP/6-31+G* method and are shown in Table 8.3. Only the z component of the property tensor has been computed.

It is observed that the effect of the position of Li on the polarizability and first hyperpolarizability of Li@C_{60} is moderate. However, the above effect on the second hyperpolarizability is remarkable, where a change of the sign is observed.

The static and some frequency dependent nuclear relaxation (nr) contributions to the (hyper)polarizabilities were also computed in addition to the static electronic properties. The method used to compute the values of the various components is based on geometry optimization in the presence of a field. A problem which may arise is the existence of multiple minima on the potential

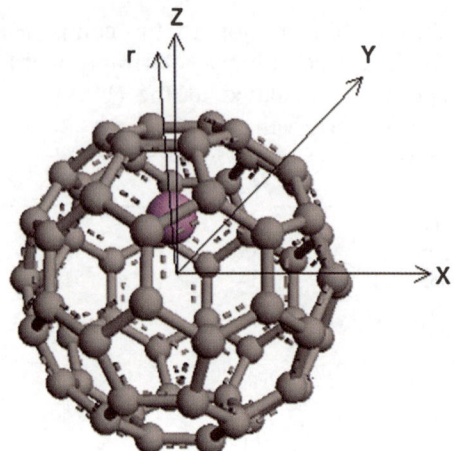

Figure 8.1   The displacement of a Li atom in the C_{60} cage.

Table 8.3   The electronic polarizability, first and second hyperpolarizability
of Li@C_{60} as a function of the distance between the Li atom and
the center of the cage (r_{Li-0}/Å). All property values were computed
with the UB3LYP/6-31+G* method and are in a.u.

r_{Li-0}	α_{zz}	β_{zzz}	$\gamma_{zzzz} \times 10^{-3}$
0	598	416	−86
0.729	594	1112	−19
0.958	593	1278	8
1.158	591	1396	32
1.358	590	1492	55
1.558	588	1569	76
2.0	584	1676	123

energy surface separated by low energy barriers. The finite-field approach
works satisfactorily as long as the field-dependent optimized structure
corresponds to the same minimum as the field-free one. An extensive study
on this subject is presented in ref. 45. However, for the case of the studied
endohedral fullerenes, it was not possible to determine the nr contribution
along the x-axis and an analytical procedure was followed.[41] The results of the
computations are summarized in Table 8.4.

It is observed that the vibrational contributions to the polarizability
components are significantly smaller than their electronic counterpart.
However, the vibrational contribution to the hyperpolarizabilities is large. In
particular, for the cation, the x and y diagonal components are larger than the
electronic ones (Table 8.2). It is also observed that the frequency-dependent nr
contributions for the first hyperpolarizability connected with the dc-Pockels
effect of Li@C_{60} are smaller compared with the corresponding electronic ones
of Table 8.2, but not negligible.

Table 8.4 The nuclear relaxation (nr) contribution to the components of the polarizability, first[a] and second hyperpolarizability of Li@C$_{60}$ and [Li@C$_{60}$]$^{+}$. All values were computed with the UB3LYP/6-31G method. The property values are given in a.u.

	Li@C$_{60}$	[Li@C$_{60}$]$^{+}$
α_{xx}^{nr}	14.7	10.4
α_{yy}^{nr}	10.3	9.4
α_{zz}^{nr}	10.2	4.5
β_{yyy}^{nr}	−126	95
β_{zzz}^{nr}	795	18
$\beta_{xxz}^{nr}\ (-\omega;\omega,0)_{\omega\to\infty}$	−199	
$\beta_{yyy}^{nr}\ (-\omega;\omega,0)_{\omega\to\infty}$	−33	
$\beta_{yyz}^{nr}\ (-\omega;\omega,0)_{\omega\to\infty}$	119	
$\beta_{zzy}^{nr}\ (-\omega;\omega,0)_{\omega\to\infty}$	−9	
$\beta_{zzz}^{nr}\ (-\omega;\omega,0)_{\omega\to\infty}$	200	
$\gamma_{xxxx}^{nr}\ \times 10^{-3}$		560
$\gamma_{yyyy}^{nr}\ \times 10^{-3}$	−90	190
$\gamma_{zzzz}^{nr}\ \times 10^{-3}$		37

[a]The frequency-dependent properties were computed employing the infinite optical frequency approximation.[41]

8.3 Sc$_2$@C$_{72}$

All the properties were computed by using the fully optimized geometry. For this purpose, the Hartree–Fock (HF) theory was used with the 6-31+G* basis set for C and the effective core potential for Sc developed by the Stuttgart–Dresden group.[46] This treats Sc as an atom with 11 valence electrons and the employed basis set involves $6s5p3d1f$. The computed structure is in agreement with that computed by Slanina *et al.*[16] (Figure 8.2).

The shortest distance between Sc and C is 2.174 Å, in agreement with ref. 16. The two Sc atoms lie along the D_2 axis, while their interatomic distance is 4.7

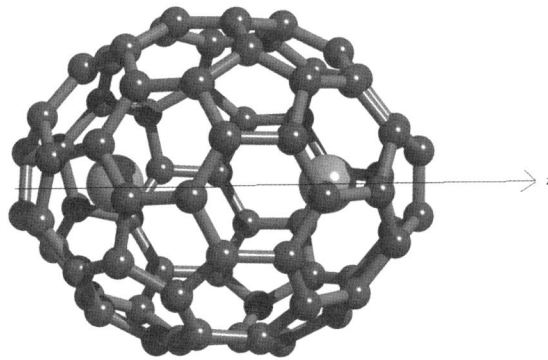

Figure 8.2 The HF optimized geometry of Sc$_2$@C$_{72}$ endohedral metallofullerene.

Å. This distance is longer than that in the isolated Sc_2 (2.28 Å, computed at the HF/6-31+G* level).

The above methodology employed for the structure optimization was used for the computation of the electronic polarizability and second hyperpolarizability of $Sc_2@C_{72}$. The usefulness of the effective core potentials for the computation of polarizabilities and hyperpolarizabilities has been discussed extensively in the literature and has been shown to provide satisfactory results for the NLO properties, compared with all-electron calculations. In order to safeguard the numerical stability of our results the Romberg fitting procedure was employed, which requires a number of computed energies, evaluated by using a number of field steps of magnitude $2^k F$, where $k = 0, 1, 2,...$ and $F = 0.001$ a.u. Since the two Sc atoms lie along the z-axis (D_2) we have restricted our computations to the z-component of the molecular property tensor.

It is known that vibrations play a key role on the reliable determination of the non-linear optical properties of molecules and clusters. Therefore, it is of interest to study their effect on the computed NLO properties of this metallofullerene. For this purpose, the field-induced coordinate (FIC) approach has been employed.[47] With this method the total nr contribution to the static polarizability, and the dynamic nr contribution of the electric-field-induced second harmonic generation (EFSHG) second hyperpolarizability, were calculated. The computational scheme involves the optimization and the evaluation of a number of necessary energy and property derivatives in a numerical fashion. Details can be found in ref. 48.

All the results of the computed properties are presented in Table 8.5. It can be seen that for the polarizability the electronic (el) contribution is dominant over the vibrational one. It was found that $\alpha_{zz}^{nr} = 30.59$ a.u., which is much smaller than its electronic counterpart ($\alpha_{zz}^{el.} = 749.13$ a.u.). Employing the infinite frequency approximation, which is known to be accurate at typical measurements we found that $\gamma_{zzzz}(-2\omega;\omega,\omega,0)^{nr}{}_{\omega \to \infty} = 98\,800$ a.u. This value is large, but it is smaller, compared with the electronic one ($\gamma_{zzzz}{}^{el} = 193\,100$ a.u.). It is observed that encapsulation of Sc_2 in C_{72} leads to a reduction of α_{zz} and γ_{zzzz} (Table 8.5).

Table 8.5 The electronic and nuclear relaxation contributions to the polarizability and second hyperpolarizability of $Sc_2@C_{72}$. All values are in a.u.

	Electronic contribution[a]	*Nuclear relaxation contribution*[b]
$\alpha_{zz}(0;0)$	749.13	30.59
	777.41[b]	
$\gamma_{zzzz}(0;0)$	193100	
	297000[b]	
$\gamma_{zzzz}(-2\omega;\omega,\omega,0)_{\omega \to \infty}$	98800	

[a] All the necessary derivatives were evaluated by employing the 6-31+G* basis function to carbon and the SDD pseudopotential to Sc. [b] C_{72} properties. The 6-31+G* basis set was used for carbon.

8.4   Ti@C$_{28}$

Skwara *et al.* performed an exhaustive computational study of the endohedral fullerene Ti@C$_{28}$.[49] They investigated the potential energy surface, as well as its bonding properties, electronic spectra, vibrational spectra, and determined its electronic and vibrational (hyper)polarizabilities, and compared several of these properties with those of the empty fullerene C$_{28}$, as well as with C$_{28}$H$_4$. The ground state of C$_{28}$ has been predicted previously to be a quintet open shell of T_d symmetry. By contrast, the endohedral fullerene Ti@C$_{28}$ as well as C$_{28}$H$_4$ are predicted to be singlets. C$_{28}$ has been found in abundance as a product of laser vaporization of C$_{60}$, but appears to be very reactive. Specifically, it forms very stable endohedral complexes with tetravalent transition metals like Ti, Zr, Hf, U.

At the RMP2/6-31G(d) level, the ground state of Ti@C$_{28}$ was found to be of C_{3v} symmetry, about 2.3 kcal mol^{-1} below a transition state of C_{2v} symmetry, in contrast to a previous report,[50] which predicted the state of C_{2v} symmetry as the ground state, using a lower level of theory. The Ti atom in the C_{3v} structure was shifted by 0.4579 Å along the C_3 axis, with respect to the origin in the center of nuclear charge. Additional higher-order saddle points of C_{3v} (second-order saddle point) and T_d symmetry (third-order saddle point) on the potential energy surface were found at 7.1 and 47.8 kcal mol^{-1}, respectively, above the global minimum. Additionally to the stationary points, the interconnection of all the different stationary points on the potential energy surface was explored. The same energetic ordering of the stationary points was found at the HF and B3LYP levels of theory, using again the 6-31G(d) basis set. The equilibrium structure of C$_{28}$ was assumed to be the $^5A_2$ state of T_d symmetry and computed at the ROMP2/6-31(d) level. At the unrestricted Møller–Plesset second-order perturbation theory (UMP2) level, considerable spin contamination was detected. Comparison of the computed C$_{28}$($^5A_2$) and Ti@C$_{28}$($^1A_1$) structures, as well as the Ti($^3F_2$) ground state, at the R(O)MP2/6-31G(d) levels, and taking into account a Boys–Bernardi counterpoise correction, yielded a binding energy of 181.3 kcal mol^{-1} for the endohedral complex, compared to literature values of 258.3 kcal mol^{-1} at the LDA/VTZP[51] and -18.4 kcal mol^{-1} at the HF/DZ level.[50] The large value computed indicates a strong bonding between Ti and the cage. The character of this bonding was further explored with an Atoms In Molecules (AIM) analysis, which revealed that although the four Ti...C interactions are essentially of closed-shell character, they are at the same time partially covalent.

As a next step in the characterization of Ti@C$_{28}$, the UV-Vis spectrum was computed at the adiabatic TDDFT level and compared with that of the exohedral counterpart C$_{28}$H$_4$ (Figure 8.3). For this purpose, the first 200 electronic states were determined with the BLYP, B3LYP and CAM-B3LYP functionals, using the 6-31+G(d) basis set and the MP2/6-31G(d) optimized geometry. BLYP was chosen according to a previous report, which indicated that non-hybrid functionals perform reasonably well in the calculation of electronic absorption spectra of fullerenes.[52] For the fullerenes investigated in

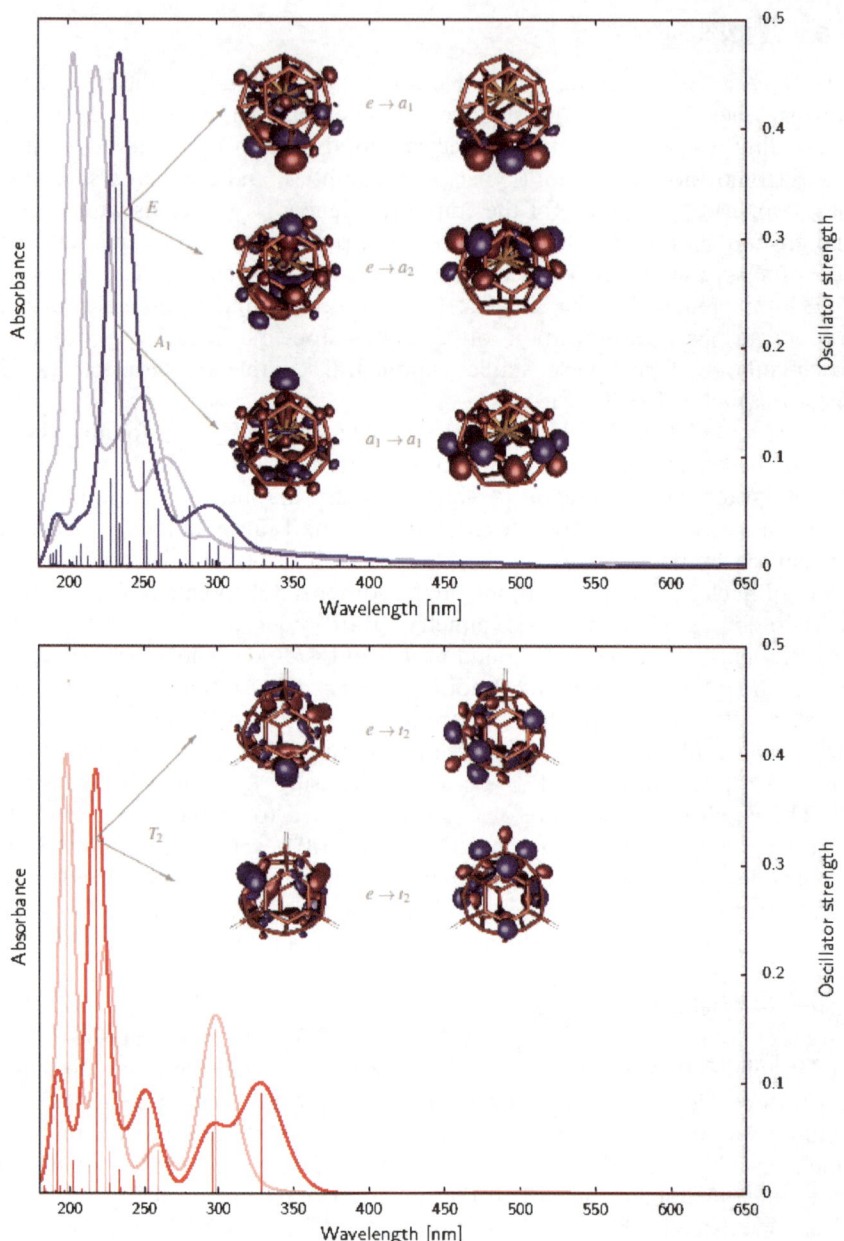

Figure 8.3   UV-Vis spectra of $Ti@C_{28}$ (top) and $C_{28}H_4$ (bottom) computed by the TDDFT method using BLYP (solid dark lines), B3LYP (solid light lines), and CAM-B3LYP (dashed lines, only on top figure) functionals and the 6-31+G(d) basis set. The line positions were convoluted with Gaussian functions with full widths at half-maximum of 3000 cm^{-1}. Reprinted with permission from ref. 49. Copyright (2011) American Chemical Society.

ref. 49, it turned out that the hybrid functional B3LYP as well as the long-range-corrected functional CAM-B3LYP predicted qualitatively similar absorption spectra as BLYP, with the same origin of the most intense lines, although slightly blue-shifted. Unfortunately, neither experimental spectra nor previous calculations exist for comparison.

In line with what has been found for other fullerenes, the first excited state of Ti@C_{28}, located at 650 nm and was dominated by the highest occupied molecular orbital (HOMO) to lowest unoccupied molecular orbital (LUMO) transition, is dipole-forbidden and does not occur in the computed spectrum (Figure 8.3). The same holds for most lines occurring at wavelengths above 300 nm. The most intense lines found with the BLYP functional occur at 232 nm and 236 nm and are of A_1 and E symmetry, respectively. The major contributing molecular orbitals involved in these transitions are also shown in Figure 8.3. With increasing exact exchange contribution, the most intense lines are blue-shifted by 20 nm (B3LYP) and 35 nm (CAM-B3LYP).

Comparison of the Ti@C_{28} spectrum with that of the exohedral $C_{28}H_4$ reveals the influence of the Ti encapsulation on the electronic absorption spectrum. In general, both spectra are quite similar, with the main difference being that the two high-intensity lines of the Ti@C_{28} spectrum merge into one single line of T_2 symmetry which, in addition, is blue-shifted (by 217 nm and 198 nm with BLYP and B3LYP, respectively). It is argued that the main effect of Ti encapsulation is the redshift of the largest peak, accompanied by an intensity increase of *ca.* 20%.

In addition to the electronic absorption spectra, IR and Raman spectra of Ti@C_{28} and $C_{28}H_4$ were also computed at the harmonic oscillator level using both the B3LYP and CAM-B3LYP functionals, together with the 6-31+G(d) basis set. Figure 8.4 shows the resulting spectra after convolution of each line with a Lorentzian with a full width at half-maximum of 10 cm^{-1}. Previous studies[53–55] reported that B3LYP in combination with a basis set containing both diffused and polarized functions performs quite well for vibrational frequencies, IR and Raman intensities, compared with experiment and/or high-level *ab initio* calculations. Thus, a reasonably good accuracy is also assumed for the corresponding properties of Ti@C_{28} and $C_{28}H_4$ reported in ref. 49, for which again no experimental values are available for comparison. The most relevant difference between the vibrational features of Ti@C_{28} and $C_{28}H_4$ with respect to the electric properties is a low-frequency doubly-degenerate vibration of the former, associated with large displacements of the Ti atom inside the cage.

The non-zero diagonal electronic contributions to the electrical properties (dipole, polarizability, first and second hyperpolarizabilities) of Ti@C_{28}, C_{28}, and $C_{28}H_4$ calculated at the R(O)HF and R(O)MP2 levels of theory with the 6-31+G(d) basis set, are shown in Table 8.6. The coordinate axes are chosen in such a way that z coincides with the C_3 symmetry axis, and $\beta_{xxx} = 0$. The effect of electron correlation on the hyperpolarizabilities appeared to be significant. The reduction of the magnitude of the hyperpolarizabilities upon encapsula-

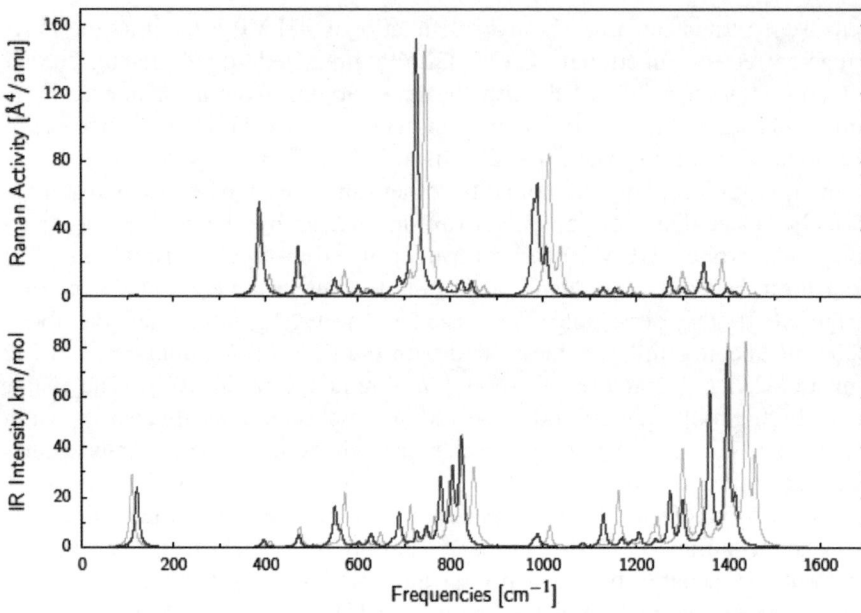

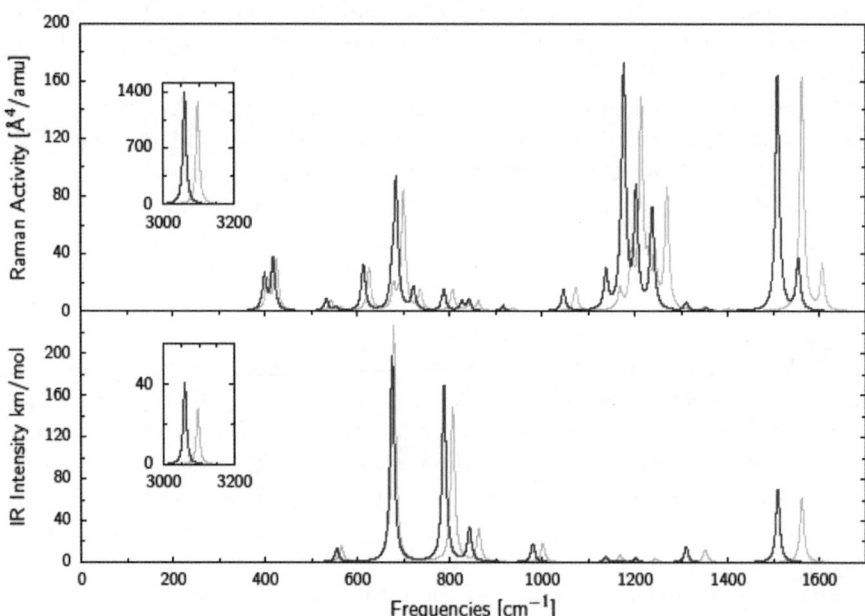

Figure 8.4 IR and Raman spectra of Ti@C$_{28}$ (top) and C$_{28}$H$_4$ (bottom) in harmonic approximation using B3LYP (dark blue lines) and CAM-B3LYP (light blue lines) functionals and the 6-31+G(d) basis set. The insets show the C–H stretching bands above 3000 cm^{-1}, which are present only in C$_{28}$H$_4$. Reprinted with permission from ref. 49. Copyright (2011) American Chemical Society.

Table 8.6   Static diagonal components of the electronic dipole and (hyper)polarizabilities of Ti@C$_{28}$, C$_{28}$H$_4$ and C$_{28}$ (with the cage geometry corresponding to the removal of Ti from Ti@C$_{28}$).

	Ti@C$_{28}$		C$_{28}$ (C$_{3v}$)		C$_{28}$H$_4$	
	RHF	*ROMP2*	*ROHF*	*ROMP2*	*RHF*	*RMP2*
μ_z	1.612	1.088	−0.0009	−0.055	0.0	0.0
α_{xx}	267.20	275.96	254.92	277.40	246.59	251.06
α_{zz}	245.81	258.83	258.94	281.04	246.59	251.06
β_{xxx}	−33.9	−126.6	−40.6	−192.0	0.0	0.0
β_{zzz}	117.3	23.1	−92.0	−588.4	0.0	0.0
$\gamma_{xxxx} \times 10^{-3}$	96.2	119.5	101.8	149.6	72.0	104.0
$\gamma_{zzzz} \times 10^{-3}$	79.3	103.7	99.9	141.2	72.0	104.0

tion of Ti into C$_{28}$ is explained in ref. 49 as a consequence of the localization of the four unpaired electrons due to the bonding between the cage and the Ti atom. This differs from the usual explanation of such effects as a "compression" effect (*e.g.* in ref. 56–59). Comparison with the properties of C$_{28}$H$_4$, which are also smaller than those of C$_{28}$, shows that the explanation advanced in ref. 49 is probably more adequate, as there can be no compression effect in this case.

In addition to the electronic contributions, static vibrational contributions to the electrical properties were also computed in ref. 49, at the so-called double harmonic level. At this level the vibrational contributions are in the Bishop–Kirtman square bracket nomenclature[60] given by eqn (8.1)–(8.3):

$$\alpha^{\mathrm{DH}} = \left[\mu^2\right]^{0,0} \tag{8.1}$$

$$\beta^{\mathrm{DH}} = \left[\mu\alpha\right]^{0,0} \tag{8.2}$$

$$\gamma^{\mathrm{DH}} = \left[\alpha^2\right]^{0,0} + \left[\mu\beta\right]^{0,0} \tag{8.3}$$

These contributions were computed at the restricted Hartree–Fock (RHF) level and at the density functional theory (DFT) level with three different functionals (BLYP, B3LYP, CAM-B3LYP) and the 6-31+G(d) basis set. The values obtained are shown in Table 8.7, where the static electronic contributions are also shown for comparison. For γ^{DH}, only the first term $[\alpha^2]^{0,0}$ was computed. As a prerequisite for the correct computation of vibrational NLO properties, the computations were performed on geometries optimized at the corresponding level/basis set combination. Thus, the electronic RHF values differ from those in Table 8.6, where the restricted Møller–Plesset second-

Table 8.7 Static electronic and double harmonic vibrational contributions to the (hyper)polarizabilities of Ti@C_{28}.

	BLYP	*B3LYP*	*CAM-B3LYP*	*RHF*
α^e_{xx}	272.0	264.4	258.7	258.4
α^e_{zz}	259.9	250.0	241.8	238.6
α^e_{av}	267.9	259.6	253.1	251.8
$[\mu^2]^{0,0}_{yy}$	21.2	27.1	38.0	58.1
$[\mu^2]^{0,0}_{zz}$	4.2	3.9	3.8	4.7
$[\mu^2]^{0,0}_{av}$	15.5	19.3	26.5	40.3
β^e_{yyy}	−71.2	−65.7	−66.6	−66.5
β^e_{zzz}	4.3	25.5	61.3	98.6
β^e_{av}	7.6	−10.4	−14.1	−34.4
$[\mu\alpha]^{0,0}_{yyy}$	−273.5	−341.4	−448.5	−673.0
$[\mu\alpha]^{0,0}_{zzz}$	6.5	27.4	43.1	69.4
$[\mu\alpha]^{0,0}_{av}$	196.2	195.3	275.9	175.4
$\gamma^e_{yyyy} \times 10^{-3}$	110	98	85	94
$\gamma^e_{zzzz} \times 10^{-3}$	92	81	70	74
$\gamma^e_{av} \times 10^{-3}$	103	91	79	87
$[\alpha^2]^{0,0}_{yyyy} \times 10^{-3}$	24	23	22	23
$[\alpha^2]^{0,0}_{zzzz} \times 10^{-3}$	18	16	14	14
$[\alpha^2]^{0,0}_{av} \times 10^{-3}$	21	19	19	19

order perturbation theory RMP2 optimized geometry was used. Additional values were computed in ref. 46 with the smaller 6-31G(d) basis set, and comparison showed that the effect of diffused functions in the basis set is large, especially on the electronic properties (*e.g.* γ^e_{av} = (79–103) × 10^3 a.u. with 6-31+G(d), but (5–7) × 10^3 a.u. with 6-31G(d)).

Table 8.7 shows that α^{DH} and β^{DH} computed with BLYP and B3LYP functionals are quite similar, but differ substantially from those properties calculated with the long-range corrected CAM-B3LYP potential, suggesting that for these properties the treatment of the exchange correlation (xc) potential at long range is important. On the other hand, this seems not to be the case for $[\alpha^2]^{0,0}$.

Comparison of electronic and vibrational (DH) contributions shows that for β, the vibrational contribution is much larger that the electronic one, whereas for α and γ the reverse appears to be the case. An analysis of the normal mode contributions to β^{DH} shows that the largest contribution comes from the lowest frequency degenerate pair of e symmetry vibration, which is also associated with a large displacement of the Ti atom with respect to the cage. An energy profile for the normal mode displacement of the mode in question shows that the energy surface for this mode is quite shallow. Therefore, it is reasonable to assume that anharmonic contributions (*i.e.*

terms beyond the double harmonic approximation) are important. To illustrate this, the so-called nr term for β_{yyy} was calculated at the B3LYP/6-31+G(d) level and was found to be 2 000 a.u., several times larger than the DH value. Thus, all the conclusions concerning the relative magnitude of vibrational and electronic contributions to the (hyper)polarizabilities of Ti@C_{28}, based on the DH approximation, should be considered preliminary until higher-level calculations are available. In addition to the nr contribution, which takes only the lowest-order anharmonicities into account, the higher-order c-zpva contribution may also be substantial. A possible treatment of the full range of anharmonicities for the vibrational NLO properties of Ti@C_{28} using an extension of a recently developed method[45] is envisaged for future work.

8.5   Analysis of the Binding Energy in Biological Systems

The recognition of different ligands by their receptors depends on interactions between them. The type of interactions arising is defined by the chemical structure and the conformations of the molecules. Thus, rational drug design identifies these interactions and opts to enhance them. The binding free energy (ΔG) is divided into enthalpic (H) and entropic (S) terms. The structural information of different complexes along with theoretical and experimental binding free energy estimations offer insight into the relationship between structure and activity. A general observation is that many drug candidates show enthalpic-driven contributions.[61] For example, tipranavir and indinavir — two inhibitors of HIV-1 protease — have very similar entropic contributions ($\sim$ -14 kcal mol^{-1}), but tipranavir has an enthalpy of -0.7 kcal mol^{-1} and indinavir 1.8 kcal mol^{-1}. This difference in the enthalpy increases the affinity of tipranavir by 70 to 19 pM.[62] One type of interaction that alters the binding of a drug candidate is hydrogen bonding (H-bond). The presence of electronegative groups (*e.g.* –OH, –NH$_2$, –COOH) creates H-bond interactions between the ligand and residues of the protein, thus increasing the enthalpic term in the binding energy.[63]

Major contributions to the binding energy in a biological system also come from van der Waals and electrostatic interactions. The presence of metal atoms in proteins (*e.g.* metalloproteases) increases the electrostatic interactions between a ligand and the macromolecule.[64] Wu *et al.*[65] developed a force field to better represent the interactions of zinc with proteins and ligands through quantum mechanics/molecular mechanics (QM/MM) calculations. Shinoda and Tsukube[66] explored the interactions between cytochrome C and different small molecules, while Villa *et al.*[67] studied the substrate specificity in meprin metalloproteinases and the type of interactions induced by its binding to the enzyme. The COX and LOX proteins are another family of enzymes where electrostatic and hydrophobic interactions are of major importance.[68] In the LOX family of enzymes, the presence of an iron atom at the catalytic site reveals the importance of electrostatic interactions as shown by a density functional study for the catalytic mechanism of the enzyme by Borowski and Broclawik.[69]

Another factor to take into account is the environment of the biological system. Hydrophobicity can play an important role in the binding of molecules. Additionally, membrane permeability has great impact in the absorption and activity of a pharmaceutical substance.[70] Politi *et al.* showed the importance of van der Waals and non-polar interactions in the binding of aliskiren to the active site of renin.[71] Thus, it is vital for researchers to explore options that increase the hydrophobic nature of a drug and combine attractive/repulsive (hydrophobic, van der Waals) components.

As mentioned above, hydrogen bonds have a major impact in the binding of a ligand to a biological macromolecule as well as in various enzymatic reactions. Lang *et al.*[72] explored the impact of hydrogen bonding in the catalysis of oxidation of L-Arg by heme proteins. They specifically focused on residues directly interacting with the heme and the effect mutations have on the catalysis. Yuzlenko and Lazaridis,[73] using molecular dynamic techniques, investigated the interactions between ionizable amino acid side chains and lipid bilayers. Their aim was to explore the importance of hydrogen bonding and salt bridges in the stability of large biological systems such as proteins and membranes. Moffett *et al.*[74] employed *in silico* methods to design optimized p38α kinase inhibitors. They used fragment-based drug design (FBDD) in order to identify new structures that increase the hydrogen bonding potential with specific residues (*e.g.* Arg70) of the enzyme. NMR spectroscopy has been used by Limbach *et al.*[75] to investigate the role of the environment in the formation of hydrogen bonds in biological systems. They worked with the enzyme cofactor pyridoxal 5′-phosphate (PLP) in different environments ranging from aqueous solutions to solid enzyme–substrate systems.

Therefore, in drug design interactions play a crucial role. It is important to design ligands so as not only to maximize the binding to the target molecule but also to optimize their interactions with different systems encountered in their path to the target. As shown by Shaw *et al.*,[76] an extensive study of biological systems can reveal the role of different interactions (*e.g.* van der Waals, hydrophobic, electrostatic) in the system's mode of action. Thus, researchers can have a clearer idea regarding the steps required for the design of a more effective drug.

Here, we shall review the analysis of the binding energy of two fullerene derivatives and C_{60} with HIV-1 PR. Using the AMBER software package,[77] molecular dynamics (MD) simulations in explicit solvent were performed in complexes of HIV-1 protease with the fullerene derivatives. Fullerene structures are shown in Figure 8.5. The protein–ligand complexes were energy-minimized and subsequently heated to 300 K. The MD simulations were run under constant pressure for 20 ns. Furthermore, the binding energies of the complexes were calculated using the MM–PBSA method.[78] With this procedure the interactions in each complex are calculated using molecular mechanics (MM) in the gas phase, and the solvation energy is estimated by

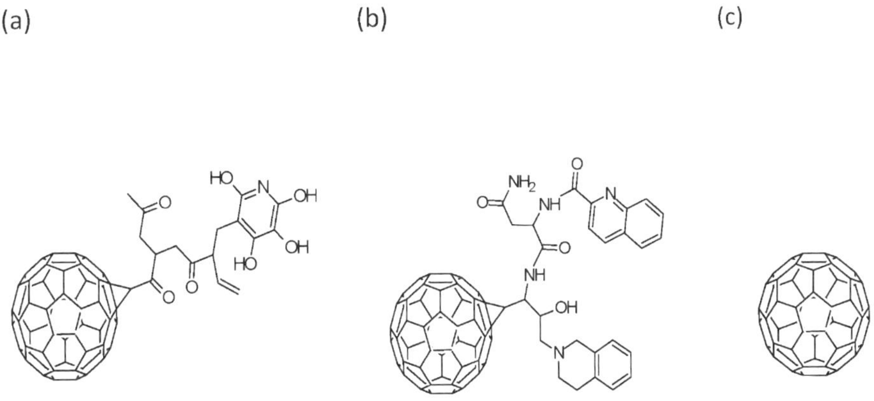

(a)   (b)   (c)

Figure 8.5   Fullerene derivatives studied in complexes with HIV-1 PR. (a) Compound **1**, (b) compound **2** and (c) fullerene core C_{60}.

solving the Poisson–Boltzmann equation. In summary, this method is expressed with the following equations:

$$\Delta G_{bind} = G_{complex} - \left(G_{protein} + G_{ligand}\right) \qquad (8.4)$$

$$= \Delta H - T\Delta S \qquad (8.5)$$

$$= \Delta E_{MM} + \Delta G_{solv} - T\Delta S \qquad (8.6)$$

where ΔH is the enthalpy and $T\Delta S$ is the conformational entropy contribution. ΔG_{MM} and ΔG_{solv} are given by:

$$\Delta E_{MM} = \Delta E_{elec} + \Delta E_{vdW} \qquad (8.7)$$

$$\Delta G_{solv} = \Delta G_{PB} + \Delta G_{NP} \qquad (8.8)$$

ΔG_{PB} is the electrostatic contribution to the solvation free energy. ΔG_{NP} is the non-polar solvation energy, computed by employing the solvent-accessible surface area (SASA), according to the equation:

$$\Delta G_{NP} = \gamma SASA + \beta \qquad (8.9)$$

As mentioned before, our aim is the investigation of the interactions between the ligand candidates and the target molecules. Therefore, the analysis in the HIV-1 PR complexes focuses on the aspects that offer insights into the type of

Table 8.8 Contributions to binding energy for three HIV-1 PR complexes.

	ΔE_{elec}	ΔE_{vdW}	ΔG_{NP}	ΔG_{PB}	ΔH	ΔG_{bind}
HIV-1 PR–compound **1**	−14.74	−83.90	−26.35	59.92	−65.08	−11.97
HIV-1 PR–compound **2**	−20.78	−76.80	−25.14	60.37	−62.35	−12.75
HIV-1 PR–C_{60}	−5.24	−65.37	−34.25	56.59	−48.27	−1.25

interactions present. An important feature observed in our simulations is the behavior of the two flaps that enclose the active site. In agreement with experimental studies, we observed that this particular region of the protein traps the ligand into the active site cavity. Further analysis revealed that certain binding-cavity residues form hydrogen bonds with the fullerene derivatives. This behavior can explain the high binding energy observed for the complexes. Unlike the natural substrate of the protease and other inhibitors, the designed fullerene derivatives do not bind directly to the active site of the protein.[79]

As shown in equations (8.4)–(8.9), with the MM–PBSA technique it is possible to decompose the binding energy to its separate components and identify similarities or differences between the derivatives studied. Table 8.8 summarizes the results from our calculations. The most important aspect of our analysis is the fact that in all three complexes the most significant contribution to the binding energy comes from van der Waals interactions and non-polar contributions to solvation. Comparing the results of the different fullerene derivatives we conclude that the fullerene core is the most important contributor to the van der Waals interactions. The substituents in the fullerenes provide the necessary chemical groups for enhanced interactions with the enzyme, thus increasing the binding energy of the complexes.

8.6 Amino Acid Fullerene Derivatives Bound to HIV-1 PR

The need for new anti-HIV drugs has stimulated a range of research approaches, including the development of drugs that act to inhibit the HIV type I aspartic protease (HIV-1 PR). In this regard, it has been shown that [60]fullerene and its analogues inhibit HIV-1 PR *via* entrapping of the C_{60} cage into the hydrophobic binding pocket of the protein, at the interface between the two PR dimers.[80,81] So far, a large number of potential fullerene derivatives has been reported by several groups for different targets; however, only a few of them have been considered for their HIV-1 PR inhibitory effects.[82] Despite the almost perfect fit of C_{60} into the HIV-1 PR binding pocket, both first and second generations of fullerene derivatives have not shown a significant inhibitory effect against HIV-1 PR compared to the standard drugs (*e.g.* amprenavir), which show sub-nanomolar level inhibition. However, their high stabilities, low cytotoxicities and reduced potential to be processed to the toxic metabolites make them interesting compounds against HIV-1 PR inhibition.[83] The synthesis of novel fullerene derivatives and their *in vitro* tests are not easy tasks in the wet-labs and usually require a great deal of effort and time. Therefore, computational techniques can be alternatively used to provide valuable information and may help to derive novel HIV-1 PR inhibitors with improved binding affinities. In this chapter, a successful *in silico* approach for the development of fullerene-based HIV-1 PR inhibitors will be reviewed.[84–86]

HIV-1 PR belongs to the family of retroviral proteases and is a C_2 symmetric, homodimeric (*i.e.* two identical segments with 99 amino acids each), aspartic protease. The binding site of HIV-1 PR is an open-ended cylindrical hydrophobic cavity of about 10 Å diameter composed of catalytic residues Asp25 and Asp25′ (Figure 8.6). Since an X-ray structure of the HIV-1 PR/fullerene analogue complex has not yet been reported, the crystal structure of HIV-1 PR complexed with a haloperidol derivative (pdb: 1AID)[87] has been used as reference structure for molecular docking and MD simulations.

Molecular docking and MD simulations have been performed with FlexX and GROMACS, respectively. The coordinates of the active-site residues of HIV-1 PR may be different for the case of the HIV-1 PR/fullerene derivative complex compared to the crystallized complex HIV-1 PR/haloperidol. Thus, the correct binding mode of the studied fullerene inhibitors can best be obtained from the MD simulations of HIV-1 PR/fullerene derivative complex.[82] The best derived docking poses were used as initial geometries for the MD simulations. Since the FlexX docking algorithm uses a rigid target conformation for the protease during the docking simulations, MD studies may assist in generating proper input coordinates for further analyses. In addition, a deeper understanding of the mechanistic events associated with HIV-1 PR binding can be also achieved by analyzing the MD trajectories.[82] Thus, together with the fullerene-bound complex, a control (HIV-1 PR

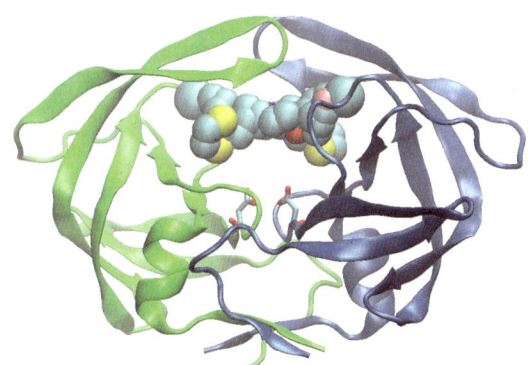

Figure 8.6   Structural representation of HIV-1 PR: C_2 symmetric, homodimeric (*i.e.* two identical segments with 99 amino acids each), aspartic protease. Figure shows haloperidol-bound HIV-1 PR.

without fullerene inhibitor) MD simulation was also performed. Consistent structural differences have been observed between the fullerene-free and fullerene-bound systems of the HIV-1 PR. In the fullerene-bound forms, the flaps are pulled in toward the dual Asp catalytic site (closed form), while the structure for the free HIV-1 PR flaps shifted away from the catalytic site (semi-open form).[82]

The coordinate file for the fullerene-bound HIV-1 PR produced from the MD simulations has been used and tested for the docking scores. The predicted pIC_{50} values of the fullerene derivatives have been compared to the actual pIC_{50} values and a high correlation between them has been found. Thus, a database that includes more than 100 fullerenes has been constructed by searching the [60]fullerene derivatives in the literature.[86] These fullerene analogues have been synthesized in the past decade and most of them were designed for other biomedical applications. Since MD simulations assisted in obtaining the correct enzyme input coordinates for docking the fullerene analogues, compounds in the database have been docked to the target and their binding scores and binding interactions were compared. One group of compounds showed particular promise: fullerene-based peptidic derivatives and their structural analogues.[86,88,89] Their high affinities are connected with the ability of pendant groups to form H-bonds and electrostatic interactions with the catalytic site of the enzyme as well as van der Waals interactions between the fullerene cage and the non-polar surface of active site amino acid residues.[86] The structures that showed better binding scores than the currently reported most potent fullerene-based HIV-1 PR inhibitor (compound **1**, Table 8.9) were collected and subjected to biological tests. These *in vitro* analyses proved the computational predictions: The bucky amino acid (compound **3S**, Table 8.9) showed a similar binding affinity ($K_i = 120$ nM) to compound **1** ($K_i = 103$ nM). Furthermore, the 9-fluorenylmethoxycarbonyl-protected bucky amino acid (compound **2S**, Table 8.9) showed about 3-fold better activity ($K_i = 36$ nM) than the most active currently available fullerene derivative **1**.[86,89]

This is considered to be an important finding because a new anti-HIV fullerene derivative is proposed, which is approximately 3-fold more active than those already reported and it is expected to stimulate further research for even more effective anti-HIV drugs. Moreover, it can be useful for directing future three-dimensional quantitative structure–activity relationships (3D-QSAR) studies. Obviously, when the number of compounds in the data set used for 3D-QSAR studies increases, the stability of the constructed QSAR model and the output from the model is expected to be increased. Therefore, each new measurement of fullerene derivatives adds valuable data for 3D-QSAR studies which help to understand the steric and electrostatic requirements of HIV-1 PR inhibition. Thus, these inhibition data of novel fullerene derivatives will assist in obtaining a diverse range of binding affinity values and will be very useful also for future 3D-QSAR studies.

Table 8.9 Comparison of experimental and computational binding energies and binding affinities of fullerene amino acid derivatives at HIV-1 PR. The binding affinity of compound **1** at HIV-1 PR was reported in ref. 90 and the binding affinities of **2S** and **3S** were reported in a recent publication.[86]

Compound no.	Structure	Experimental binding affinities, K_i/nm	Experimental binding energies/kJ mol^{-1}	Computational (docking) binding energies/kJ mol^{-1}
1		103	−40.13	−36.66
2S		36	−42.75	−43.54 (*C1, S) −39.40 (*C1, R)
3S		120	−39.75	−37.60 (*C1, S) −36.56 (*C1, R)

8.7   MMK16 into COX-2/LOX-5 Enzymes

Enzymes cyclooxygenase-2 (COX-2)[90] and lipoxygenase-5 (LOX-5)[91] have been implicated in inflammation responses and in the progression of neoplasia.[90,92–94] Toward the development of potent anti-inflammatory compounds, MMK pyrrolidinone-based analogues have been synthesized to inhibit both enzymes.[95,96] 4-[(2S)-2-(1H-imidazol-1-ylmethyl)-5-oxotetrahydro-1H-pyrrol-1-yl]methylbenzenecarboxylic acid (MMK16) is a promising candidate inhibitor since it possesses pyrrolidinone and carboxylate groups, which are both associated with anti-inflammatory action.[97,98]

Molecular dynamics (MD) and molecular docking techniques provided useful information regarding principal interactions between MMK16 and COX-2/LOX-5 as well as a dynamic description of the systems by monitoring conformational changes induced upon binding.[98] Aspirin and caffeic acid are known inhibitors of COX-2 and LOX-5, respectively. Thus, comparative docking calculations for MMK16 into COX-2 and for aspirin into COX-2 suggested that MMK16 is a more potent inhibitor of the enzyme than aspirin, mainly due to the orientation of MMK16 in the binding cavity and the stabilization of its structure via electrostatic interactions. Similarly, electrostatic interactions between the catalytic Fe(III) in LOX-5 and the imidazole ring of MMK16 favor binding of MMK16 (compared to caffeic acid) into LOX-5. Hydrogen bonding (HB) and van der Waals interactions further stabilize MMK16 inside COX-2/LOX-5.

Additionally, a 12 ns all-atom MD simulation in an implicit solvent environment[99] at 300 K for the MMK16–COX-2 system resulted in a stable structure for the enzyme, after an initial conformational change, which was induced mainly by the increased flexibility of residues belonging to the membrane-binding domain and active-site region (Figure 8.7a). This implied that the initial, docked structure of MMK16 may be unstable and rather equilibrates toward different conformations. Indeed, HB analysis showed the rearrangement of interactions during the course of the simulation: the initial HB interaction between the imidazole ring of MMK16 and active-site residue Arg120 was replaced by another interaction between the pyrrolidinone oxygen atom of MMK16 and Arg120. This HB rearrangement resulted in a displacement of MMK16 to a new position that favors enhanced interactions with the binding cavity. Although the active site (His90, Arg120, Gln192, Tyr355, Arg513, Val523, Glu524) appeared overall stable, it underwent a substantial structural change to account for the new arrangement of MMK16. Additional interactions such as MMK16–Ser530 and MMK16–Asp362 stabilized further MMK16 into COX-2 (Figure 8.7b).

A representation of MMK16 into the binding cavity of LOX-5 is given in Figure 8.7c. A noticeable conformational change was also observed for LOX-5 upon beginning the 12 ns MD simulation of the MMK16–LOX-5 complex. However, structural changes on this enzyme were considered irrelevant to the active site since the latter appeared practically stable throughout the simulation and it did not participate in any HB interactions with MMK16.

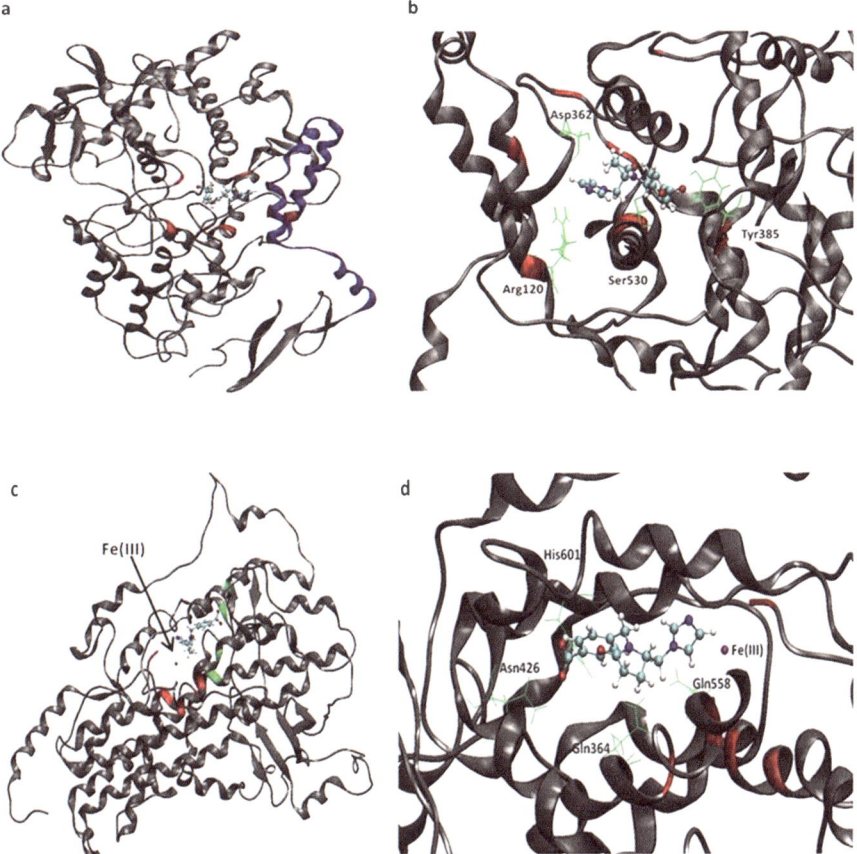

Figure 8.7   (a) Schematic representation of MMK16 into COX-2. Active-site residues (His90, Arg120/513, Gln192, Tyr355, Val523, Glu524, in red) form a tunnel through which inhibitors bind to the catalytic cavity of the enzyme. The flexible membrane-binding domain (residues 73–123, in blue) induces structural changes to COX-2. (b) Close-up view of MMK16 into COX-2 binding cavity. Residues forming hydrogen bonds with MMK16 are shown in green. (c) Schematic representation of MMK16 into LOX-5. Residues involved in hydrogen bonds with MMK16 (Gln364/558, Asn426, His601, in green) do not belong to the active-site region (His368/373/551, Asn555, Ile674, in red). Catalytic Fe(III) is in constant interaction with MMK16. (d) Close-up view of MMK16 into LOX-5 binding cavity. Residues forming hydrogen bonds with MMK16 are shown in green and Fe(III) in purple.

Additionally, atomic fluctuation calculations for active-site residues (His368, His373, His551, Asn555, Ile674) yielded values well below 1 Å, thus indicating the very low mobility of this region. In accordance with the docking results, the distance between the imidazole ring and the catalytic Fe(III) in the active site remained very stable during the simulation around the average 2.15 Å. HB analysis revealed the main reason for the structural change induced to LOX-5 during the beginning of the simulation: the initially unstable MMK16–His433 interaction has been gradually replaced by a strong HB network involving interactions between MMK16 and non-active-site residues Gln364, Asn426, Gln558 and His601 (Figure 8.7d).

Finally, the ability of MMK16 to exclude water from the the enzymes' binding cavities was explored to reveal that the solvent-accessible surface area (SASA) of the compound inside COX-2 appeared almost 2-fold greater than inside LOX-5. Since it is anticipated that a strong hydrophobic environment may act favorably toward binding in COX/LOX systems, the exposure of MMK16 to the solvent may play a crucial role.

8.8   Aliskiren in Solution and Bound to Renin

Renin inhibition is of paramount importance toward effective therapeutic strategies against hypertension.[100,101] Renin is a 340-amino acid, highly selective aspartic protease, which cleaves angiotensinogen to produce the peptide angiotensin I.[102] An increase in blood pressure occurs when angiotensin I is converted to angiotensin II by the angiotensin-converting enzyme (ACE). The active site of renin is a deep cavity to the bottom of which a pair of catalytic triads (Asp32/215, Thr33/216, Gly34/217) lies.[102–104] In 2007, aliskiren was approved by the US Food and Drug Administration (FDA) as the first oral renin inhibitor for the treatment of hypertension.[105–107] It possesses four chiral centers and prevents the conversion of angiotensinogen to angiotensin I upon direct binding to the catalytic cavity of renin. A highly potent inhibitor for human renin, aliskiren has $IC_{50} = 0.6$ nM and biological half life ≈ 24 h.[108]

In a recent study, Politi *et al.* investigated the conformational properties of aliskiren in dimethylsulfoxide (DMSO) using MD calculations and NMR spectroscopy.[71] The X-ray crystal structure of aliskiren was solvated in DMSO and heated at 300 K before being subjected to a lengthy MD simulation for 200 ns. The trajectory obtained after the simulation was clustered to group aliskiren structures based on conformational similarities as defined by a 2.5 Å cutoff among all conformations. In agreement with the NMR results, clustering analysis revealed that aliskiren appears flexible and adopts two main conformations in DMSO solution. The first representative structure was dominant during the simulation (73% of time present) and suggested an extended conformation for aliskiren, while the second representative structure appeared for 27% of the simulation and denoted a "bent" structure. Of major significance was the observation that the extended conformation of aliskiren bore a striking resemblance to the crystal structure of the drug when bound to renin (Figure 8.8a). Therefore the use of the crystal structure as the initial structure to investigate the dynamic properties and binding modes of aliskiren in renin could be well rationalized.

Thus, the same authors were the first to perform MD simulations and MM–PBSA free energy calculations for the aliskiren–renin complex to explore

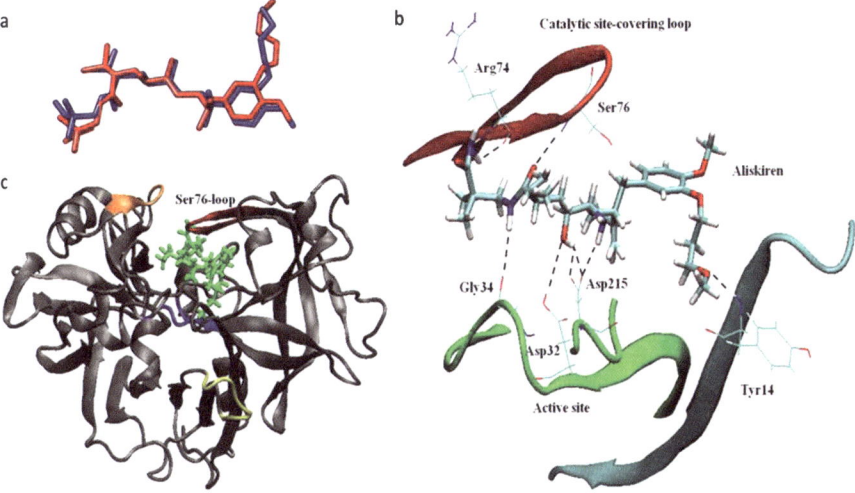

Figure 8.8 (a) The representative structure of aliskiren in DMSO (obtained after clustering calculations, red) resembles the crystal structure of aliskiren (blue). (b) Principal hydrogen bonds between aliskiren and renin: eight interactions involving active-site and Ser76-loop residues stabilize the drug inside the protease. (c) The very mobile Ser76 loop in the *apo* form of renin is stabilized in a closed conformation upon aliskiren entrapping into the binding cavity. Aliskiren (green) lies on top of the active site (blue) in an extended conformation.

binding patterns, principal interactions and conformational changes associated with aliskiren binding to renin.[71] A 15 ns MD simulation at 300 K and in explicit water was initiated from the X-ray crystal structure of aliskiren–renin to observe that the complex presented a high degree of stability, further suggesting that the simulated system equilibrates toward conformations that resemble the crystal structure. HB analysis attributed the increased stability of the complex to several HB interactions between aliskiren and active-site residues of renin. The major HB interactions between aliskiren and renin are presented in Figure 8.8b. In particular, Asp32/215 and Gly34 at the active site are involved in stabilizing the drug inside the cavity of renin, while residues such as Arg74, Ser76 and Tyr14 also participate in HB interactions with aliskiren.

Although the structure of the complex remained stable during the simulation, a striking feature was observed when the flexibility of the *apo* (ligand-free) form of renin was considered. Amino acid residues in the vicinity of the "tip" residue Ser76 (Arg74–Tyr75–Ser76–Thr77–Gly78) form a loop that belongs to the outer region of the protease, lying on top of the active site. The loop appeared increasingly flexible in the *apo* renin, with "open" and "closed" conformations being in dynamic equilibrium. It was then suggested that ligand access to the cavity may depend on the sufficient opening of this loop. This was a direct implication that this region serves as a modulator for substrate entrance to the binding cavity of renin. Interestingly, upon aliskiren binding to renin, the loop stabilized permanently via two HB among Arg74, Ser76 and aliskiren. Ser76 remained attached to aliskiren in a closed conformation that entrapped the drug inside renin to yield a compact structure stabilized by strong interactions involving active-site residues, the loop and aliskiren (Figure 8.8b and c).

Finally, MM–PBSA free-energy calculations were applied to estimate the binding energy in the aliskiren–renin complex to be -12.0 kcal mol^{-1}, in agreement with the experimental value (-12.6 kcal mol^{-1}).[109] Further energy decomposition revealed that the formation of the complex is mainly driven by the non-polar contribution to solvation (-28.1 kcal mol^{-1}) and by van der Waals (-35.7 kcal mol^{-1}) interactions. As this trend has been already verified in other protein complexes by independent studies, it could be used as an effective tool for compound optimization in drug design.[110,111]

8.9 Drug–Biosurface Interactions

Recent reviews point out the significance of drug–membrane interactions.[112–115] This is because of two major reasons. First, some classes of drug molecules act directly at the lipid bilayer core, perturbing the lipid component or triggering the modulation of the proteins, and some others still have to interact with the membranes in order to reach their interior site of action. It is for these reasons that intense studies have been applied in recent decades to study the efficiency of drugs when interacting with the membranes and correlate it with their pharmacological effects. At the moment, such a correlation is not straightforward and to analyze the fingerprint of the drug in the membranes is a tedious work that involves the application of many biophysical methodologies. Thanks to advances in the capabilities of analytical instrumentation and methodologies developed, detailed information on drug-membrane interactions[116–146] has been acquired.

Membrane bilayers are complex entities composed of different proteins and lipids. Both components are very important for a drug to exert its biological effect. In our studies, we have focused on the drug action at the lipid component of the membranes. Here we will give an example from our recent work to illustrate the importance of the drug–membrane interactions. It has been proposed that antihypertensive drugs, and more particularly the AT$_1$ antagonists which act through blocking vasoconstrictive hormone angiotensin II, are first embedded into the lipid bilayers and then through lateral diffusion reach the active site of their receptor. The molecular basis of AT$_1$ antagonist action is shown in Figure 8.9.[147–154]

It is evident that lipid bilayers play an important role in AT$_1$ antagonism. For this reason, we found it very interesting to study the membrane–AT$_1$ interactions. Thus, it appears very important to relate the fingerprint of an AT$_1$ antagonist and its pharmacological action (Figure 8.10). *This would lead to the synthesis of drugs possessing more specific and better pharmacokinetic properties.*

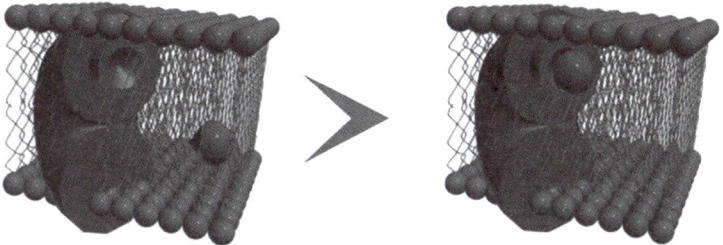

Figure 8.9 The mechanism of action for the prototype of AT_1 antagonists, losartan shown as a sphere.

First, we defined the properties of the drugs in the lipid bilayers that determine their fingerprints and then we attempted to quantify them using various physical and chemical methodologies. Such properties are: (a) topography (localization); (b) perturbation as it is depicted by *gauche–trans* isomerization and chain mobility.

To show the importance of drug–membrane interactions, we provide an example of a comparative study between the two drugs losartan and candesartan cv (Figure 8.11 and ref. 151).

Losartan (Cozaar) is the first commercially developed competitive/surmountable antihypertensive AT_1 antagonist of the sartans class. Candesartan (CV-11974) is a non-competitive/insurmountable antagonist and is the active metabolite of candesartan cilexetil (Atacand), which belongs to the sartan class. Compared to losartan, it exerts a longer duration of action and shows the highest receptor affinity among the AT_1 sartan antagonists.[155,156]

At the given experimental conditions (pH $\sim$ 7), both the acidic tetrazole and carboxylate groups of candesartan are mostly deprotonated ($pK_a \sim 6$ for the terazole ring and $pK_a \sim 3$–4 for the carboxylic acid). Losartan is in a potassium salt form and bears only the acidic group of tetrazole with a negative charge. Dipalmitoylphosphatidylcholine (DPPC) is a neutral zwitterionic molecule at physiological pH, bearing the positively charged headgroup choline and the negatively charged phosphate moiety. It is anticipated that the two agents would exhibit different electrostatic interactions with the bilayer interface due to the difference in their negative charges, implying a

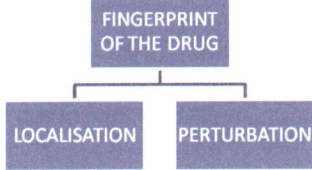

Figure 8.10 The fingerprint of the drug is determined by its orientation, localization and a general perturbation term which includes its thermal, dynamic and other effects that can be quantified.

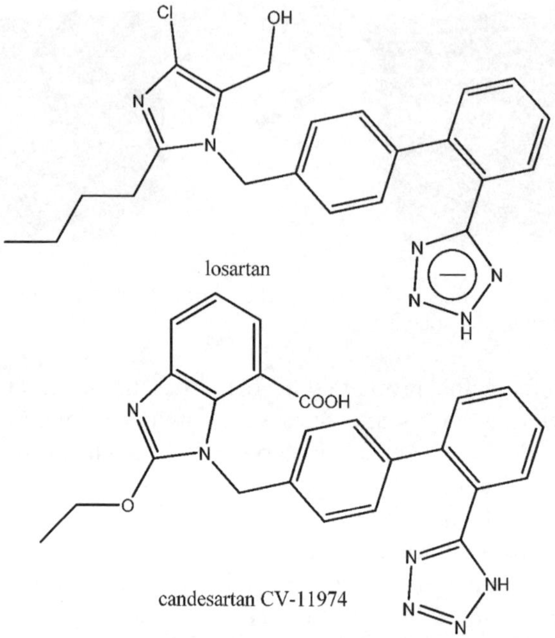

Figure 8.11   Chemical structures of losartan and candesartan CV-11974.

differentiated affinity to the membrane surface as well as a different immersion of the drugs in the model membranes.

The interactions of the two molecules were studied through the application of differential scanning calorimetry (DSC), Raman, and solid-state ^{31}P NMR spectroscopies. A ^{31}P CP NMR broadline fitting methodology in combination with *ab initio* computations was implemented and, in conjunction with DSC and Raman results, provided valuable information regarding the perturbation, localization, orientation, and dynamic properties of the drugs in membrane models.

In particular, losartan is localized at the interface which covers the polar region and upper segment of the lipophilic region to maximize its amphipathic interactions. Such a localization of the drug could induce a local curvature and enlarge the space between the adjacent alkyl chains. This could allow the tails of the alkyl chains of the next layer to entangle, introducing tail interdigitation. On the other hand, candesartan at low concentrations affects only the headgroup, probably spanning between the water interface and headgroup region. This may be attributed to attractive electrostatic interactions between the two anions of candesartan and the positively charged nitrogen atom of the choline group, and repulsive interactions with the phosphate group, thus leading to its higher affinity for aggregation with the water interface as it adopts a more accessible area to the hydrophilic environment. At higher concentrations, candesartan strengthens the polar interactions and also affects

Table 8.10 Comparison of interface activity of candesartan CV-11974 and losartan in phosphatidylcholine bilayers.

Methodology	Effect	Candesartan CV-11974	Losartan
Differential scanning calorimetry	Inhibition of pre-transition	No inhibition	Inhibition
Raman spectroscopy	Shift of 714 cm^{-1} peak attributed to C–N stretch vibration	Shift to higher values	Shift to higher values
^{31}P NMR spectroscopy	Effect on σ_{iso} values	Lowers σ_{iso} values	More significant lowering of the σ_{iso} values

the packing of the alkyl chains, probably due to partial penetration into the hydrocarbon region.

Both sartan molecules decrease the mobilization of the phospholipid alkyl chains. Losartan exerts stronger interactions in comparison to candesartan, as depicted by the more prominent thermal, structural, and dipolar ^{1}H–^{31}P changes that are caused in the lipid bilayers. At higher concentrations, candesartan strengthens the polar interactions and induces increased order at the bilayer surface. At the highest concentration used (20 mol%), only losartan induces the formation of microdomains attributed to the flexibility of its alkyl chain. The comparative perturbing effects and interface activity of the molecules are shown in Tables 8.10 and 8.11.

The above observed variations in membrane interactions might relate to the differentiated pharmacological profile of the two studied AT$_1$ antagonists and plausibly in part explain the more potent biological profile of candesartan. In addition, the obtained results suggest a relationship between the diffusion efficacy and the pharmacological potency of the studied sartan agents. Thus, losartan's tendency to form domains in the lipid bilayers could presumably retard its diffusion toward the active site of the AT$_1$ receptor. Moreover, the

Table 8.11 Comparison of the perturbing effect of candesartan CV-11974 and losartan in phosphatidylcholine bilayers.

Methodology	Effect	Candesartan CV-11974	Losartan
Differential scanning calorimetry	Abolish the pretransition, effect on the value of the main phase transition temperature and cause of broadening	No abolition, small effect	Abolition, stronger lowering of the main phase transition and broadening
Raman spectroscopy	$I_{2935/2880}$ ratio	Lowering	More drastic lowering

diffusion may be retarded by its stronger binding to the headgroup region as well as the induction of the interdigitation effect. On the other hand, candesartan at higher concentration is not inhibited by such effects toward its diffusion trip at the AT_1 receptor. Its membrane perturbation effects are milder and, in contrast to losartan, its incorporation does not induce interdigitation to the lipid matrix.

Acknowledgments

H. Tzoupis acknowledges the following: This research has been co-financed by the European Union (European Social Fund – ESF) and Greek national funds through the Operational Program "Education and Lifelong Learning" of the National Strategic Reference Framework (NSRF) – Research Funding Program: Heracleitus II. Investing in knowledge society through the European Social Fund. G. Leonis acknowledges the funding provided by the European Commission for the FP7-REGPOT-2009-1 Project 'ARCADE' (Grant Agreement No. 245866). M. G. Papadopoulos acknowledges the High-Performance Computing Infrastructure for South East Europe's Research Communities (HP-SEE), a project co-funded by the European Committee (under Contract No. 261499) through the Seventh Framework Programme. Also we acknowledge the CINECA award under the ISCRA initiative, for the availability of high-performance computing resources and support.

References

1. H. W. Kroto, J. R. Heath, S. C. O'Brien, R. F. Curl and R. E. Smalley, *Nature*,1985, **318**, 162.
2. P. Avouris, *Acc. Chem. Res.*, 2002, **35** 1026.
3. P. Delaney and J. C. Greer, *Appl. Phys. Lett.*, 2004, **84**, 431.
4. A. Gromov, N. Krawez, A. Lassesson, D. I. Ostrovskii and E.E.B. Campbell, *Curr. Appl. Phys.*,2002, **2**, 51.
5. H. Shinohara, *Rep. Prog. Phys.*, 2002, **63**, 843.
6. L. Dunsch and S. Yang, *Electrochem. Soc. Interface*, 2006, **15**, 34.
7. Y. Chai, T. Guo, C. Jin, R. E. Haufler, L. P. F. Chibante, J. Fure, L. Wang, J. M. Alford and R. E. Smalley, *J. Phys. Chem.*, 1991, **95**, 7564.
8. R. Beyers, C.-H. Kiang, R. D. Johnson, J. R. Salem, M. S. de Vries, C. S. Yannoni, D. S. Bethune, H. C. Dorn, P. Burbank, K. Harich and S. Stevenson, *Nature*, 1994, **370**, 196.
9. D. S. Bethune, R. D. Johnson, J. R. Salem, M. S. de Vries and C. S. Yannoni, *Nature*, 1993, **366**, 123.
10. D. E. Manolopoulos and P. W. Fowler, *Chem. Phys. Lett.*, 1991, **187**, 12.
11. P. W. Fowler and D. E. Manoloupoulos, *An Atlas of Fullerenes*, Oxford University Press, Oxford, 1995.
12. K. Kobayashi and S. Nagase, *Chem. Phys. Lett.*, 1996, **262**, 227.
13. S. Guha and K. Nakamoto, *Coord. Chem. Rev.*, 2005, **249**, 1111.

14. I. Infante, L. Gagliardi and G. E. Scuseria, *J. Am. Chem. Soc.*, 2008, **130**, 7459.

15. A. Rehaman, L. Gagliardi and P. Pyykkö, *Int. J. Quantum Chem.*, 2007, **107**, 1162.

16. Z. Slanina, Z. Chen, P.v.R. Schleyer, F. Uhlik, X. Lu and S. Nagase, *J. Phys. Chem. A*, 2006, **110**, 2231.

17. L. Dunch, A. Bartl, P. Georgi and P. Kuran, *Synth. Met.*, 2001, **121**,1113.

18. K. Ohno, Y. Maruyama, K. Esfarjani, Y. Kawazoe, N. Sato, R. Hatakeyama, T. Hirata and M. Niwano, *Phys. Rev. Lett.*, 1996, **76**, 3590.

19. M. S. Syamala, R. J. Gross and M. Saunders, *J. Am. Chem. Soc.*, 2002, **124**, 6216.

20. T. Cai, L. Xu, H. W. Gibson, H. C. Dorn, C. J. Chancellor, M. M. Olmstead and A. L. Balch, *J. Am. Chem. Soc.*, 2007, **129**,10795.

21. C. R. Wang, T. Kai, T. Tomiyama, T. Yoshida, Y. Kobayashi, E. Nishibori, M. Takata, M. Sakata and H. Shinohara, *Nature*, 2000, **408**, 426.

22. S. Stevenson, P. W. Fowler, T. Heine, J. C. Duchamp, G. Rice, T. Glass, K. Harich, E. Hajdu, R. Bible and H. C. Dorn, *Nature*, 2000, **408**, 427.

23. Z. Q. Shi, X. Wu, C. R. Wang, X. Lu and H. Shinohara, *Angew. Chem., Int. Ed.*, 2006, **45**, 2107.

24. S. Yang, A. A. Popov and L. Dunsch, *Angew. Chem., Int. Ed.*, 2007, **46**, 1256.

25. H. Kato, A. Taninaka, T. Sugai and H. Shinohara, *J. Am. Chem. Soc.*, 2003, **125**, 7782.

26. T. Wakahara, H. Nikawa, T. Kikuchi, T. Nakahodo, G. M. A. Rahman, T. Tsuchiya, Y. Maeda, T. Akasaka, K. Yoza, E. Horn, K. Yamamoto, N. Mizorogi, Z. Slanina and S. Nagase, *J. Am. Chem. Soc.*, 2006, **128**, 14228.

27. Z. Slanina, F. Uhlık, S.-L. Lee, L. Adamowicz and S. Nagase, *Int. J. Quantum Chem.*, 2007, **107**, 2494.

28. S. Yang, A. A. Popov and L. Dunsch, *J. Phys. Chem. B*, 2007, **111**, 13659.

29. C. M. Beavers, T. Zuo, J. C. Duchamp, K. Harich, H. C. Dorn, M. Olmstead and A. L. Balch, *J. Am. Chem. Soc.*, 2006, **128**, 11352.

30. M. M. Olmstead, H. M. Lee, J. C. Duchamp, S. Stevenson, D. Marciu, H. C. Dorn and A. L. Balch, *Angew. Chem., Int. Ed.*, 2003, **42**, 900.

31. K. Kobayashi and S. Nagase, in *Endofullerenes: A New Family of Carbon Clusters*, ed. T. Akasaka and S. Nagase, Kluwer Academic Publishers, Dordrecht, The Netherlands, 2002, p. 99.

32. T. S. M. Wan, H. W. Zhang, T. Nakane, Z. D. Xu, M. Inakuma, H. Shinohara, K. Kobayashi and S. Nagase, *J. Am. Chem. Soc.*, 1998, **120**, 6806.

33. Y.-H. Cui, W. Q. Tian, J.-K. Feng and D.-L. Chen, *J. Comput. Chem.*, 2008, **29**, 2623.

34. R. Tellgmann, N. Krawez, S.-H. Lin, I. V. Hertel and E. E. B. Campbell, *Nature*, 1996, **382**, 407.
35. D. B Whitehouse and A. D. Buckingham, *Chem. Phys. Lett.*, 1993, **207**, 332.
36. M. R. Pederson, T. Baruah, P. B. Allen and C. Schmidt, *J. Chem. Theory Comput.*, 2005, **1**, 590.
37. E. E. B. Campbell, M. Fanti, I. V. Hertel, R. Mitzmer and F. Zerbetto, *Chem. Phys. Lett.*, 1998, **288**, 131.
38. E. E. B. Campbell, S. Couris, M. Fanti, E. Koudoumas, N. Krawez and F. Zerbetto, *Adv. Mater.*, 1999, **11**, 405.
39. M. Yaghobi, R. Rafie and A. Koohi, *J. Mol. Struct.: THEOCHEM*, 2009, **905**, 48.
40. M. Yaghobi and A. Koohi, *Mol. Phys.*, 2010, **108**, 119.
41. H. Reis, O. Loboda, A. Avramopoulos, M. G. Papadopoulos, B. Kirtman, J. M. Luis and R. Zalesny, *J. Comput. Chem.*, 2011, **32**, 908.
42. M. Zhang, L. B. Harding, S. K. Gray and S. A. Rice, *J. Phys. Chem. A*, 2008, **112**, 5478.
43. J. Hernandez-Rojas, J. Breton and J. M. Gomez-Lllorente, *Chem. Phys. Lett.*, 1995, **235**, 160.
44. P. W Fowler and P. A Madden, *Phys. Rev. B*, 1984, **30**, 6131.
45. J. M. Luis, H. Reis, M. G. Papadopoulos and B. Kirtman, *J. Chem. Phys.*, 2009, **131**, 34116.
46. http://www.theochem.uni-stuttgart.de/pseudopotentials/clickpse.en.html.
47. J. M. Luis, M. Duran, B. Champagne and B. Kirtman, *J. Chem. Phys.*, 2000, **113**, 5203.
48. O. Loboda, R. Zalesny, A. Avramopoulos, J. M. Luis, B. Kirtman, N. Tagmatarchis, H. Reis and M. G. Papadopoulos, *J. Phys. Chem. A*, 2009, **113**, 1159.
49. B. Skwara, R. W. Gora, R. Zalesny, P. Lipkowski, W. Bartkowiak, H. Reis, M. G. Papadopoulos, J. M. Luis and B. Kirtman, *J. Phys. Chem. A*, 2011, **115**, 10370.
50. T. Guo, R. E. Smalley and G. E. Scuseria, *J. Chem. Phys.*, 1993, **99**, 352.
51. B. L. Dunlap, O. D. Haberlen and N. J. Rosch, *J. Phys. Chem.*, 1992, **96**, 9095.
52. R. Bauernschmitt, R. Ahlrichs, F. Hennerich and M. Kappes, *J. Am. Chem. Soc.*, 1998, **120**, 5052.
53. C. A. Jimenez-Hoyos, B. G. Janesko and G. E. Scuseria, *Phys. Chem. Chem. Phys.*, 2008, **10**, 6621.
54. M. D. Halls and H. B. Schlegel, *J. Chem. Phys.*, 1998, **109**, 10587.
55. M. D. Halls and H. B. Schlegel, *J. Chem. Phys.* 1999, **111**, 8819.
56. M. G. Papadopoulos and A. J. Sadlej, *Chem. Phys. Lett.*, 1998, **288**, 377.
57. P. W. Fowler and P. A. Madden, *Phys. Rev. B*, 1984, **29**, 1035.
58. P. W. Fowler, *J. Phys. Chem.*, 1985, **89**, 2581.
59. A. Kaczmarek and W. Bartkowiak, *Phys. Chem. Chem. Phys.*, 2009, **11**, 2885.

60. B. Kirtman, B. Champagne and J.M. Luis, *J. Comput. Chem.*, 2000, **16**, 1572.
61. C. Bissantz, B. Kuhn and M. Stahl, *J. Med. Chem.*, 2010, **53**, 5061.
62. E. Freire, *Drug Discovery Today*, 2008, **13**, 869.
63. C. A. Lipinski, F. Lombardo, B. W. Dominy and P. J. Feeney, *Adv. Drug Delivery Rev.*, 1997, **23**, 3.
64. P. Caccin, O. Rossetto, M. Rigoni, E. Johnson, G. Schiavo and C. Montecucco, *FEBS Lett.*, 2003, **542**, 132.
65. R. Wu, Z. Lu, Z. Cao and Y. Zhang, *J. Chem. Theory Comput.*, 2011, 7, 433
66. S. Shinoda and H. Tzukube, *Chem. Sci.*, 2011, **2**, 2301.
67. J. P. Villa, G. P. Bertenshaw and J. S. Bond, *J. Biol. Chem.*, 2003, **278**(43), 42545.
68. E. Pontiki, D. Hadjipavlou-Litina, K. Litinas, O. Nicolotti and A. Carotti, *Eur. J. Med. Chem.*, 2011, **46**(1), 191.
69. T. Borowski and E. Broclawik, *J. Phys. Chem. B*, 2003, **107**, 4639.
70. E. H. Mojumbar and A. P. Lyubartsev, *Biophys. Chem.*, 2010, **153**, 27.
71. A. Politi, G. Leonis, H. Tzoupis, D. Ntountaniotis, M. G. Papadopoulos, S. G. Grdadolnik and T. Mavromoustakos, *Mol. Inf.*, 2011, **30**, 973.
72. J. Lang, J. Santolini and M. Couture, *Biochemistry*, 2011, **50**, 10069.
73. O. Yuzlenko and T. Lazaridis, *J. Phys. Chem. B*, 2011, **115**, 13674.
74. K. Moffett, Z. Konteatis, D. Nguyen, R. Shetty, J. Ludington, T. Fujimoto, K. J. Lee, X. Chai, H. Namboodiri, M. Karpusas, B. Dorsey, F. Guarnieri, M. Bukhtiyarova, E. Springman and E. Michelotti, *Bioorg. Med. Chem. Lett.*, 2011, **21**, 7155.
75. H. H. Limbach, M. Chan-Huot, S. Sharif, P. M. Tolstoy, I. G. Shenderovich, G. S. Denisov and M. D. Toney, *Biochim. Biophys, Acta*, 2011, **1814**, 1426.
76. B. F. Shaw, G. F. Schneider, H. Arthanari, M. Narovlyansky, D. Moustakas, A. Durazo, G. Wagner and G. M. Whitesides, *J. Am. Chem. Soc.*, 2011, **133**, 17681.
77. D. A. Case, T. A. Darden, T. E. Cheatham III, C. L. Simmerling, J. Wang, R. E. Duke, R. Luo, R. C. Walker, W. Zhang, K. M. Merz, B. P. Roberts, B. Wang, S. Hayik, A. Roitberg, G. Seabra, I. Kolossváry, K. F. Wong, F. Paesani, J. Vanicek, J. Liu, X. Wu, S. R. Brozell, T. Steinbrecher, H. Gohlke, Q. Cai, X. Ye, J. Wang, M.-J. Hsieh, G. Cui, D. R. Roe, D. H. Mathews, M. G. Seetin, C. Sagui, V. Babin, T. Luchko, S. Gusarov, A. Kovalenko and P. A. Kollman, *AMBER 11*, University of California, San Francisco, 2010.
78. P. A. Kollman, I. Massova, C. Reyes, B. Kuhn, S. Huo, L. Chong, M. Lee, T. Lee, Y. Duan, W. Wang, O. Donini, P. Cieplak, J. Srinivasan, D . A. Case and T. E. Cheatham III, *Acc. Chem. Res.*, 2000, **33**(12), 889.
79. R. Sijbesma, G. Srdanov, F. Wudl, J. A. Castoro, W. Q. Charles, S. H. Friedman, D. L. Decamp and G. L. Kenyon, *J. Am. Chem. Soc.*, 1993, **115**, 6506.

80. R. Sijbesma, G. Srdanov, F. Wudl, J. A. Castoro, W. Q. Charles, S. H. Friedman, D. L. Decamp and G. L. Kenyon, *J. Am. Chem. Soc.*, 1993, **115**, 6510.

81. S. Durdagi, T. Mavromoustakos, N. Chronakis and M. G. Papadopoulos, *Bioorg. Med. Chem.*, 2008, **16**, 9957.

82. D. Martin and M. Karelson, *Lett. Drug Des. Discovery*, 2010, **7**, 587.

83. H. Tzoupis, G. Leonis, S. Durdagi, M. G. Papadopoulos, V. Bouhlis and T. Mavromoustakos, *J. Comput.-Aided Mol. Des.*, 2011, **25**, 959.

84. S. Durdagi, T. Mavromoustakos and M. G. Papadopoulos, *Bioorg. Med. Chem. Lett.*, 2008, **18**, 6283.

85. S. Durdagi, C. T. Supuran, A. T. Strom, N. Doostdar, M. K. Kumar, A. R. Barron, T. Mavromoustakos and M. G. Papadopoulos, *J. Chem. Inf. Model.*, 2009, **49**, 1139.

86. E. Rutenber, E. B. Fauman, R. J. Keenan, S. Fong, P. S. Furth, P. R. Ortiz de Montellano, E. Meng, I. D. Kuntz, D. L. DeCamp, R. Salto, J. R. Ros, C. S. Craik and R. M. Stroud, *J. Biol. Chem.*, 1993, **268**, 15343

87. J. Yang and A. R. Barron, *Chem. Commun.*, 2004, 2884.

88. J. Yang, L. B. Alemany, J. Driver, J. Yang and A. R. Barron, *Chem.–Eur. J.*, 2007, **13**, 2530.

89. S. H. Friedman, P. S. Ganapathi, Y. Rubin and G. L. Kenyon, *J. Med. Chem.*, 1998, **41**, 2424.

90. I. Morita, *Prostaglandins Other Lipid Mediators*, 2002, **68–69**, 165–175.

91. B. Samuelsson, *Am. J. Respir. Crit. Care Med.*, 2000, **161**(2), S2–S6.

92. H. Sheng, J. Shao, J. D. Morrow, R. D. Beauchamp and R. N. DuBois, *Cancer Res*, 1998, **58**(2), 362.

93. M. Cuendet and J . M. Pezzuto, *Drug Metabol. Drug Interact.*, 2000, **17**, 109.

94. A. J. Dannenberg and K. Subbaramaiah, *Cancer Cell.*, 2003, **4**, 431.

95. T. Mavromoustakos, P. Moutevelis-Minakakis, C. G. Kokotos, P. Kontogianni, A. Politi, P. Zoumpoulakis, J. Findlay, A. Cox, A. Balmforth, A. Zoga and E.Iliodromitis, *Bioorg. Med. Chem.*, 2006, **14**, 4353.

96. W. Sneader, in *Drug Discovery: A History*, John Wiley & Sons, West Sussex, 2005, p. 68.

97. M . G . Malkowski, S . L . Ginnell, W . L . Smith and R . M. Garavito, *Science*, 2000, **289**, 1933.

98. N. Neophytou, G. Leonis, N. Stavrinoudakis, M. Simčič, S. Golič Grdadolnik, E. Papavassilopoulou, G. Michas, P. Moutevelis-Minakakis, M. G. Papadopoulos, M. Zing and T. Mavromoustakos, *Mol. Inf.*, 2011, **30**, 473.

99. G. D. Hawkins, C. J. Cramer and D. G. Truhlar, *J. Phys. Chem.*, 1996, **100**, 19824.

100. J. M. Wood, J. L. Stanton and K. G. Hofbauer, *J. Enzyme Inhib. Med. Chem.*, 1987, **1**, 169.

101. N. A. Powell, F. L. Ciske, C. Cai, D. D. Holsworth, K. Mennen, C. A. Van Huis, M. Jalaie, J. Day, M. Mastronardi, P. McConnell, I. Mochalkin, E.

Zhang, M. J. Ryan, J. Bryant, W. Collard, S. Ferreira, C. Gu, R. Collins and J. J. Edmunds, *Bioorg. Med. Chem.*, 2007, **15**, 5912.
102. M. Paul, A. Poyan Mehr and R. Kreutz, *Physiol. Rev.*, 2006, **86**, 747.
103. A. P. Politi, S. Durdagi, P. M. Minakakis, T. Mavromoustakos and G. Kokotos, *J. Mol. Graph. Model.*, 2010, **29**, 425.
104. A. R. Leach, in *Molecular Modelling: Principles and Applications*, Pearson Education Ltd, Essex, 2nd edn, 2001.
105. J. A. Staessen, Y. Li and T. Richart, *Lancet*, 2006, **368**(9545), 1449.
106. M. J. Brown, *Circulation*, 2008, **118**, 773.
107. A. Gradman, R. Schmieder, R. Lins, J. Nussberger, Y. Chiang and M. Bedigian, *Circulation*, 2005, **111**(8), 1012.
108. J. M. Wood, J. Maibaum, J. Rahuel, M. G. Grütter, N. C. Cohen, V. Rasetti, H. Rüger, R. Göschke, S. Stutz, W. Fuhrer, W. Schilling, P. Rigollier, Y. Yamaguchi, F. Cumin, H. P. Baum, C. R. Schnell, P. Herold, R. Mah, C. Jensen, E. O'Brien, A. Stanton and M. P. Bedigian, *Biochem. Biophys. Res. Commun.*, 2003, **308**(4), 698.
109. J. Nussberger, G. Wuerzner, C. Jensen and H. R. Brunner, *Hypertension*, 2002, **39**, e1.
110. I. Stoica, S. K. Sadiq and P. V. Coveney, *J. Am. Chem. Soc.*, 2008, **130**, 2639.
111. T. Hou, J. Wang, Y. Li and W. Wang, *J. Chem. Inf. Model.*, 2011, **51**, 69.
112. M. Lucio, J.L.F.C. Lima and S. Reis. *Curr. Med. Chem.*, 2010, **17**,1795.
113. A. M. Seddon, D. Casey, R. V. Law, A. Gee and R. H. Templer, O. Ces, *Chem. Soc. Rev.*, 2009, **38**, 2509.
114. T. Mavromoustakos, P. Zoumpoulakis, I. Kyrikou, A. Zoga, E. Siapi, M. Zervou, I. Daliani, D. Dimitriou, A. Pitsas, C. Kamoutsis and P. Laggner, *Curr. Top. Med. Chem.*, 2004, **4**, 445.
115. J. K. Seydel, M. A. Velasco, E. A. Coats, H. P. Cordes, B. Kunz and M. Wiese, *Quant. Struct.-Act. Relat.*, 1992, **11**, 205.
116. D. A. Middleton, *Annu. Rep. NMR Spectrosc.*, 2007, **60**, 39.
117. J. K. Seydel and M. Wiese, *Drug–Membrane Interactions: Analysis, Drug Distribution, Modeling*, Wiley-VCH Verlag GmbH & Co., Weinheim, 2002.
118. T. Mavromoustakos, *The Use of Differential Scanning Calorimetry to Study Drug-Membrane Interactions*, Methods in Molecular Biology, Methods in Membrane Lipids, ed. A. M. Dopico, Humana Press Inc., Tuttowas, NJ, 2008, vol. 400, ch. 39, pp. 587–600 (other articles are related to other methodologies).
119. X. L. Liu, P. Fan, M. Chen, H. Hefesha, G. K. E. Scriba, D. Gabel and A. Fahr, *Helv. Chim. Acta*, 2010, **93**, 203.
120. M. Carini, G. Aldini, G. M. Orioli and M. R. M. Facino, *Curr. Pharm. Anal.*, 2006, **2**, 141.
121. T. Mavromoustakos, D. P. Yang, A. Charalambous, L. G. Herbette and A. Makriyannis, *Biochim. Biophys. Acta*, 1990, **1024**, 336.
122. T. Mavromoustakos, D. P. Yang, W. Broderick, D. Fournier, L. G. Herbette and A. Makriyannis, *Pharmacol. Biochem. Behav.*, 1991, **40**, 547.

123. D. W. Chester, V. Skita, H.S. Young, T. Mavromoustakos and P. Shrittmatter, *Biophys. J.*, 1992, **61**, 1224.

124. D. P. Yang, T. Mavromoustakos, K. Beshah and A. Makriyannis, *Biochim. Biophys. Acta*, 1992, **1103**, 25.

125. D. P. Yang, T. Mavromoustakos and A. Makriyannis, *Life Sci.*, 1993, **53**, 117.

126. M.A. Makriyannis, T. Mavromoustakos, K. Kelly and K. R. Jeffrey, *Biochim. Biophys. Acta*, 1993, **1151**, 51.

127. T. Mavromoustakos, D. P. Yang and A. Makriyannis, *Biochim. Biophys. Acta*, 1994, **1194**(1), 69.

128. T. Mavromoustakos, D. P. Yang and A. Makriyannis, *Biochim. Biophys. Acta*, 1995, **1237**, 183.

129. T. Mavromoustakos, D. P. Yang and A. Makriyannis, *Biochim. Biophys. Acta*, 1995, **1239**(2), 257.

130. T. Mavromoustakos, E. Theodoropoulou, D. Papahatjis, T. Kourouli, D. P. Yang, M. Trumbore and A. Makriyannis, *Biochim. Biophys. Acta*, 1996, **1281**(2), 235.

131. M. Koufaki, T. Calogeropoulou, T. Mavromoustakos, E. Theodoropoulou, A. Tsotinis and A. Makriyannis, *J. Heterocycl. Chem.*, 1996, **33**, 619.

132. T. Mavromoustakos, D. P. Yang and A. Makriyannis, *Life Sci.*, 1996, **59**(23), 1969.

133. T. Mavromoustakos, E. Theodoropoulou, C. Dimitriou, J. Matsoukas, D. Panagiotopoulos and A. Makriyannis, *Lett. Pept. Sci.*, 1996, **3**(4), 175.

134. T. Mavromoustakos, E. Theodoropoulou and D. P. Yang, *Biochim. Biophys. Acta*, 1997, **1328**(1), 65.

135. T. Mavromoustakos, A. Papadopoulos, E. Theodoropoulou, C. Dimitriou and E. A. Byza, *Life Sci.*, 1998, **62**(20), 1901.

136. T. Mavromoustakos and E. Theodoropoulou. *Chem. Phys. Lipids*, 1998, **92**(1), 37.

137. T. Mavromoustakos, D. Papahatjis and P. Laggner, *Biochim. Biophys. Acta*, 2001, **1512**(2), 183.

138. H. Maswadeh, C. Demetzos, I. Daliani, I. Kyrikou, T. Mavromoustakos, A. Tsortos and G. Nounesis, *Biochim. Biophys. Acta*, 2002, **1567**, 49.

139. I. Kyrikou, I. Daliani, T. Mavromoustakos, H. Maswadeh, C. Demetzos, S. Xatziantoniou, S. Giatrellis and G. Nounesis, *Biochim. Biophys. Acta*, 2004, **1661**, 1.

140. I. Kyrikou, S. Xadjikakou, D. Kovala-Demertzi, K. Viras and T. Mavromoustakos, *Chem. Phys. Lipids*, 2004, **132**, 157.

141. I. Kyrikou, T. Mavromoustakos, S. Xatziantoniou, A. Georgopoulos and C. Demetzos, *Chem. Phys. Lipids*, 2005, **133**, 125.

142. N.-P. Benetis, I. Kyrikou, T. Mavromoustakos and M. Zervou, *Chem. Phys.*, 2005, **314**(1–3), 57.

143. V. Saroglou, S. Xatziantoniou, M. Smyrniotakis, I. Kyrikou, T. Mavromoustakos, A. Zompra, V. Magafa, P. Cordopatis and C. Demetzos, *J. Pept. Sci.*, 2006, **12**, 43.

144. C. Koukoulitsa, I. Kyrikou, C. Demetzos and T. Mavromoustakos, *Chem. Phys. Lipids*, 2006, **144**, 85.
145. I. Kyrikou, C. Poulos, N. Benetis, K. Viras, M. Zervou and T. Mavromoustakos, *Biochim. Biophys. Acta*, 2008, **1778**, 113.
146. C. Koukoulitsa, S. Durdagi, E. Siapi, C. Villalonga-Barber, X. Alexi, B. R. Steele, M. Micha-Screttas, M. N. Alexis, A. Kakoulidou and T. Mavromoustakos, *Eur. Biophys. J.*, 2011, **40**, 865.
147. P. Zoumpoulakis, I. Daliani, M. Zervou, I. Kyrikou, E. Siapi, G. Lamprinidis, E. Mikros and T. Mavromoustakos, *Chem. Phys. Lipids*, 2003, **125**, 13.
148. C. Fotakis, S. Gega, E. Siapi, C. Potamitis, K. Viras, P. Moutevelis-Minakakis, G. Kokotos, S. Durdagi, S. Grdadolnik, B. Sartori, M. Rappolt and T. Mavromoustakos, *Biochim. Biophys. Acta*, 2010, **1798**(3), 422.
149. T. Mavromoustakos, P. Chatzigeorgiou, C. Koukoulitsa and S. Durdagi, *Int. J. Quantum Chem.*, 2011, **6**, 1172.
150. C. Potamitis, P. Chatzigeorgiou, E. Siapi, T. Mavromoustakos, A. Hodzic, F. Cacho-Nerin, P. Laggner and M. Rappolt, *Biochim. Biophys. Acta*, 2011, **1808**, 1753.
151. C. Fotakis, D. Christodouleas, P. Zoumpoulakis, A. Gili, E. Kritsi, N.-P. Benetis, M. Zervou, H. Reis, M. G. Papadopoulos and T. Mavromoustakos, *J. Chem. Phys. B*, 2011, **115**, 6180.
152. D. Ntountaniotis, G. Mali, S. G. Grdadolnik, M. Halabalaki, A.-L. Skaltsounis, C. Potamitis, E. Siapi, P. Chatzigeorgiou, M. Rappolt and T. Mavromoustakos, *Biochim. Biophys. Acta*, 2011, **1808**, 2995.
153. T. Mavromoustakos, S. Durdagi, C. Koukoulitsa, M. Simcic, M. G. Papadopoulos, M. Hodoscek and S. G. Grdadolnik, *Curr. Med. Chem.*, 2011, **18**, 2517.
154. T. Mavromoustakos, P. Moutevelis-Minakakis, G. Kokotos, E. Papavassilopoulou, C. Potamitis, C. Fotakis, P. Chatzigeorgiou, K. Vyras, C. Koukoulitsa, E. Kalatzis and S. Durdagi, Peptide mimetics drugs and their interdigitation with lipid bilayers, in *Essays on Contemporary Peptide Science*, ed. P. Cordopatis, Research SignPost (India), 2011, ch. 6, pp. 95–113. ISBN: 978-81-308-0428-6.
155. L. J.Wagenaar, A. A. Voors, H. Buikema, A. van Buiten, R. H. Luubeck, P. W. Boonstra, D. J. van Veldhuisen and W. H. van Gilst, *Cardiovasc. Drugs Ther.*, 2002, **16**, 311.
156. T. Unger, *Eur. Heart J. Suppl.*, 2004, **6**, H11.

CHAPTER 9

Thermodynamic Cartography and Structure-Property Mapping of Potential Nano-Hazards

AMANDA S. BARNARD

CSIRO Materials Science & Engineering, Parkville, 3052, Australia
E-mail: amanda.barnard@csiro.au

9.1 Introduction

The synthesis and application of a range of functional nanoscale materials (nanomaterials) promises to provide us with new tools in the fight against disease, climate change and the generation and storage of renewable energy. Reading the chapters in this book, and the associated literature, we can see that the *nanoscience* is maturing into *nanotechnology*, and the development of a variety of anticipated applications, such as sensors (including biosensors), tips for scanning probe microscopy, electrochemical actuators and batteries, is now beginning to be realized. While many of these devices are already familiar to us, the diversity of nanomaterial properties (that are dissimilar to their macroscopic counterparts) means that many materials are being used in non-traditional situations. That is to say, nanomaterials of a particular substance are finding their way into products and places in which is it not normally used, and entering a life-cycle that is uncharacteristic of the bulk (macroscale) version. Scientists and engineers are therefore required to consider the consequences and implications of this type of re-tasking, and navigate a delicate balance between technological advancement and social awareness.

RSC Nanoscience & Nanotechnology No. 25
Towards Efficient Designing of Safe Nanomaterials: Innovative Merge of Computational Approaches and Experimental Techniques
Edited by Tomasz Puzyn and Jerzy Leszczynski
© The Royal Society of Chemistry 2012
Published by the Royal Society of Chemistry, www.rsc.org

When balancing these two issues, it is helpful to think about the problem from both the technological *and* the toxicological perspectives. The nanoparticles we engineer do have a particular end-use in mind. They are not random, and their intended function tells us as much about their properties as their structural specifications. The way that nanomaterials are being applied is also important. In many cases they will be incorporated into macroscale products, and we may be tempted to think that this renders them innocuous. However, considering the current move to recyclable or biodegradable products, it is still more prudent to assume that the introduction of these engineered nanoparticles into the natural world is an inevitability, rather than a "worst case scenario". Therefore, in parallel to studies aimed at optimizing industrial performance, the toxicology of all types of nanomaterials is also under scrutiny,[1–3] along with the consideration of possible environmental impacts[4–7] and issues surrounding workplace safety.[8,9] So, if we step back and take a broad view of the field, we can see that both the technological and the toxicological fields are focussed on the same goal: ensuring sustainability for the future of nanotechnology.

In this context, sustainability encapsulates a number of aspects. Firstly, there is an expectation that this new technology be economically viable. We all want our new devices to be cost effective, so this can mean generally using less, and striving to develop nanomaterials which are not based on rare commodities or resources that are not currently available *via* reliable trade agreements. In additional to this, there is increasing demand that new technologies be environmentally friendly, so that they do not damage the environment when discarded, when in use, or when in production (*i.e.* low carbon emissions). Ideally we also want the new nanotech-products to be recyclable and/or biodegradable, and of course, we also want nanotechnology to be safe. This means that the next generation of "in-demand" smart products must be simultaneously efficient, safe and environmentally friendly and, above all else, they must be reliable and perform their function in a predictable way. This is a demanding resumé, and in reality not all nanomaterials will meet these requirements. In such cases we must decide either to mitigate (avoid the unmanageable), or adapt (manage the unavoidable) to preserve our quality of life.

While challenging, these issues are not entirely unfamiliar. In the past many nanoparticles were simply referred to as ultra-fine particles, which were known to either occur naturally,[10] or to be introduced unintentionally through human activities or as industrial products.[11] The natural sources include gas-to-particle conversions, forest fires, volcanoes, viruses, biogenic sources, and ferritin; the incidental sources include internal combustion engines, power plants, incinerators, jet engines, fumes (metal polymer, *etc.*), heated surfaces (such as in frying, boiling or grilling) and electric motors. In general, much of the study of hazards associated with nano-sized particles is based on established research on these airborne ultra-fine particles, and so are the assumptions.

The main difference between this field and the emerging field of nanotoxicology is the issue of the nanomaterials being *engineered*. The nanomaterials engineered for the tasks mentioned above typically have no natural analogue, and are often artificially modified to induce a specific functionality. This introduces a degree of unpredictability, since we are presented with a range of largely untested materials that are (literally) unique on an atomic scale. Unlike the nanomaterials from the natural and unintentional sources, intentionally produced nanoparticles (made in laboratories) have not been around long enough for us to gather sufficient data to gauge their long-term potential for causing adverse effects. In time we could easily accumulate the necessary information and knowledge, but economic and social pressures call for immediate solutions, particularly in regard to potential toxicity.

It is also important to remember that nanomaterials do not exist in isolation, nor in ultra-high vacuum (UHV) or at zero temperature, and their structure and properties are subject to change. We must take into consideration the issue of the stability of nanomaterials, and their relationship to the immediate environment. While we are accustomed to thinking about what nanomaterials may do to the environment, it is often forgotten that the environment can also have an important impact on nanomaterials. There is a type of cyclic causality at work here, where environmental variations (such as changes in temperature, pressure, humidity or electrostatic charge) can alter the stability and properties of nanomaterials, thereby changing the way they interact with the environment. This has the unfortunate consequence of reducing our dynamic control over properties, and our ability to anticipate nano-hazards in a given situation.

Our only defence against the technological and environmental ambiguity introduced by this cyclic causality is to increase our knowledge of the various types of instabilities native to nanomaterials, along with their respective causes. This includes structural instability, but also more subtle instabilities that may be more difficult to detect, such as conformational instabilities of impurities and functional defects or ligands adsorbed at the surfaces. Consider the situation where (given a suitable perturbation such as a significant increase in temperature) impurities thought to be stable within the core of a nanoparticle were to diffuse to the surface, thereby changing the surface reactivity or causing stabilizing surfactants to desorb. Similarly, changes in the pH of a storage solution could also induce surfactants to desorb, and alter the degree of agglomeration of collections of nanoparticles. As nanomaterials have a high surface area, maintaining stable surfaces both in devices and in storage media is of critical importance.

So how do we approach these issues? One of the motivations of this book is to describe ways that modellers can engage in this area, and to provide a broad overview of methods for predicting risks and hazards associated with nanomaterials. In this particular chapter we will examine some analytical methods for predicting the structure and stability of nanomaterials, and how these may be used as a guide to their properties – including those perceived to be hazardous. We will begin by outlining how this aspect features into the bigger picture of nano-hazard prediction (and, ultimately, prevention), followed by discussion of the stability of titania photocatalysts. This will be followed by a short case study, highlighting how these techniques can be combined with other modelling methods to create complementary structure-property maps to facilitate informed decision making.

9.1.1 Strategic Approaches to Predicting Nano-hazards

To understand how studies of nanoparticle structure and stability relate to the topic at hand it is convenient to break this complex problem down into a scheme of interacting nodes,[12] each responsible for a particular piece of the puzzle. An example is shown in Figure 9.1. The first node (upper left) in this scheme is entitled *Nano-hazards*. Throughout this book there are a multitude of descriptions of nano-hazards, as they pertain to undesirable interactions with life forms or with natural ecosystems (including the introduction of air- or water-borne pollution, contamination of soils, detriment to the food chain and reactions leading to increases in salinity or imbalances in natural resources that were not present before). These issues are all combined in this node.

The ultimate goal of this scheme is to make a link between the *Nano-hazards*, and *Prevention*, shown in the node at the lower left corner of Figure 9.1. This can mean preventing exposure or setting exposure limits (dosimetry), prescribing modifications to the nanomaterials that will reduce or eliminate the danger, setting appropriate safety guidelines and labelling, designing containment and storage facilities, designing disposal systems, or introducing

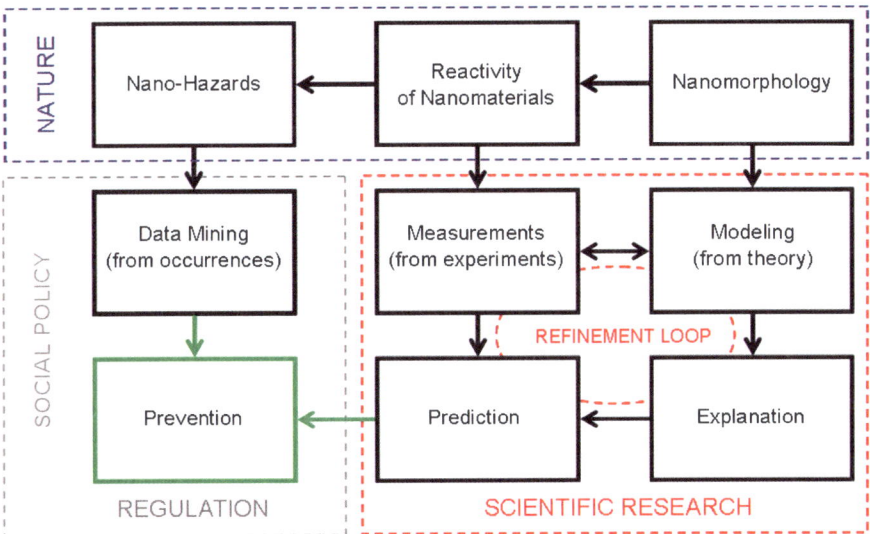

Figure 9.1   Schematic representation of the multi-disciplinary landscape that is navigated by those working on various aspects of (potential) nano-hazards, showing the interaction between different aspects of the problem and the actors involved in formulating solutions. Reproduced with permission from ref. 12, Royal Society of Chemistry © 2009.

regulation. This is predominantly the role of the social sciences, laws and politics, which are engaged in activities designed to protect the interests (and members) of the population and the environment. Although scientists may be in a position to advise colleagues in this space, the mechanisms are not in place for us to act on this advice autonomously. Currently, within this space, the only reliable link between *Nano-hazards* and *Prevention* is termed *Data mining*, which is based on experiences of actual events.

While observations and data collected from occurrences are irrefutable, this route is highly undesirable since it requires that something dangerous has already happened. One does not need to look far into the past to see where mistakes have been made (such as asbestos and DDT) and recall the detrimental effects these mistakes have had upon human health and the economy. While it is true that we were able to learn from these mistakes and identify prevention methods going forward, clearly the cost was too big and the damage too great to ever let this type of error in judgement happen again. Therefore the route given in the first column must be considered socially unacceptable, and we need to find a better way.

To identify an alternative path from *Nano-hazards* to *Prevention* we must first consider what we (as scientists) know about nano-hazards at a fundamental level. How do natural phenomena give rise to a property that is hazardous? Based on the current empirical evidence and limited scientific studies that have been reported, we find that there is a strong link between nano-hazards and the reactivity of nanoparticle surfaces (although there are other sources). This includes chemical reactions and bio–nano interactions occurring at individual surface facets, and in particular with individual "active sites" such as edges, corners, kinks and steps on surface facets, and surface defects. An example of this link is the production of reactive oxygen species (ROS) on the surfaces of metal oxide nanoparticles, which can ultimately give rise to oxidative damage of tissue when the two come in contact.[13] However, although the reactivity of nanoparticles is observable, it is a natural phenomenon, and ultimately beyond our control.

From extensive observation, we are, however, aware that the reactivity of nanoparticles can exhibit a high degree of selectivity which depends sensitively on the material (both composition and solid phase), the size (surface-to-volume ratio), and on the shape, *i.e.* on the nanomorphology. In addition to size, shape and phase this term also encompasses aspects such as the topology, symmetry and crystallographic forms presented, the growth direction of quasi-one-dimensional nanostructures such as nanorods and nanowires, and the chirality of nanotubes. Many fundamental properties of nanomaterials have already been shown to have a strong dependence on their nanomorphology, and indeed this is the very essence of structure-property relationships.

Although the link between *Nanomorphology* and *Reactivity* is also a natural phenomenon, there has been some success in engineering the morphology of nanoparticles so as to minimize or maximize certain interactions or reactive properties. Variations in nanomorphology are either thermodynamically or kinetically driven, and include materials parameters such as size, coatings and surfactants, supports and substrates, and composition (dopants); and environmental parameters such as temperature, chemical environment, and exposure to external fields. Since so many of the influencing factors have been identified, a detailed understanding of the relationship between each parameter and particle shape may be used to tailor the properties of nanomaterials on an individual and collective basis. This link is often exploited by those designing nanomaterials for specific applications, who routinely modify the size, shape and surface chemistry. Hence, we already have a reasonable understanding of how these two areas relate to one another, and how they relate to the local environment.

Together the *Reactivity* and *Nanomorphology* provide points in the scheme where we can access *Nano-hazards*, and it is from here that an alternative route to *Prevention* can be found. Given the fact that a definitive characterization of *Reactivity* can (at this stage) only be achieved after samples have been produced, it is suggested that this link is best made to (specific) experimental *Measurements* (see Figure 9.1). This ensures that the characterization of *Reactivity* takes account of the natural distribution of possible values resulting from the polydispersivity of sizes, shapes and surface chemistries exhibited by real samples, and the possibility of an unpredictable number (and type) of structural defects. Conversely, given the large number of parameters that influence *Nanomorphology*, it is suggested that this link is best made to *Modeling* using theory and computer simulations. Using this approach it is possible to test one parameter in isolation from all others in a way that cannot be done efficiently using experimental methods (where samples typically change in response to being probed).

9.1.2 Combining Theory, Simulation and Experiment

Both *Measurements* and *Modeling* fall squarely within our sphere of influence, but since these complementary areas are not unrelated, a "refinement loop" is formed as highlighted in Figure 9.2. The usefulness of theory and computer simulations lies in their ability to accurately reproduce experiment, and this can only be verified *via* a structured system of comparison. Analytic theories, computational algorithms and simulation methods must be constantly compared to physical measurements to improve their accuracy and ensure reliability. Similarly, a systematic examination of all possible nanoparticle/environmental combinations using experimental methods has already become unfeasible, and the targeted application of these resources is more scientifically (and economically) responsible. It is in this regard that modeling can complement the experimental effort by identifying the most important areas to focus on, and reveal the underlying mechanisms to be tested.

One of the main benefits in combining these different approaches is that they provide very different, but complementary, information. In the vast majority of cases the type of results accessible to experiments is at the mercy of the type of sample. Real samples contain an ensemble of different structures, since a

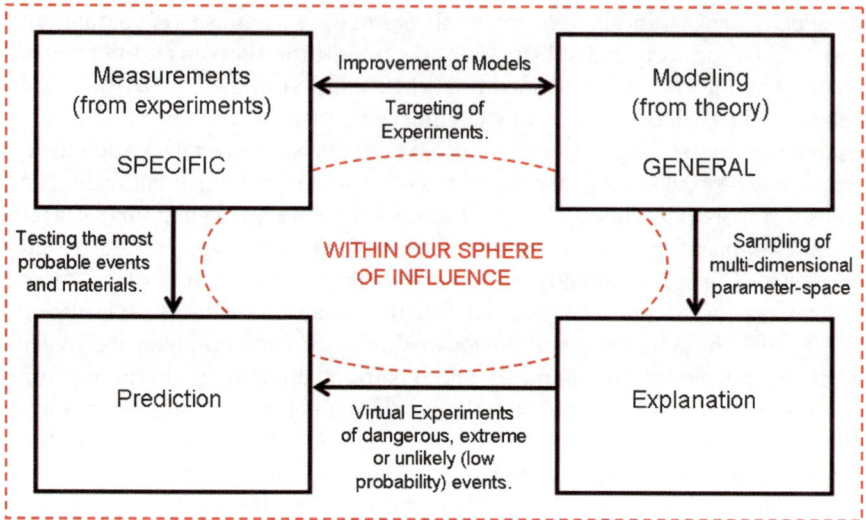

Figure 9.2   Schematic representation of the refinement loop where complementary approaches can combine to produce reliable predictions. Reproduced with permission from ref. 12, Royal Society of Chemistry © 2009.

natural degree of polydispersity is inevitable. This means that the results of experimental measurements are effectively an average over all these different structures, and ultimately provide an "approximate solution" (or "statistical solution"), albeit to an "exact problem".

In contrast, theory and simulation have the opposite perspective. In the virtual world of modeling it is very straight forward to treat isolated, singular structures, and to define that structure explicitly and with exquisite control. The disadvantage in this case is that the complexity inherent in real samples is usually out of reach of most models, which typically include a sub-set of the physio-chemical parameters associated with the nanostructure and the surroundings. This means that modeling provides an "exact solution" to an "approximate problem".

Another advantage of including modeling in this strategy is that (once experimentally validated) it can rapidly sample a very large multi-dimensional parameter space. Since modeling is capable of doing this in a systematic manner it can also provide a much needed *Explanation* of the underlying mechanisms responsible for nano-hazards. It should be noted that *Nano-hazards* and *Explanation* are represented at opposite extremes of the diagram in Figure 9.1, serving to highlight the long way we have to go to bring these two concepts together. Once both the *Measurements* and *Modeling* approaches have been validated, and a suitable partnership has been achieved, tasks may be selectively assigned to each of these paths, so as to maximize both scientific impact and the responsible use of available resources. By activating both paths in the refinement loop (Figure 9.2) we can generate a complete scientific picture which provides a fundamental basis for *Prediction*. Both paths feed into this node, and both are equally important, because they provide unique input.

However, this being said, just as measurements are not a "one-size fits all" solution, it is not immediately clear which is the right modeling technique for a given situation. In general, answering this question is the aim of this book, and we will see that there are a range of different techniques designed to address different areas of need. When it comes to the modeling of the structure of nanomaterials we are fortunate in that a hierarchy of computational techniques exists from which to choose.[14] In order of decreasing complexity we have quantum chemical methods (born of Hartree–Fock theory), electronic structure methods such as the various flavours of density functional theory (DFT), and semi-empirical methods; all beginning with *ab initio* assumptions based on nuclear and electronic interactions (though each with a different Hamiltonian). These are highly sophisticated methods that, due mainly to computational restrictions, are typically used to examine simple and idealized systems at a specific point in time.[15] At a reduced level of sophistication (or conversely, a higher order of approximation) we have classical and empirical methods, which allow for the explicit examination of more realistic systems and the consideration of dynamical properties. Already we can see how these techniques can provide unique and complementary information in their own right.

As the degree of complexity of the system increases, such as when we introduce a broader and diverse morphological ensemble, or a more rich description of the surrounding environment is required, we must typically turn to analytic and continuum methods. Analytic methods offer the advantage of capturing the phenomenology while only needing a limited set of parameters to describe a large variety of configurations. However, this comes at the expense of coarse-graining over much of the explicit detail. Such methods can be very powerful for screening or rapidly sampling a multi-dimensional parameter space, and identifying important combinations and dependencies that will require more meticulous attention.

One such method is the mapping of the structure of nanomaterials in a space defined by a set of environmental parameters, a field known as *thermodynamic cartography*. In principle, there are no limits to the number of structural or environmental parameters one may include (often referred to as "dimensions"), but it is advisable to restrict this type of modeling to those dimensions with the greatest influence, so that the calculations remain tractable and statistically significant. Thermodynamic cartography has proved to be very useful in describing the equilibrium and non-equilibrium structure of nanomaterials such as oxides,[16–21] sulphides[22–27] and metals.[28–34] In the following sections the thermodynamic cartography of titania nanoparticles will be described and used in the construction of structure-property maps as they relate to the use of this material in modern sunscreens.

9.2 Thermodynamic Cartography of Nanoscale Titania

Nanostructures of titania (TiO_2) are an ideal exemplary system, as they are of considerable scientific interest due to their superior performance in a range of advanced photochemical applications. The optical properties of titania are known to be size-dependent, and it has also been shown that the crystalline phase (anatase or rutile) and the shape of individual nanocrystals are critical parameters in determining their photocatalytic efficiency[35,36] and overall suitability for particular applications.[37–39] Recent studies have demonstrated remarkable control over the morphology of titania nanostructures;[40] however, this requires very specific synthesis conditions which are dissimilar to the environment in which these materials will ultimately be used, and the (greater) natural environment with which they are likely to interact. Predicting the effect which changing the thermal and chemical environment will have upon titania nanostructures is highly desirable, but the incorporation of environmentally relevant parameters into theoretical descriptions of nanomaterials is not trivial.

In general it is well known that rutile is the thermodynamically stable phase of titania under ambient conditions at the macroscale,[41] and that anatase is the thermodynamically stable phase at the nanoscale.[42,43] A number of authors[44–47] have shown that the synthesis of nanocrystalline titania consistently resulted in anatase nanocrystals which transform to rutile upon reaching a particular size.[48] The size range of the anatase-to-rutile phase transition for hydrothermal samples (at ~ 650 to ~ 800 K) has been predicted to be in the range 11.4 nm to 17.6 nm.[44] However, anatase nanocrystals are often observed over this size[35,43,49,50] depending upon the temperature and surface chemistry.

To understand why this reversal in phase stability occurs, and how it relates to size, shape, surface chemistry and temperature, we may use an analytic shape-dependent thermodynamic model based on a geometric summation of the Gibbs free energy.[14,51,52] The simplified version of the model used here is applicable specifically to isolated, defect-free structures over ~ 3 nm in diameter, and is written as:

$$G_x(\mu,T) = \Delta_f G_x^0(T) + \frac{M_x}{\rho_x}\left[1 - \frac{2\sum_i f_i \sigma_{xi}(\mu,T)}{\langle R \rangle B_0} + \frac{P_{ext}}{B_0}\right]\left(q\sum_i f_i \gamma_{xi}(\mu,T)\right) \quad (9.1)$$

where M is the molar mass and ρ is the density of a titania in phase x.

Unlike other models used to describe the free energy of materials, this model specifically accounts for the polyhedral shapes of nanocrystals (as well as the finite size) *via* the sum over the individual contribution from the exposed surface facets, i. The model includes the surface-to-volume ratio, q, which is both size- and shape-dependent; and the specific surface energy, $\gamma_{xi}(\mu,T)$, of facet i exposed to the specific surface chemistry described by μ at a temperature T. The volume dilation induced by the isotropic surface stresses σ_{xi} and external pressure P_{ext} is also included *via* the Laplace–Young formalism,[52] using the bulk modulus B_0 (of phase x), and the average particle radius $<R>$ calculated using a spherical approximation. In all cases described in latter sections, atmospheric external pressure has been assumed, and the standard free energy of formation $\Delta_f G^0_x(T)$ (with x = anatase or rutile) is described using a regression representation widely used in phase diagram calculations, and data from the JANAF tables.[53]

One of the great advantages of using this technique is that the model provided in eqn (9.1) may be parameterized using *ab initio* electronic structure methods. This means that one can access the accuracy and sensitivity of first principles methods, and apply them to sizes that are traditionally inaccessible to these simulation techniques. This is particularly helpful when it is necessary to distinguish between the different types of reactions occurring on surfaces, since subtle differences are often associated with electron or proton exchange.

To demonstrate the use of this technique, let us consider a simple example where we calculate the size-dependent phase transformation between anatase and rutile, in the absence of surface passivation, at low temperature. In this example, DFT has been used to calculate the lattice parameters (needed to find ρ), B_0, and γ_{xi} and σ_{xi} for $i = \{100\}$, $\{110\}$ and $\{011\}$ when $x =$ rutile, and $i = \{100\}$, $\{101\}$ and $\{001\}$ when $x =$ anatase. The orientation of these facets and an example of the morphologies that they enclose are shown in Figure 9.3.

Using the DFT parameters (provided in ref. 54), it is possible to optimized the particle shape, and find the crossover in free energy, as shown in Figure 9.4. At this point it is pertinent to point out that since there is a volumetric disparity between anatase and rutile (due to the differences in ρ_x), identification of phase transformation sizes should be performed as a function of the number of titania formula units (N_{TiO2}). Once the critical number of formula units has been determined it may then be converted to the equivalent nanocrystal diameter, in this case ~ 9.6 nm for anatase. This is characteristic of the phase transformation size if the particles are at equilibrium, but in actual fact samples may contain a range of shapes that depart from equilibrium, or exhibit specific shapes may have been engineered to increase the photocatalytic efficiency.

It has been previously reported that the photocatalytic efficiency is greater on the $\{001\}$ facet of anatase,[55] and the $\{110\}$ facets or rutile,[56] so considerable effort is being directed to making titania nanoparticles that maximize the fractional area in these crystallographic orientations.[55] If we artificially alter the shapes to minimize and/or maximize the fraction of anatase $\{001\}$ and rutile $\{110\}$ there will be consequences for the phase stability. We can see from Figure 9.5 that the transformation size can fall anywhere from ~ 8 to ~ 18 nm,

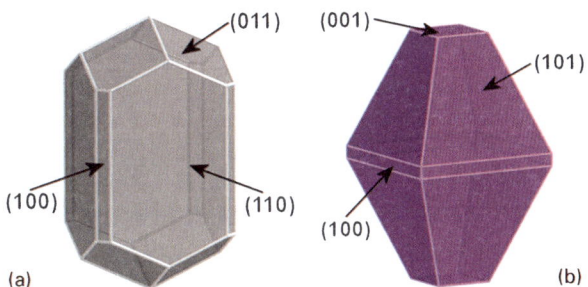

Figure 9.3   Illustrations of the faceted morphology of rutile (a) and anatase (b), with the crystallographic (*hkl*) orientation of the facets marked.

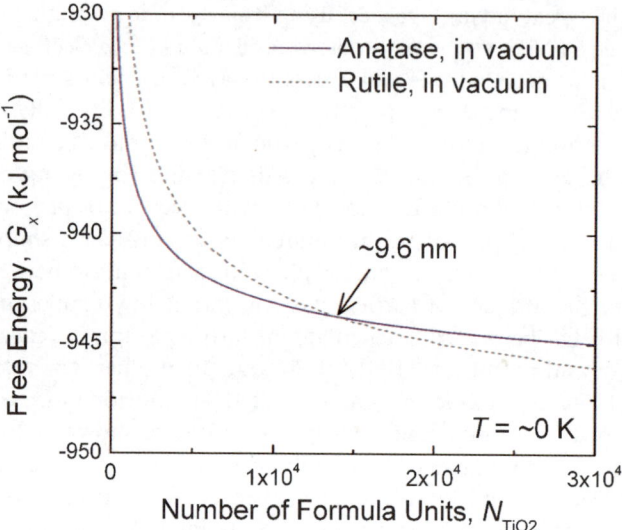

Figure 9.4   Free energy of formation for anatase (solid line) and rutile (dashed line) as a function of size (in number of TiO_2 formula units). The average diameter of anatase nanoparticles at the phase transformation crossover is marked.

depending on the shapes involved. Size, shape and phase are all intrinsically linked, and we in fact have a shape-dependent phase transformation size, as opposed to a size-dependent phase transformation.

Transition Size* (nm)					
	17.85	17.9	17.7	17.3	15.7
	12.75	12.8	12.6	12.3	10.7
	11.05	11.1	11.0	10.5	9.0
	10.35	10.4	10.3	9.8	8.3
	9.95	10.0	9.9	9.5	7.9

Figure 9.5   Matrix of the shape-dependent phase transformation size, as the shape of anatase (columns) and rutile (rows) is artificially altered to increase or decrease the fractional area of the reactive anatase {001} and rutile {110} facets. * refers to the equivalent average diameter of anatase. Reproduced with permission from ref. 14, IOP Publishing © 2010.

Although these results are in excellent agreement with the size-dependent phase boundary between anatase and rutile that has been calculated a number of times over the past decade, like past studies it fails to capture the dynamic variations in nanomorphology as a function of temperature,[45,57,58] and does not include the important influence of surface adsorbates.[57] From experiment, it is usually found that titania surfaces include TiOH surface hydroxyl groups,[49,59-64] and under neutral pH conditions the surfaces are terminated with H_2O (either as molecular H_2O or as dissociated $OH^- + H^+$) or OH.[43,59,60] These terminations are more commonly observed than alternative combinations of O and H,[57] and are most likely to be present on the surface when titania particles are exposed to pH neutral atmospheric conditions.[59,60]

These issues may also be probed using the same thermodynamic modelling approach used above, by applying monolayers of different types of surface groups to represent different types of chemical conditions when undertaking the DFT simulations of the surface properties. It is during the DFT parameterization that the chemical complexity can be introduced, as we can see by reviewing the work on H_2O- and OH-terminated surfaces. The structure and energy of the H_2O- and OH-terminated surfaces of anatase and rutile have been described in great detail elsewhere,[65] but briefly the modified specific surface energies may be calculated from the chemical potential of the bulk material, μ_{TiO2}, and the total energy of the surface slabs, $E_{xi}(N_{TiO2})$, using the expression:

$$\gamma_{xi,y}(\mu) = \frac{1}{2A_{xi}}\left[E_{xi}\left(N_{TiO_2}\right) - N_{TiO_2}\mu_{TiO_2} - N_y\mu_y\right] \quad (9.2)$$

where A is the area of the surface and N_{TiO2} is the number of titania units in the (stoichiometric) cell. Here, μ_y is the chemical potential of the adsorbates of species y (where y can be any surface group, but in the present context is H_2O or OH), and N_y is the total number of adsorbate molecules per unit A.[65] The chemical potential μ_y must be defined with respect to a chosen frame of reference, and in order to be directly comparable, must be in mutual equilibrium with the underlying material, and any molecular alternatives. In this example the μ_y have been constructed from the chemical potentials of water and hydrogen, such that:

$$\mu_{H_2O} = E_{H_2O} + \frac{h\nu_{H_2O}}{2} + k_B T \ln\left(\frac{PV}{k_B T}\right) \quad (9.3)$$

$$\mu_{OH} = \mu_{H_2O} - \frac{1}{2}\left[E_{H_2} + \frac{h\nu_{H_2}}{2} + k_B T \ln\left(\frac{PV}{k_B T}\right)\right] \quad (9.4)$$

where E_y is the total electronic energy of molecule y. This means that water and hydroxyl are in mutual equilibrium with each other within the stability fields of the chemical potential reservoir. As usual, k_B is Boltzmann's constant, T, P and v are the temperature, pressure and sum of the vibrational frequencies in the reservoir, and V is the quantum volume:

$$V = \left(\frac{h^2}{2\pi m_x k_B T} \right)^{3/2},$$
(9.5)

where m_x is the molecular mass of x, and h is Planck's constant.[66] Since σ is proportional to $\partial G/\partial A$ the temperature dependence of $\sigma_{xi,y}$ may be assumed to be consistent with $\gamma_{xi,y}$, and the same slope applied. The energetic terms in the chemical potentials, and the $T = 0$ K coefficients for the surface energies and isotropic surfaces stresses, have been previously calculated from first principles and are listed in ref. 65.

Based on these results, Figures 9.6(a) and (b) shows the free energy comparisons for OH- and H₂O-terminated anatase and rutile respectively, along with the optimized nanomorphologies (shown as insets). Here we can see that both the shape and the phase transformation sizes differ, depending on the type of surface chemistry, and they are both different from the *in vacuo* results for clean surfaces (shown in Figure 9.4). In both cases anatase is favoured more at larger sizes (than *in vacuo*), but the passivation of the surfaces with OH groups also affects the equilibrium morphology, increasing the relative area of the reactive facets and the aspect ratio. The anatase nanoparticles will also be expected to develop {100} facets, in addition to {101} and {001} facets, forming a "belt" about the centro-symmetric plane.

These results have been plotted with a value of $T = 300$ K, and by repeating the modeling for a range of temperatures the temperature dependence of the transformation size can be obtained. By mapping the position of the crossover points in $<D,T>$ space, while using the lowest energy surface adsorbate

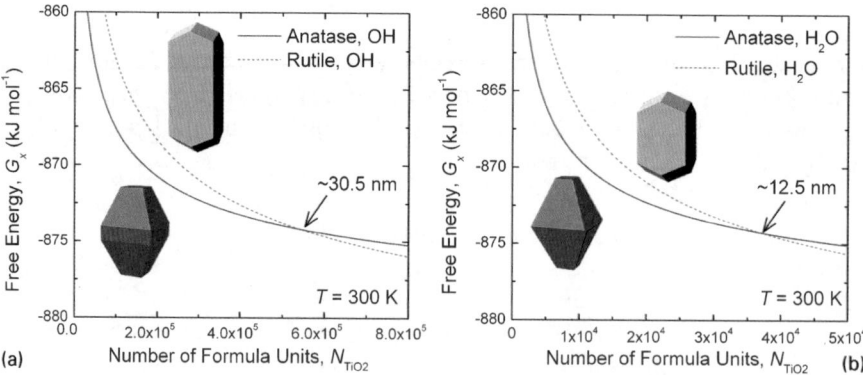

Figure 9.6 Free energy of formation for anatase and rutile nanoparticles with (a) OH-terminated surfaces, and (b) H₂O-terminated surfaces, at $T = 300$ K. The optimized morphology for each titania/adsorbate combination is shown as an inset.

species and optimizing the shape at each point, we can map the morphology of nanoscale titania. However, in this definition of the chemical potential, T is the temperature of the surrounding reservoir, and cannot describe material-specific temperature-dependent behaviour such as nanoparticle melting.

Therefore, to complete the thermodynamic cartography we must extend these results to the construction of a nanoscale phase diagram, so we need another expression to model the high temperature solid–liquid phase transformation. The classical definition of the size-dependent melting temperature $T_m(D)$, is expressed in terms of the bulk melting temperature $T_m(\infty)$, such that:

$$\frac{T_m(D)}{T_m(\infty)} = 1 + \frac{A_t(\gamma_l - \gamma_s)}{V_t \Delta H_m(\infty)} \tag{9.6}$$

where A_t is the total surface area, V_t is the total volume, γ_l is the surface energy of the liquid, and $\Delta H_{m,\infty}$ is the melting enthalpy. This formalism has recently been shown to be applicable to titania nanocrystals over ~ 2 nm by Guisbiers *et al.*[58] but by recognizing that A_t/V_t is the definition of q (which changes with both size and temperature during the morphological optimization), a more general description of the size- and shape-dependent melting temperature for nanocrystals in equilibrium is given by:

$$T_{mx}(D) = T_{mx}(\infty)\left[1 + q(T,D)\frac{\gamma_l - \sum_i f_i \gamma_{xi}(T)}{\Delta H_{mx}(\infty)}\right] \tag{9.7}$$

Note that we have dropped the subscript for the surface groups y, since we are concerned with temperatures that are greater than the characteristic desorption temperatures. Before melting occurs the surfaces will likely become "clean".

$T_m(D)$ may be calculated for any shape (enclosed by the surfaces *i*) using the known bulk melting temperatures, the surface energy of the liquid[67] and the enthalpy of melting from experiment.[68] It is important to point out that since q is now expressed as $q(T,D)$, $T_{mx}(D)$ is obtained by self-consistently solving for the average diameter D when $T = T_{mx}(D)$ during each optimization of $q(T,D)$ (*i.e.* during the morphological optimization to minimize the total free energy of the system).

From each optimization the values of $T_{mx}(D)$ and $G_x(\mu, T)$ must be compared for anatase with OH and H_2O, and rutile with OH and H_2O, and the lowest energy phase is mapped in $<D,T>$ space. Since the model is analytic, this can be done quite rapidly. The results of these comparisons are shown in Figure 9.7, where we can see that the position of the anatase-to-rutile solid–solid phase transition line differs depending on the type of adsorbed groups.[69] The phase boundary is no longer abrupt, and a reaction zone is formed where each combination is in an effective state of thermodynamic coexistence. At these sizes and temperatures the relative stability of anatase

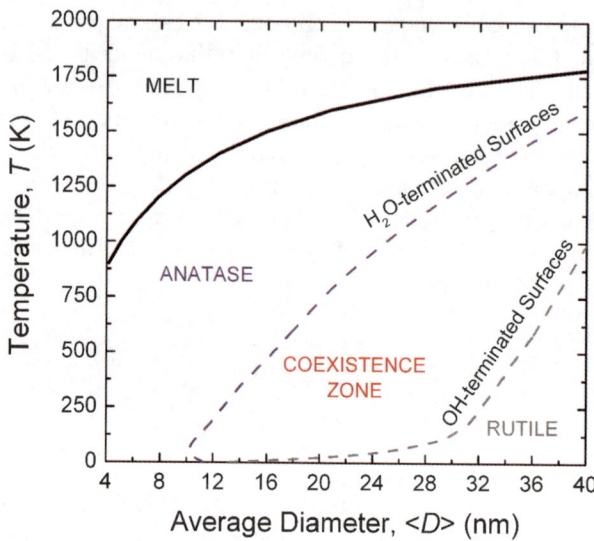

Figure 9.7 The <*D,T*> phase map of titania, based on first principles calculations and melting enthalpies from experiment. Shown are the solid–solid phase transition lines for nanocrystals with H₂O- and OH-terminated surfaces, and the coexistence region where the relative stability depends on the type of adsorbed groups. Reproduced with permission from ref. 69, American Chemical Society © 2008.

and rutile will be sensitive to the surface chemistry, and solid–solid phase transitions will be coupled with adsorption–desorption reactions, as well as shape-transitions.

This is an example of the type of results that could occupy the *Modeling* and *Explanation* nodes of the introductory scheme.

9.2.1 Comparison with Experiment

As mentioned above in the Introduction, the modeling of *Nanomorphology* can be related to *Measurements*, as shown in the feedback loop in Figure 9.2. This will be particularly useful in the reaction zone, where the modeling is unable to provide more detail. Combining this important feature of the titania nanoscale phase diagram with appropriate experiments began with verifying the surface chemistry present on real samples (shown in Figure 9.8(a) and (b)) using Fourier transform infrared (FTIR) spectroscopy (shown in Figure 9.8(c)).[69] The size of small rutile crystals in these experiments was measured at approximately 14 to 18 nm, and in this case we can see broad peaks from ~ 3000 to $3700\ \text{cm}^{-1}$ due to the stretching mode between O and H. These peaks could be associated with both H_2O and OH groups. The peak at ~ 1650 to $1630\ \text{cm}^{-1}$ is due to the bending mode between two H atoms of H_2O (in pure water) or the H_2O of the titania surface group. The large dispersion (~ 20 cm^{-1}) of the peak indicates the presence of surface H_2O groups, in addition to

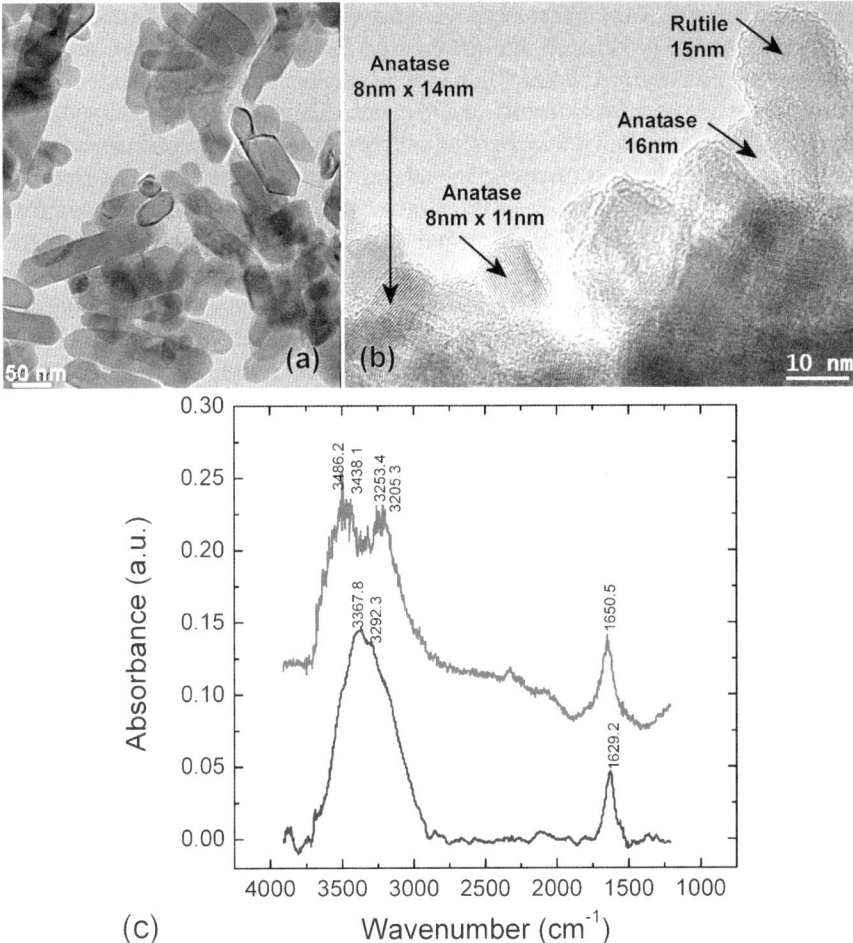

Figure 9.8 (a) TEM image showing rutile nanocrystals formed in 0.5 M HCl solution at 180 °C. Small rutile crystals have diameters of 14 to 18 nm. (b) TEM image showing rutile and anatase nanocrystals formed in 0.2 M HCl solution at 180 °C. (c) FTIR spectra from pure water (upper spectrum) and anatase-dominated titania (lower spectrum) in neutral condition. The FTIR spectra indicate that both OH and H_2O groups occur on the titania surface. Reproduced with permission from ref. 69, American Chemical Society © 2008.

neutral water, which have a slightly different bending frequency. The large absorbance peak for the anatase-dominated titania sample indicates that there are OH surface groups in addition to H_2O surface groups, which is consistent with the observations of others.[70]

In general, an examination of the modest list of references included here reveals that that anatase-to-rutile transformation is routinely found to occupy the reaction zone of the phase map, at least, when grown using hydrothermal synthesis. However, because the crystal sizes are often determined based on bulk X-ray powder diffraction (not individual nanocrystals measured using transmission electron microscopy (TEM), as in Figure 9.8), it is very difficult to quantify the size boundary or size overlap zone. Direct comparison with theory and experiment is further complicated by the fact that studies that use TEM show that titania samples present a variation in size and shape, some of which are beyond those predicted during the morphological optimizations that preceded the construction of the phase map. In Figure 9.8(a) for instance, the length of the rutile crystals ranges from 50 to 200 nm, and the shapes are more consistent with those guided by kinetic considerations.[71,72] Nevertheless, based on a careful series of experimental observations, a simple and successful comparison was reported in ref. 69, as shown in Figure 9.9.

The solid–liquid transition line is more difficult to validate. Obtaining higher temperatures (over 1000 K) and true equilibrium conditions simultaneously in the laboratory is challenging, but is achieved easily in nature and preserved geologically. With the benefit of time, natural nanocrystals are able to adopt their equilibrium structure subject to environmental conditions. Therefore, for this comparison the authors reported on naturally occurring anatase nanocrystal clusters from a young and fresh basalt glass layer in the McKinney Butte Basalt of Snake River Group Basalt in the Snake River

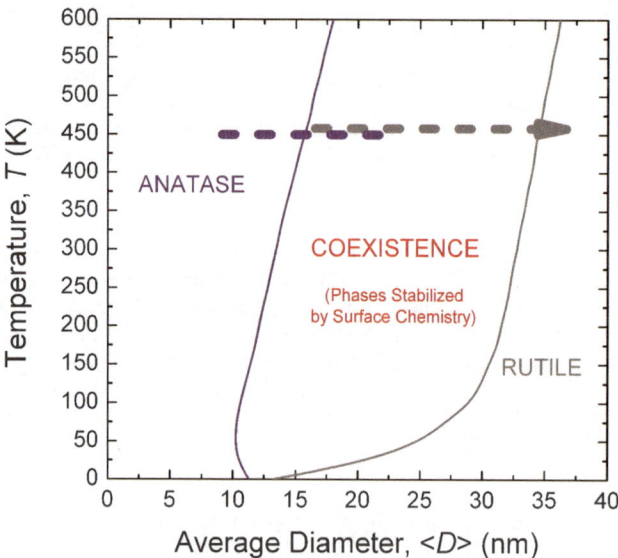

Figure 9.9   The low temperature and size regime with the size range of experimental observations at 180 °C marked. The identification of individual anatase and rutile nanocrystals is based on their periodicities of lattice fringes in high-resolution TEM images. The diameter refers to the anatase phase. Reproduced with permission from ref. 69, American Chemical Society © 2008.

area of Idaho.[73] Although the exact condition for the anatase formation in these layers is not certain, it was explained that these anatase nanocrystals formed in the hot environment of basalt rock due to the presence of Fe(III) in solid solution. This supports the idea that anatase is the thermodynamically stable phase for titania nanocrystals at high temperatures. In general, the solubility of Fe(III) in either rutile or anatase is very low, and one expects that such a small amount of Fe in these natural anatase nanocrystals will not influence the anatase stability with respect to rutile.

9.3 Structure-Property Mapping of Photocatalysis

It has been well established that the design of modern high-performance technologies often begins with the creation of robust structure-property maps. However, as we have seen above, unlike molecules and materials at other length scales, the structure of nanomaterials is not static and may alter significantly (and permanently) in response to changes in their surroundings. Titania is no different, and since the size, crystalline phase and particle morphology (shape) have all been found to be critical parameters[35,36,55,74–77] in determining the overall suitability for specific applications, the thermodynamic cartography outlined here precedes the development of reliable structure-property relationships.

This is particularly important in the case of titania, since the photoactive surfaces produce ROS[78] or *free radicals*. Free radicals are generated by photocatalytic reactions at active cation sites, and include the hydroxyl radical, $OH^{\bullet}$, the superoxide radical, $O^{\bullet-}$, and hydrogen peroxide, H_2O_2.[79–82] Among them, the $OH^{\bullet}$ radical is extremely important, as it has been suggested to be a product from the photocatalytic oxidation of water.[80–82] While a detailed understanding of the photocatalytic reactions on titania surface remains elusive, a basic understanding may be constructed based on the work by a number of research groups.[83–86]

For example, it has been established that the addition of H_2O_2 to the titania photocatalytic system accelerates $OH^{\bullet}$ formation, thereby improving the reaction activity.[78,87–89] The $OH^{\bullet}$ is formed on reduction of H_2O_2 with a conduction band electron, e_{cb}^-:

$$H_2O_2 + e_{cb}^- \rightarrow OH^{\bullet} + OH^- \qquad \text{(Scheme 9.1)}$$

which is the primary reaction for the production of $OH^{\bullet}$, or by $O_2^{\bullet-}$:

$$H_2O_2 + O_2^{\bullet-} \rightarrow OH^{\bullet} + OH^- + O_2 \qquad \text{(Scheme 9.2)}$$

Alternatively, H_2O_2 can be oxidized to $O_2^{\bullet-}$ by a valence band hole, h_{vb}^+:

$$H_2O_2 + h_{vb}^+ + 2OH^- \rightarrow O_2^{\bullet-} + 2H_2O \qquad \text{(Scheme 9.3)}$$

which is the primary reaction for $O_2^{\bullet-}$ production, or by $OH^{\bullet}$:

$$H_2O_2 + OH^{\bullet} + OH^- \rightarrow O_2^{\bullet-} + 2H_2O \qquad \text{(Scheme 9.4)}$$

We can see in each case that H_2O_2 is an important species in these reactions, and provides a very useful measure of ROS production from titania. However, the important point in the present context is that each of these reactions occur at under-coordinated cationic Ti surface sites, and that the reaction efficiency is directly proportional to the number of active sites present in a given sample.[78,80]

These reactive species are useful in water purification or self-cleaning surfaces, but can potentially cause oxidative stress in tissue or even damage to DNA.[13,90] Since many applications require the photoactivity to be maximized, identification of the optimal balance between the performance and the safety of titania photocatalysts can be assisted by a structure-property map relating the production of free radicals to the physical characteristics of the nanoparticles as a function of the surrounding environment. While this could be done systematically (at some expense) using exclusively experimental methods, producing and characterizing all possible material and environmental combinations guarantees that the most hazardous variations will be produced, if only for the purposes of testing. By combining the thermodynamic cartography of titania outlined above with calculations of the number of under-coordinated Ti atoms on specific surface facets, a structure-property map of the catalytic efficiency and potential toxicity from free radicals may be systematically obtained.

In addition to the nanoscale phase diagram, which also predicts the shape at each point in the $<D,T>$ manifold, it is also important to understand how these reactions relate more generally to the overall nanomorphology. Rutile has been shown to stabilize the formation of $O_2^{\bullet-}$, but anatase surpassed rutile in terms of production efficiency (from H_2O_2).[78] The consistently higher activity of anatase with respect to that of rutile has previously been interpreted using photo-conductance measurements, which evidence the higher aptitude of anatase to photoadsorb and photodesorb oxygen, and to have a lower relative electron–hole recombination rate.[74] Also, as mentioned above, it has been found that the reaction efficiency is maximized on the anatase {001} facets, and on rutile {110}. Therefore, the number and density of active under-coordinated Ti surface sites available to participate in photocatalytic reactions within a given sample is a function of the particle phase and shape, as well as size.

Based on the assumption that calculations of the free radical production efficiency provide a measure of activity and relative ROS mediated nanotoxicity, we may use the prefactor f_i in eqn (9.1) to quantify the fraction of the total surface area that is associated with the photoactive facets at each and every point in the phase diagram. In general, the anatase {001} facet is a minority facet ($f_{\{001\}} < f_{\{101\}}$), the fractional area of which is almost constant with respect to size,[57] and increases slowly with increasing temperature.[69] In contrast, the rutile {110} facet is a majority facet ($f_{\{110\}} > f_{\{011\}}$), the fractional area of which is slowly increasing or decreasing with respect to size (depending on the surface groups),[57] and decreases with increasing temperature.[69] To determine the number of potentially reactive sites on anatase {001}, we may simply use the known lattice constants and the in-plane density of cations. If all under-coordinated Ti atoms on the anatase {001} facets are available to produce free radicals, the maximum number of active sites is 6.99×10^{18} m^{-2}. Repeating this procedure for the rutile {110} facets gives the maximum number of active sites of 5.12×10^{18} m^{-2}. These numbers may be thought of as "active-site population numbers".

Combining these active-site population numbers with the predicted shapes under different conditions one can generate a set of equilibrium "activity maps" for anatase or rutile in $<D,T>$ space, passivated with H_2O or OH, as shown in Figures 9.10(a) to (d) respectively.[91] Here we can see that smaller

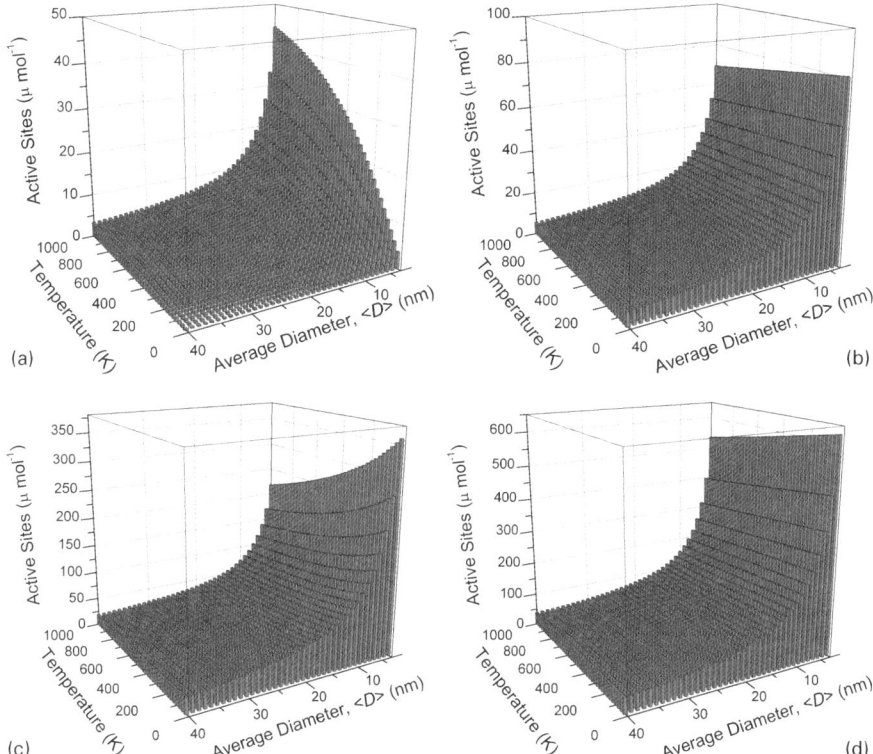

Figure 9.10 Number of active sites for free radical production associated with (a) {001} surface facets on polyhedral anatase passivated with H_2O, (b) {001} surface facets on polyhedral anatase passivated with OH, (c) {110} surface facets on polyhedral rutile passivated with H_2O, and (d) {110} surface facets on polyhedral rutile passivated with OH, as a function of temperature and nanoparticle size. We can see here how the different environmental conditions determine the preferred shape, and therefore the number of reactive cationic sites. Reproduced with permission from ref. 91, Royal Society of Chemistry © 2011.

nanoparticles are more active (and hence potentially more toxic), as a direct result of the higher q. These results also indicate that if titania nanoparticles were annealed (below the desorption temperature) then morphological transformations are likely to reduce the photocatalytic activity of rutile and hydroxylated anatase nanoparticles (but potentially make these particles safer), but increase the activity of anatase nanoparticles with hydrated surfaces (potentially making them more toxic).

If we now project these activity maps onto the phase diagram, then an environmentally-sensitive structure-property map is produced. It is a simple matter to combine the results of Figure 9.10(a) and (d) and multiply them by a relative fraction of each polymorph within a sample. However, the phase diagram predicts that an anatase nanoparticle terminated with H_2O is the thermodynamically preferred combination at small sizes, and OH-terminated rutile is the thermodynamically preferred combination at larger particle sizes, so it is little more complicated to incorporate the reaction zone. The simplest way of doing this is to apply a weighting function that reflects the location and width of the reaction zone (with 100% anatase at sizes below this zone, and 100% rutile at sizes beyond it). One appropriate weighting function is a Fermi–Dirac distribution, where the adsorbed water is replaced by adsorbed hydroxyl groups as we cross the reaction zone (from left to right), accompanied by the transformation from anatase to rutile to preserve equilibrium. This produces the three dimensional activity/ROS-toxicity map for titania in water shown in Figure 9.11.[91] Figure 9.11(a) predicts the activity/toxicity for a sample where complete phase transformation has occurred, and there is 100% anatase or 100% rutile to the left and the right of the reaction zone respectively. Figure 9.11(b) predicts the activity/toxicity for a mixed phase sample where an

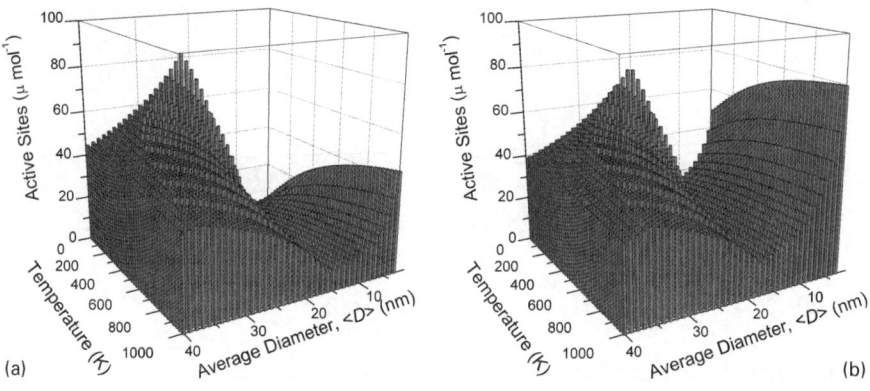

Figure 9.11   Maximum number of free radicals generated by titania nanoparticles under ambient temperature and pressure, (a) with complete size/temperature-dependent phase transformation and equilibrium surface chemistry, and (b) with equilibrium surface chemistry and size/temperature-dependent phase transformation with 10% phase contamination on each side of the reaction zone. Reproduced with permission from ref. 91, Royal Society of Chemistry © 2011.

incomplete phase transformation has occurred, and there is 10% phase contamination on each side of the reaction zone (10% rutile in the anatase zone and *vice versa*).

9.3.1 Comparison with Experiment

This analysis predicts an increase in active sites up to a size of 33 nm (at room temperature), and then a $\sim 1/R$ decrease over this size. Although unusual, this profile is in good agreement with measurements comparing the physicochemical properties of titania nanoparticles (*e.g.*, size, surface area and crystal phase) on their oxidant generating capacity by Jiang *et al.*[92] To facilitate a direct comparison with the results in Figure 3a of ref. 92, the predicted number of active sites (in µM) is shown (as a function of size, at 300 K) in Figure 9.12. In the case of the experimental results there is a small distribution in the sizes of the particles included in the measurement, although they are very close to being monodispersed. Using the same standard deviation in the particle size as reported in ref. 92, and translating this to a variation in the number of active sites associated with the same variation in size, the predicted results also include the uncertainties that would be expected in the experiments.

This provides another useful tool in the preliminary assessment of tailor-made or modified specimens, since the optimization of the photocatalytic properties of hydrothermal titania has been previously shown (through systematic experimental investigation) to be largely controlled by selective combination of temperature, pH, and duration of the hydrothermal treatment which influences phase composition, particle size and shape.[35]

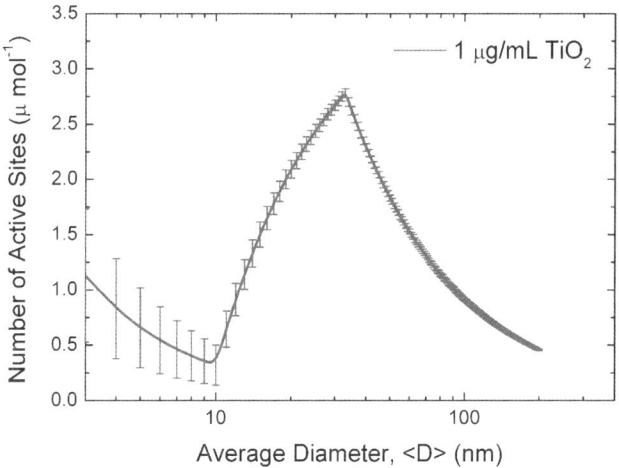

Figure 9.12 Comparison of the size-dependent potential for ROS generation (a) calculated from first principles for 1 µg ml^{-1}. Reproduced with permission from ref. 91, Royal Society of Chemistry © 2011.

The results presented here assume all active sites have an (equal) probability of producing free radicals, although differential cytotoxic responses in human dermal fibroblasts (HDF) and A549 cells observed by Sayes *et al.*[93] suggests the rutile surface can be up to 100 times less active than anatase. If a lower efficiency of free radical generation were assumed on a rutile {110} facets (with less active Ti sites producing free radicals) the magnitude of the maxima at $\sim$33 nm would diminish, but the location of the maximum remains stationary since it is related to the nanoscale phase diagram. When a ratio of $100 \times$ is applied, the activity of rutile is negligible with respect to anatase, and Figure 9.11(a) and (b) appear identical to Figure 9.10(a). The location ($<D,T>$ coordinates) of the maxima is, however, sensitive to phase contamination.

9.4   Case Study: Sunscreen

Although we can see how the creation of structure-property maps can aid in the design of more efficient or safer nanomaterials, it is often necessary to combine groups of complementary structure-property maps to form a basis for decision making. Nanomaterials are typically multi-functional (they do more than one thing at a time), which is one of the reasons that they are preferred over their macroscale counterparts for a variety of applications.

One example of an application of titania nanoparticles where the consideration of potential toxicity is an important factor is sunscreen. In sunscreens nanoparticles are included as individual entities suspended in the host medium[94] and, although the form of the nanoparticle used in the product formulation is not necessarily the form to which environmental and biological stressors are exposed, these products have recently raised some concerns. To date *ex vivo* testing indicates that the nanoparticles in sunscreen remain on the surface of the skin and in the stratum corneum among keratinized cells, and do not reach viable cells beneath.[95] However, this barrier is not impenetrable, since human skin is always marred by imperfections (whether visible or not), is easily abraded, and is perforated by hair follicles and pores. Studies focussing on other types of nanoparticles (fluorescent quantum dots) have previously shown that hair follicles and abrasions increase penetration.[96,97] However, the fate of nanoparticles does not necessarily determine the fate of the ROS, which may or may not penetrate the superficial dead layer of skin, irrespective of where the nanoparticles reside. Moreover, although very different to the interaction of sunscreens with the skin, the issue of ROS generation by sunscreen preparations has also caught the attention of other areas of research. A recent study has cited the ROS generated by nanoparticles in sunscreens used by workers as the cause of unsightly hand- and finger-shaped defects on pre-painted steel roofing.[98]

The reason titania (or zinc oxide) nanoparticles are introduced into sunscreens is that their optical properties provide superior protection against harmful ultraviolet (UV) light, and small nanoparticles are transparent. It is these optical properties that increase both the efficacy and aesthetic appeal of the product, and the photocatalytic property (while active) is extraneous. However, all of these properties are intrinsically linked, and each must be considered when making informed decisions regarding the size, type and quantity of nanoparticles to be used.

9.4.1 Potential Toxicity from ROS

As we have seen above, by counting the number of potentially reactive sites on anatase {001} and rutile {110}, while accounting for the changes in particle shape with size and temperature, it is possible to estimate the potential ROS production at any size, D.[15,69] If we assume that the volumetric concentration of the nanoparticles in the sunscreen host medium is sufficiently low as to exclude secondary reactions, then the number of active sites per mole (as shown in Figure 9.13(a)) will be linear with an increase in the number concentration (N) of titania nanoparticles, C (*i.e.* it is additive). This will be denoted by $N(D,C)$, and adding this new dimension provides a 2-D prediction of ROS generation as a function of these basic product configurations, as shown in Figure 9.13 for $10^9 < C < 10^{11}$.

Readers should note that on the *x*-axis a "preparation-independent" value of C has been used, referred to as *particle loading*, which could be dispersed in any medium to achieve a traditional *concentration*.

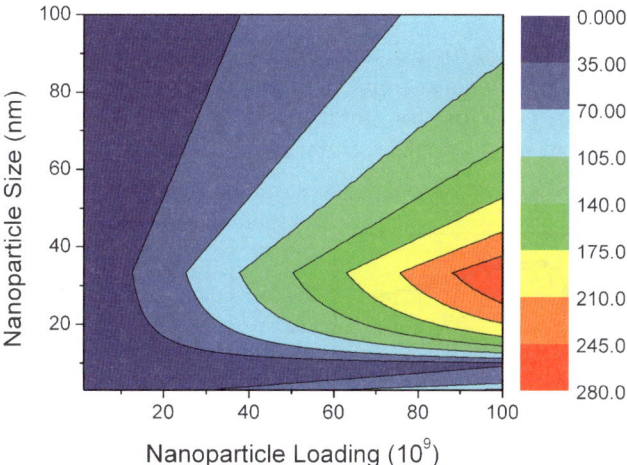

Figure 9.13 The potential for generation of free radicals attributed to a particle loading of $10^9 < C < 10^{11}$ titania nanoparticles per unit volume, based on the size- and temperature-dependent thermodynamic cartography. The coloured scale bar represents the absolute number of active sites for ROS generation per μM. Reproduced with permission from ref. 105, Nature Publishing Group © 2010.

9.4.2 Efficacy

The traditionally accepted way of assessing the efficacy of a sunscreen product is based on a measure known as the Sun Protection Factor (SPF). This value provides a guide as to amount of protection a particular sunscreen preparation provides against damaging UV irradiation, and is given by:

$$SPF(D,C) = \frac{\int S(\lambda)E(\lambda)d\lambda}{\int S(\lambda)E(\lambda)T(\lambda,D,C)d\lambda}. \tag{9.8}$$

In this expression $S(\lambda)$ is the solar spectrum (see Figure 9.14(a)), $E(\lambda)$ is the erythemal action spectrum (see Figure 14b) and $T(\lambda,D,C)$ is the transmittance. The product of $S(\lambda)$ and $E(\lambda)$ form an effective spectrum that can be obtained from the ISO/CIE Standard.[99] According to the Beer–Lambert law the value of $T(\lambda,D,C) = 10^{-\varepsilon(\lambda,D)CL}$ is a function of the molar extinction coefficient $\varepsilon(\lambda,D)$, the concentration of nanoparticles C (per unit volume), and the path length L, which can be assumed to be $L = 2$ µm (based on the recommended applied dose of 2 mg cm^{-2} in compliance with FDA and COLIPA methodologies).

Because the size of the nanoparticles is smaller than the wavelengths of light in both relevant parts of the electromagnetic spectrum (UV and visible), the calculations of the size-dependent molar extinction coefficient $\varepsilon(\lambda,D)$ must be performed using Lorenz–Mie–Debye theory:[100]

$$\varepsilon(\lambda,D) = C\left(\frac{\pi D^2}{4}\right)Q_{ext}(\lambda,D) \tag{9.9}$$

where D is the average particle diameter (based on a spherical approximation), C is the particle loading (in units of nanoparticles per unit volume), and $Q_{ext}(\lambda,D)$ is the Q-factor for total extinction. $Q_{ext}(\lambda,D)$ is a sum of contributions from single-particle scattering and absorption:

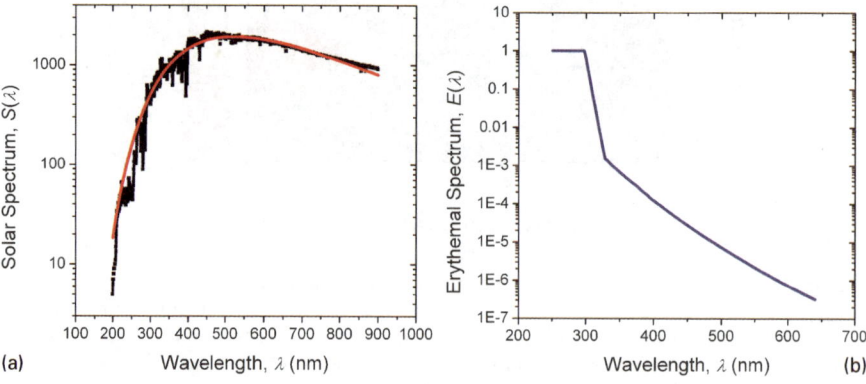

(a) Wavelength, λ (nm) (b) Wavelength, λ (nm)

Figure 9.14 (a) The solar spectrum, $S(\lambda)$, and (b) the erythemal action spectrum, $E(\lambda)$.

$$Q_{ext}(\lambda,D) = Q_{sc}(\lambda,D) + Q_{ab}(\lambda,D) \qquad (9.10)$$

where

$$Q_{sc}(\lambda,D) = \frac{8}{3}\left(\frac{\pi D}{\lambda}\right)^4 \mathrm{Re}\left[\left(\frac{\tilde{n}^2-1}{\tilde{n}^2+2}\right)^2\right] \qquad (9.11)$$

and

$$Q_{ab}(\lambda,D) = \frac{4\pi D}{\lambda}\,\mathrm{Im}\left[\frac{\tilde{n}^2-1}{\tilde{n}^2+2}\right]. \qquad (9.12)$$

In both cases, $\tilde{n} = n - in'$ is the complex refractive index, as provided by Popov *et al.*[101] This technique provides rigorous analytical solutions to Maxwell's equations for light scattering by an isotropic sphere embedded in a homogeneous medium, and has previously been successfully applied to titania.[102] The surrounding medium may be assumed to be skin, with the refractive index of the horny layer of 1.53 being available in ref. 103. The calculated Q-factors are shown in Figure 9.15.

Based on these results, the $SPF(D,C)$ has been determined for wavelengths over the UV spectrum (270 nm $< \lambda <$ 400 nm) for nanoparticles between 3 nm

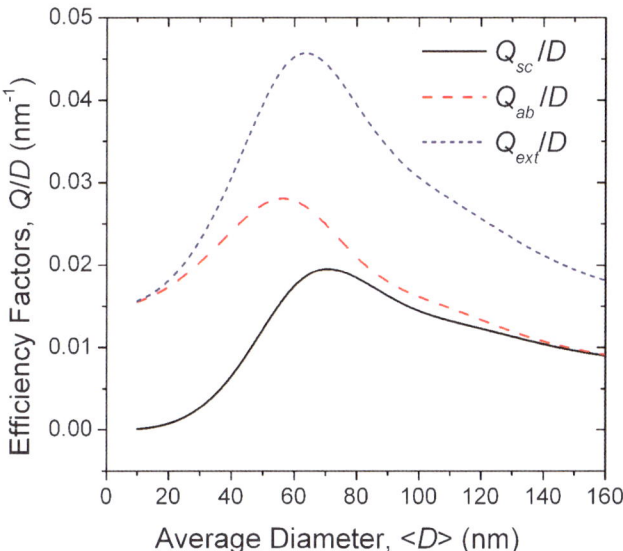

Figure 9.15 The optical efficiency factors (Q-factor) divided by the particle size, for total extinction ($Q_{ext}(\lambda)/D$), scattering $Q_{sc}(\lambda)/D$), and absorption ($Q_{ab}(\lambda)/D$), at a wavelength of $\lambda =$ 310 nm. The maxima in each value occur at 70.8 nm, 56.2 nm, and 63.7 nm, respectively.

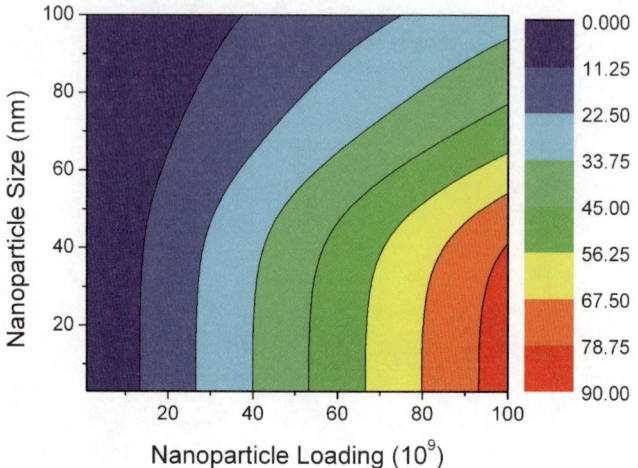

Figure 9.16   The SPF attributed to a particle loading of $10^9 < C < 10^{11}$ titania nanoparticles per unit volume, based on Lorenz–Mie–Debye theory calculations for 270 nm $< \lambda <$ 400 nm, and $L = 2$ μm. The coloured scale bar represents the absolute SPF value. Reproduced with permission from ref. 105, Nature Publishing Group © 2010.

to 100 nm in average diameter, and the results are shown for $10^9 < C < 10^{11}$ titania nanoparticles per unit volume in Figure 9.16. The results show at this loading, the SPF attributed to the nanoparticles ranges from ~90 at small particle sizes, to less than 10 when the particle occupy the sub-micron regime (~100 nm). These results are consistent with available experimental observations.[104]

9.4.3   Aesthetics

As mentioned above, another advantage of using nanoscale particles in sunscreens is that they are optically transparent to visible light. In the past sunscreens containing metal oxides particles were relatively unattractive, appearing as an opaque white topical cream, as the particles within were typically in the micron regime.

The transparency of a material at a given wavelength (as opposed to the transmittance) is defined by:

$$T(\lambda,D,C) = \exp[-\varepsilon(\lambda,D)CL] \tag{9.13}$$

where (as above) $\varepsilon(\lambda,D)$ is the molar extinction coefficient. In this case however, as we are interested in the interactions with the visible spectrum, $\varepsilon(\lambda,D)$ is calculated for wavelengths 400 nm $< \lambda <$ 700 nm. The size-dependent transparency for the entire visible range is then determined using:

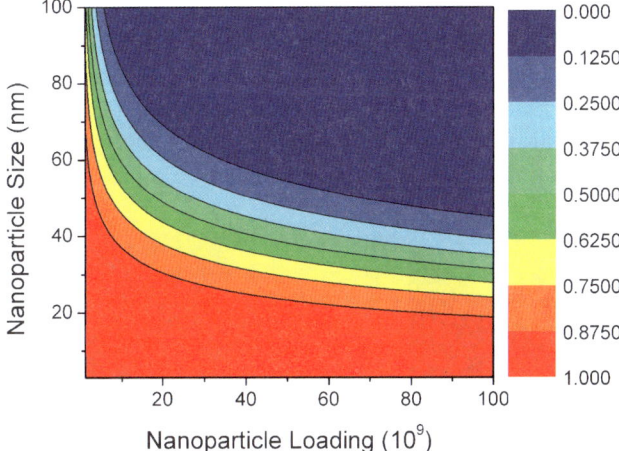

Figure 9.17 The transparency of titania nanoparticles with a particle loading of 10^9 < C < 10^{11} nanoparticles per unit volume, based on Lorenz–Mie–Debye theory calculations for 400 nm < λ < 700 nm, and L = 2 µm. The coloured scale bar represents the absolute fraction of transmitted light (0 = opaque, 1 = transparent). Reproduced with permission from ref. 105, Nature Publishing Group © 2010.

$$T(D,C) = \int \exp[-\varepsilon(\lambda,D)CL]d\lambda. \qquad (9.14)$$

These results are provided for 10^9 < C < 10^{11} nanoparticles per unit volume in Figure 9.17, showing that ~10 nm particles are almost completely transparent, $T(D,C) \approx 1$, but that they rapidly become more absorbing as the size increases. At $D \approx 85$ nm the particles are opaque, $T(D,C) = 0$, in the visible range of the spectrum.

9.4.4 Cross-comparison and Relationship to Regulations

Since each of these predictions is a function of D and C, we can conveniently map each property using these same physical parameters. Figures 9.13–9.15 are directly comparable in terms of product configurations. By comparing these results we can see that the only particle size and concentration that will remain functional (delivering superior efficacy and high transparency) while offering low relative toxicity is a large amount of very small nanoparticles.[105]

Results such as these may prove useful for manufacturers, and/or for those involved in setting the standards that regulate the industry. Currently there are no specific regulations relating to the size of nanoparticles that may be used in sunscreens, but there are existing regulations that restrict the concentrations[106] or the claimed protection.[107,108] In the latter case consumers see these regulations in the form of approximate maximal SPFs, such as 30+ or 50+, depending on the regulations involved. To determine how these restrictions relate to the sunscreen structure-property maps, the regions disallowed under the SPF 30+ and SPF 50+ regulations have been blocked out in the results shown as labelled in Figure 9.18. In both cases we can see that the existing regulations do a remarkably good job of simultaneously allowing attractive transparent preparations and excluding product configurations with the greatest potential toxicity.

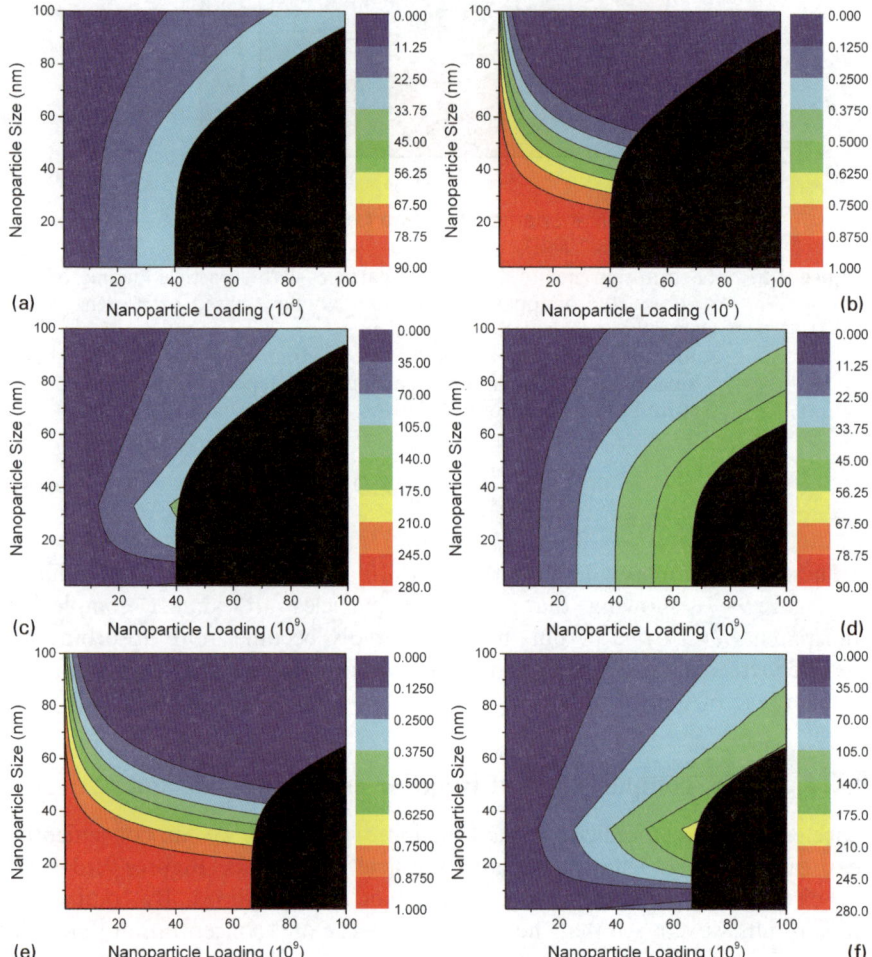

Figure 9.18   Quantifying the effect of current regulations on the efficacy, transparency and potential toxicity from free radicals produced by titania nanoparticles in sunscreen: (a) accessible SPF size/concentration combinations, (b) transparency combinations, and (c) potential toxicity combinations under a 30+ regulation; and the (d) accessible SPF combinations, (e) transparency combinations, and (f) potential toxicity combinations under a 50+ regulation. In all cases the blackened areas are inaccessible under the regulations. Modified with permission from ref. 105, Nature Publishing Group © 2010.

When using this information, one must always remember that the only measure of toxicity that is included here is possible ROS generation, and small nanoparticles may present a health risk for a variety of other reasons that are not included in this work.[109] Small (10 to 20 nm) particles have also been shown to induce oxidative DNA damage, lipid peroxidation, and micronuclei formation, and increase hydrogen peroxide and nitric oxide production in a human bronchial epithelial cell line (BEAS-2B cells), even in the absence of photoactivation.[110] In more recent studies, has been found that ultra-fine titania nanoparticles can also affect the cell–matrix adhesion in keratinocytes for extracellular matrix remodeling (again in the absence of illumination),[111] and fibre-shaped particles can induce inflammasome activation and release of inflammatory cytokines through a cathepsin B-mediated mechanism.[112] Long titania nanowires, nanofibres, nanobelts or nanotubes can exhibit increased pathogenic potential, as phagocytic cells have difficulty processing these materials, resulting in lysosomal disruption and interaction with lung macrophages akin to those of asbestos or silica.[112] Although additional toxic influences could be incorporated into these analyses (if they could be related to the particle surface-to-volume ratio) such results have yet to be reported.

9.5 Conclusions

Based on this type of knowledge and information, the methods used to take this forward and translate this data into usable *Predictions* may vary. While scientists and engineers are experienced in making predictions that are testable and applicable within the realm of science, it is important in this case that such predictions are accessible to (and assessable by) those outside the field. These predictions need to be in a format that fits in with the existing risk prevention and regulatory frameworks, as well as being intelligible by a non-specialist. If this is not the case, the link between *Predictions* and *Prevention* will fail.

The examples described in this chapter also demonstrate that the route from *Nano-hazards* to *Prevention*, and the ultimate goal of making this journey without harm occurring, is more than just a multi-disciplinary problem. It is a multi-field problem. Finding an efficient strategy which covers all the bases will be as much an exercise in knowledge sharing and personal relationships as it will be in scientific discovery, and traditional competition will have to be put aside in the interest of the greater good.

References

1. H. Dumortier, S. Lacotte, G. Pastorin, R. Marega, W. Wu, D. Bonifazi, J.-P. Briand, M. Prato, S. Muller and A. Bianco, *Nano Lett.*, 2006, **6**, 1522.
2. A. Magrez, S. Kasas, V. Salicio, N. Pasquier, J. W. Seo, M. Celio, S. Catsicas, B. Schwaller and L. Forró, *Nano Lett.*, 2006, **6**, 1121.
3. K. Pulskamp, S. Diabaté and H. F. Krug, *Toxicol. Lett.*, 2007, **168**, 58.

4. K. Donaldson, V. Stone, C. L. Tran, W. Kreyling and P. J. A. Borm, *Occup. Environ. Med.*, 2004, **61**, 727.

5. C. W. Lam, J. T. James, R. McCluskey, S. Arepalli and R. L. Hunter, *Crit. Rev. Toxicol.*, 2006, **36**, 189.

6. A. S. Barnard, *Nat. Mater.*, 2006, **5**, 245.

7. H. Hyung, J. D. Fortner, J. B. Hughes and J.-H. Kim, *Environ. Sci. Technol.*, 2007, **41**, 179.

8. K. Donaldson, R. Aitken, L. Tran, V. Stone, R. Duffin, G. Forrest and A. Alexander, *Toxicol. Sci.*, 2006, **92**, 5.

9. J. Bartis and E. Landree, Nanomaterials in the Workplace: Policy and Planning Workshop on Occupational Safety and Health, RAND, Santa Monica, CA, 2006.

10. *Nature's Nanostructures*, ed. H. B. Guo and A. S. Barnard, Pan Stanford Publishing, Singapore, 2012.

11. G. Oberdörster, E. Oberdörster and J. Oberdörster, *Environ. Health Perspect.*, 2005, **113**, 823.

12. A. S. Barnard, *Nanoscale*, 2009, **1**, 89.

13. A. Nel, T. Xia, L. Mädler and N. Li, *Science*, 2006, **311**, 622.

14. A. S. Barnard, *Rep. Prog. Phys.*, 2010, **73**, 086502.

15. A. S. Barnard, *Nat. Nanotechnol.*, 2009, **4**, 332.

16. A. S. Barnard and P. Zapol, *J. Phys. Chem. B*, 2004, **108**, 18435.

17. A. S. Barnard, P. Zapol and L. A. Curtiss, *J. Chem. Theory Comput.*, 2005, **1**, 107.

18. A. S. Barnard, R. R. Yeredla and H. Xu, *Nanotechnology*, 2006, **17**, 3039.

19. A. S. Barnard and A. I. Kirkland, *Chem. Mater.*, 2008, **20**, 5460.

20. H. Guo and A. S. Barnard, *J. Mater. Chem.*, 2011, **21**, 11566.

21. H. Guo and A. S. Barnard, *J. Mater. Chem.*, 2012, **22**, 161.

22. A. S. Barnard and S. P. Russo, *J. Phys. Chem. C*, 2007, **111**, 11742.

23. A. S. Barnard and H. Xu, *J. Phys. Chem. C*, 2007, **111**, 18112.

24. A. S. Barnard and S. P Russo, *Nanotechnology*, 2009, **20**, 115702.

25. A. S. Barnard and S. P Russo, *J. Phys. Chem. C*, 2009, **113**, 5376.

26. A. S. Barnard and S. P. Russo, *J. Mater. Chem.*, 2009, **19**, 3389.

27. A. S. Barnard, C. A. Feigl and S. P. Russo, *Nanoscale*, 2010, **2**, 2294.

28. A. S. Barnard, X. M. Lin and L. A. Curtiss, *J. Phys. Chem. B*, 2005, **109**, 24465.

29. A.S. Barnard and L.A. Curtiss, *J. Mater. Chem.*, 2007, **17**, 3315.

30. A.S. Barnard, *J. Phys. Chem. C*, 2008, **112**, 1385.

31. A. S. Barnard, N. Young, A. I. Kirkland, M. A. van Huis and H. Xu, *ACS Nano*, 2009, **3**, 1431.

32. L. Y. Chang, A. S. Barnard, L. C. Gontard and R. Dunin-Borkowski, *Nano Lett.*, 2010, **10**, 3073.

33. A.S. Barnard and L.Y. Chang, *ACS Catal.*, 2011, **1**, 76.

34. A.S. Barnard, H. Konishi and H. Xu, *Catal. Sci. Technol.*, 2011, **1**, 1440.

35. A. Testino, I. R. Bellobono, V. Buscaglia, C. Canevali, M. D'Arienzo, S. Polizzi, R. Scotti and F. Morazzoni, *J. Am. Chem. Soc.*, 2007, **129**, 3564.

36. N. Balázs, K. Mogyorósi, D. F. Srankó, A. Pallagi, T. Alapi, A. Oszkó, A. Dombi and P. Sipos, *Appl. Catal., B*, 2008, **84**, 356.
37. T. Rajh, J. M. Nedeljkovic, L. C. Chen, O. Poluektov and M. C. Thurnauer, *J. Phys. Chem. B*, 1999, **103**, 3515.
38. H. Zhang, R. L. Penn, R. J. Hamers and J. F. Banfield, *J. Phys. Chem. B*, 1999, **103**, 4656.
39. H. Bullen and S. Garrett, *Nano Lett.*, 2002, **2**, 739.
40. N. Satoh, T. Nakashima, K. Kamikura and K. Yamamoto, *Nat. Nanotechnol.*, 2008, **3**, 106.
41. J. Muscat, V. Swamy and N. M. Harrison, *Phys. Rev. B*, 2002, **65**, 224112.
42. M. R Ranade, A. Navrotsky, H. Zhang, J. F. Banfield, S. H. Elder, A. Zaban, P. H. Borse, S. K. Kulkarni, G. S. Doran and H. J. Whitfield, *Proc. Natl. Acad. Sci. U. S. A.*, 2002, **99**, 6476.
43. A. A. Levchenko, G. Li, J. Boerio-Goates, B. F. Woodfield and A. Navrotsky, *Chem. Mater.*, 2006, **18**, 6324.
44. A. A. Gribb and J. F. Banfield, *Am. Mineral.*, 1997, **82**, 717.
45. H. Zhang and J. F. Banfield, *J. Mater. Chem.*, 1998, **8**, 2073.
46. A. Navrotsky in *Nanoparticles and the Environment*, *Reviews in Mineralogy Geochemistry*, ed. J. F. Banfield and A. Navrotsky, Mineralogical Society of America, Washington DC, USA, 2001, vol. 44.
47. H. Zhang, M. Finnegan and J. F. Banfield, *Nano Lett.*, 2001, **1**, 81.
48. A. Navrotsky, *Geochem. Trans.*, 2003, **4**, 34.
49. Z. V. Saponjic, N. Dimitrijevic, D. Tiede, A. Goshe, X. Zuo, L. Chen, A. S. Barnard, P. Zapol, L. A. Curtiss and T. Rajh, *Adv. Mater.*, 2005, **17**, 965.
50. W. Li, C. Ni, H. Lin, C. P. Huang and S. Ismat Shaha, *J. Appl. Phys.*, 2004, **96**, 6663.
51. A. S. Barnard and P. Zapol, *J. Chem. Phys.*, 2004, **121**, 4276.
52. A. S. Barnard, *J. Phys. Chem. B*, 2006, **110**, 24498.
53. M. W. Chase, C. A. Davies, J. R. Downey, D. J. Frurip, R. A. McDonald and A. N. Syverud, *J. Phys. Chem. Ref. Data*, 1985, **14**, 1680.
54. A. S. Barnard and P. Zapol, *Phys. Rev. B*, 2004, **70**, 235403.
55. H. G. Yang, C. H. Sun, S. Z. Qiao, J. Zou, G. Liu, S. C. Smith, H. M. Cheng and G. Q. Lu, *Nature*, 2008, **453**, 638.
56. T. Ohno, K. Sarukawa and M. Matsumura, *New J. Chem.*, 2002, **26**, 1167.
57. A. S. Barnard and L. A. Curtiss, *Nano Lett.*, 2005, **5**, 1261.
58. G. Guisbiers, O. Van Overschelde and M. Wautelet, *Appl. Phys. Lett.*, 2008, **92**, 103121.
59. U. Diebold, *Surf. Sci. Rep.*, 2003, **48**, 53.
60. M. R. Hoffmann, S. T. Martin, W. Choi and D. W. Bahnemann, *Chem. Rev.*, 1995, **95**, 69.
61. M. C. Thurnauer, T. Rajh and D. M. Tiede, *Acta. Chem. Scand.*, 1997, **51**, 610.
62. Y. Nakaoka and Y. Nosaka, *J. Photochem. Photobiol.*, *A*, 1997, **110**, 299.
63. T. Kasuga, M. Hiramatsu, A. Hoson, T. Sekino and K. Niihara, *Adv. Mater.*, 1999, **11**, 1307.

64. S. H. Szczepankiewicz, J. A. Moss and M. R. Hoffmann, *J. Phys. Chem. B*, 2002, **106**, 7654.
65. A. S. Barnard, P. Zapol and L. A. Curtiss, *Surf. Sci.*, 2005, **582**, 173.
66. G. S. Rushbrook, *Introduction to Statistical Mechanics*, Clarendon Press, Oxford, 1957.
67. Y. L. Li and T. Ishigaki, *J. Cryst. Growth*, 2002, **242**, 511.
68. W. Martinssen and H. Walimont, *Springer Handbook of Condensed Matter and Materials Data*, Springer, Berlin, 2005.
69. A. S. Barnard and H. Xu, *ACS Nano*, 2008, **2**, 2237.
70. J. Soria, J. Sanz, I. Sobrados, J. M. Coronado, A. J. Maira, M. D. Hernandez-Alonso and F. Fresno, *J. Phys. Chem. C*, 2007, **111** 10590.
71. H. Zhang and J. F. Banfield, *Am. Mineral.*, 1999, **84**, 528.
72. H. Zhang and J. F. Banfield, *Chem. Mater.*, 2005, **17**, 3421.
73. V. S. Gilleman, J. D. Kauffman and K. L. Othberg, *Thousand Springs Quadrangle, Gooding and Twin Falls Counties, Idaho*, in Idaho Geological Survey, 2005.
74. A. Sclafani and J. M. Herrmann, *J. Phys. Chem.*, 1996, **100**, 13655.
75. K. D. Jang, S.-K. Kim and S.-J. Kim, *J. Nanoparticle Res.*, 2001, **3**, 141.
76. T. Paunesku, T. Rajh, G. Wiederrecht, J. Maser, S. Vogt, N. Stojićvić, M. Protić, B. Lai, J. Oryhon, M. Thurnauer and G. Woloschak, *Nat. Mater.*, 2003, **2**, 343.
77. N. Balázs, D. F. Srankó, A. Dombi, P. Sipos and K. Mogyorósi, *Appl. Catal., B*, 2010, **96**, 569.
78. T. Hirakawa, K. Yawata and Y. Nosaka, *Appl. Catal., A*, 2007, **325**, 105.
79. T. L. Thompson and J. T. Yates Jr., *Chem. Rev.*, 2006, **106**, 4428.
80. M. R. Hoffmann, S. T. Martin, W. Choi and D. W. Bahnemann, *Chem. Rev.*, 1995, **95**, 69.
81. L. Sun and J. R. Bolton, *J. Phys. Chem.*, 1996, **100**, 4127.
82. Y. Nosaka, M. Kishimoto and J. Nishino, *J. Phys. Chem. B*, 1998, **102**, 10279.
83. Y. Nosaka, H. Natsui, M. Sasagawa and A. Y. Nosaka, *J. Phys. Chem. B*, 2006, **110**, 12993.
84. K. Ishibashi, A. Fujishima, T. Watanabe and K. Hashimoto, *J. Photochem. Photobiol., A*, 2000, **134**, 139.
85. R. Nakamura and Y. Nakato, *J. Am. Chem. Soc.*, 2004, **126**, 1290.
86. Y. Murakami, K. Endo, A. Y. Nosaka and Y. Nosaka, *J. Phys. Chem. B*, 2006, **110**, 16808.
87. L. X. Chen and C. Zhao, *Langmuir*, 2001, **17**, 4118.
88. N. San, A. Hatipoglu, G. Kocturk and Z. Cinar, *J. Photochem. Photobiol., A*, 2001, **139**, 225.
89. A. V. Vorontsov, E. V. Sainov, L. Davydov and P. G. Smirniotis, *Appl. Catal., B*, 2001, **32**, 95.
90. H. Wiseman and B. Halliwell, *Biochem J.*, 1996, **313**, 17.
91. A. S. Barnard, *Energy Environ. Sci.*, 2011, **4**, 439.
92. J. Jiang, G. Oberdörster, E. Elder, R. Gelein, P. Mercer and P. Biswas, *Nanotoxicology*, 2008, **2**, 33.

93. C. M. Sayes, R. Wahi, P. A. Kurian, Y. Liu, J. L. West, K. D. Ausman, D. B. Warheit and V. L. Colvin, *Toxicol. Sci.*, 2006, **92**, 174.
94. K. M. Tyner, A. M. Wokovich, W. H. Doub, L. F. Buhse, L.-P. Sung, S. S. Watson and N. Sadrieh, *Nanomedicine*, 2009, **4**, 145.
95. Zs. Kertesz, Z. Szikszai and A. Z. Kiss, *Quality of skin as a barrier to ultra-fine particles.* Contribution of the IBA Group to the NANODERM EU-5 Project, 2003–2004.
96. A. Vogt, B. Combadiere, S. Hadam, K. Stieler, J. Lademann, H. Schaefer, B. Autran, W. Sterry and U. Blume-Peytavi, *J. Invest. Dermatol.*, 2006, **126**, 1316.
97. L. W. Zhang and N. A. Monteiro-Riviere, *Skin Pharmacol. Physiol.*, 2008, **21**, 166.
98. P. J. Barker and A. Branch, *Prog. Org. Coatings*, 2008, **62**, 313.
99. Erthema Reference Action spectrum and Standard Erythemal Dose, Joint 99 CIE Standard, in *Human Exposure to Ultraviolet Radiation: Risks and Regulations*, Elsevier, Amsterdam, 1987, pp 83–87.
100. G. Mie, *Ann. Phys.*, 1908, **330**, 377.
101. A. P. Popov, A. V. Priezzhev, J. Lademann and R. Myllylä, *J. Phys. D: Appl. Phys.*, 2005, **38**, 2564.
102. E. S. Thiele and R. H. French, *J. Am. Ceram. Soc.*, 1998, **81**, 469.
103. V. V. Tuchin, *Tissue Optics*, SPIE Optical Engineering Press, Bellingham, 2000.
104. J. R. Villalobos-Hernández and C. C. Müller-Goymann, *Int. J. Pharm.*, 2006. **322**, 161.
105. A. S. Barnard, *Nat. Nanotech.*, 2010, **5**, 271.
106. *Sunscreening agents permitted as active ingredients in listed Products*, Therapeutic Goods Administration, Department of Health and Aging, Australian Government, http://www.tga.gov.au/pdf/archive/otc-argom-amendment-060816-sunscreens.pdf
107. *Commission Recommendation of 22 September 2006 on the efficacy of sunscreen products and the claims made relating thereto.* Official Journal of the European Union. 2006-09-22.
108. *UV Resource Guide - Sunscreens.* Arpansa. 2008-12-20.
109. M. J. Osmond and M. J. McCall, *Nanotoxicology*, 2010, **4**, 15; B. Gulson, M. McCall, M. Korsch, L. Gomez, P. Casey, Y. Oytam, A. Taylor, M. McCulloch, J. Trotter, L. Kinsley and G. Greenoak, *Toxicol. Sci.*, 2010, **118**, 140.
110. J. R. Gurr, A. S. Wang, C. H. Chen and K. Y. Jan, *Toxicology*, 2005, **213**, 66.
111. K. Fujita, M. Horie, H. Kato, S. Endoh, M. Suzuki, A. Nakamura, A. Miyauchi, K. Yamamoto, S. Kinugasa, K. Nishio, Y. Yoshida, H. Iwahashi and J. Nakanishi, *Toxicol. Lett.*, 2009, **191**, 109.
112. R. F. Hamilton, N. Wu, D. Porter, M. Buford, M. Wolfarth and A. Holian, *Part. Fibre Toxicol.*, 2009, **6**, 35.

CHAPTER 10

Nano-QSAR: Advances and Challenges

B. RASULEV[a], A. GAJEWICZ[b], T. PUZYN[b], D. LESZCZYNSKA[c] AND J. LESZCZYNSKI*[a]

[a] Interdisciplinary Nanotoxicity Center, Department of Chemistry and Biochemistry, Jackson State University, Jackson, MS, USA; [b] Laboratory of Environmental Chemometrics, Faculty of Chemistry, University of Gdansk, Gdansk, Poland; [c] Interdisciplinary Nanotoxicity Center, Department of Civil and Environmental Engineering, Jackson State University, Jackson, MS, USA
*E-mail: jerzy@icnanotox.org

10.1 Introduction

During the last three decades a dramatic increase in research and applications involving the chemistry and technology of nanomaterials has been observed. This statement can be supported by the results obtained using the Google search engine (Ngram viewer) to search publications for the period 1900–2008. An enormous growth in the frequency use of words "nano" and "nanoparticle" is noticed, starting from the decade 1980–1990 (Figure 10.1).

In this context, "nano" doesn't only mean "1000 times smaller than micro", and nanotechnologies are not just an extension of microtechnologies to a smaller scale. It is an entirely new paradigm that opens entirely novel scientific and technological opportunities.

What is a nanoparticle? Nanoparticles are building blocks for nanotechnology, and are defined as particles (structures) with at least one dimension of less than 100 nm.[1] In fact, particles in these size ranges have been used by humankind for thousands of years. However, there has been a recent

RSC Nanoscience & Nanotechnology No. 25
Towards Efficient Designing of Safe Nanomaterials: Innovative Merge of Computational Approaches and Experimental Techniques
Edited by Tomasz Puzyn and Jerzy Leszczynski
© The Royal Society of Chemistry 2012
Published by the Royal Society of Chemistry, www.rsc.org

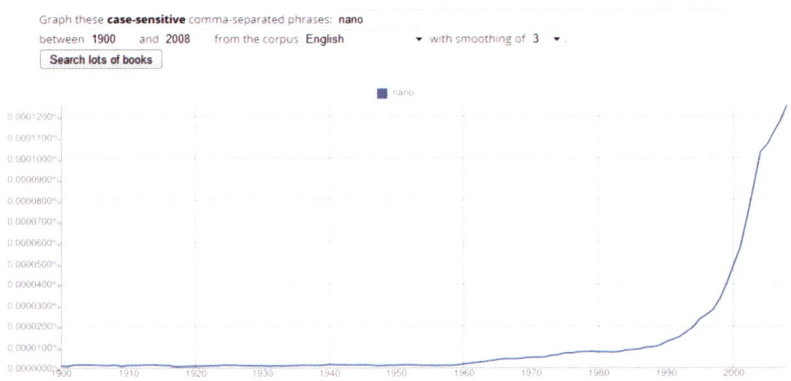

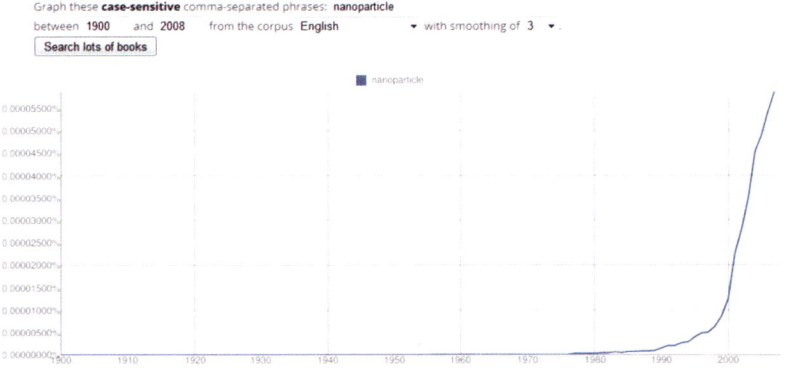

Figure 10.1 The Google Ngram viewer's results through publications for the period 1900–2008 for the frequency of use of "nano" and "nanoparticle".

renaissance in this area because of technological progress which brings an ability to synthesize and manipulate such materials.[2] Nanomaterials are being currently used as electronics, optoelectronics, in biomedical-, environmental-, material- and energy-related areas, as cosmetics, pharmaceuticals, and catalysts. Nanomaterials exhibit unique physico-chemical properties and impart enhancements to engineered materials, including better magnetic properties, improved electrical activity, and increased optical properties. Because of the potential output of this technology, applications and investment in nanotechnology research are on the rise worldwide.[3–7] Moreover, the use of nanomaterials in various industries is projected to increase dramatically in the future and, as a consequence, the contamination of the environment by these materials is expected, or at least such a possibility cannot be disregarded. In fact, nanotechnology could lead to serious environmental problems. This is because it is still largely unknown how nanoparticles will impact the environment. Besides, notwithstanding this, there is a need to comprehensively investigate all physico-chemical and also biological characteristics of nanoparticles to predict their possible impact on environment.[1,6,8–11]

There is a clear need for short-term testing of the potential hazards of nanomaterials in order to gain information necessary to evaluate their risk.[6,8,10,12–17] However, the large number of nanoparticles and the variety of their characteristics including sizes and coatings suggest that the only rational approach which avoids testing every single nanoparticle is to find a relationship between the physico-chemical characteristics of nanoparticles and their toxicity. Current risk-assessment paradigms, particularly in regulatory submissions for drugs and chemicals, generally depend on standardized methodologies. There is the potential for computational tools to refine, improve, and, in some cases, even replace existing tests. Computational tools are essential for increasing throughput, reducing the burden of animal testing, providing details of the toxicity mechanisms, and generating novel hypotheses for risk assessment.[8,17,18]

For the purpose of nanoparticle assessment such approaches as the quantitative structure–activity relationship (QSAR) can be applied.[10,19,20] Computational tools are useful not only in making predictions, but also in refining existing risk-assessment paradigms. For example, QSAR approaches applied to assessing risk may facilitate the placement of chemicals with incomplete data sets in appropriate risk categories. Computational modeling includes physiologically-based pharmacokinetic (PBPK) models, as well as modeling dose-responses.[18] If a QSAR model is then developed, ideally the dose-response toxicity of untested nanoparticle can be predicted on the basis of its physico-chemistry.[10,11,19,21] Actually, there is a strong need to extend the traditional QSAR paradigm to nanoparticles, and selected results obtained using this approach will be shown here.

In fact, to date, there is very limited information related to systematic data on experimentally measured toxic effects for various groups of nanoparticles. Only accurate and diverse experimental data can be useful for QSAR modeling purposes. However, some isolated and limited experiments have been published in the last few years.[22–39] The available toxicity data for nanoparticles at this time are mostly in vitro, for bacteria cultures.[26,28,32,33,39–42] A few studies in vivo were published for mammals, particularly for rats or mice.[23–26,28–30,35,38,39,41–44] This quite limited toxicity information makes complicated the study of structure–property and structure–toxicity relationships of nanoparticles.

In this chapter we discuss recent works related to the development of quantitative structure–property relationship (QSPR) and QSAR approaches which are feasible to apply to the physico-chemical and toxicity studies of various groups of nanoparticles. The current status of such developments assures us that computational techniques are essential for the evaluation of new nanomaterials and predictions of their possible environmental effects.

10.2 What Makes a Nanoparticle Unique?

The intrinsic properties of nanoparticles make them very relevant objects for the development of new materials with unusual and desired properties that could be very small in size.[3,9] But why are nanomaterials so special? There are several factors that contribute towards their unique characteristics. First, at the nanometre scale, the behavior of the properties of matter, such as energy, is unusual and of great consequence. This is a direct effect of the small size of nanomaterials, and the main explanation attributes these unusual properties of nanomaterials to quantum size effects. As a consequence, when a material (*e.g.*, a metal, metal oxide or carbon nanotube) is in a nano-sized form it can assume properties which are very different from those of the same material in a bulk form. For example, bulk silver is non-toxic, whereas silver nanoparticles are capable of killing bacteria upon contact.[45–47] Properties like reactivity, electrical conductivity, color, strength, and solubility change when the nanoscale level is reached.[3,48] The same material can become a semiconductor or an insulator at the nanoscale level.[49,50] Finally, nanomaterials have an increased surface-to-volume ratio compared to bulk materials.[3,9,51] This has important consequences for all those processes which occur at a material surface, such as catalysis, binding to biological species, and detection.

Physico-chemical properties that may be important in understanding the peculiar effects of nanomaterials include particle size and size distribution, agglomeration state, particle shape, crystal structure, chemical composition, surface area, surface chemistry, surface charge, electronic properties (reactivity, conductivity, interaction energies, *etc.*), and porosity.[6,9,11,52,53] All these properties are vital and have to be considered step-by-step in nanoparticle research, separately or, better if possible, in combination.

10.3 Modeling Nanoparticle Properties

From the experimental point of view, the fundamental problem in the nanoscale region is that the units are too small to see and manipulate. However, they are too large for single-pot synthesis from chemical precursors. Because it is difficult to see what takes place at the nanoscale, it is essential to develop theoretical and computational approaches sufficiently fast and accurate that the structure and properties of materials can be predicted, especially for various conditions (temperature, pressure, concentrations).[54–60] A particular advantage of using theory for such purpose is that the properties of new materials might be predicted in advance of experiments. This allows the system to be adjusted and refined (designed) so as to obtain the optimal properties before the arduous experimental task of synthesis and characterization. However, there are significant challenges in using theory to predict accurate properties for nanoscale materials. Thus, a cubic gold nanoparticle 100 nm on a side would have millions of atoms, being too large for standard classical molecular dynamics (MD) and exceedingly too large for quantum mechanics (QM).

There are many groups of nanoparticles. They may include fullerenes, carbon nanotubes, metals, quantum dots, dendrimers, ceramics, zeolites, semiconductors, polymers, and liquid crystals. The nanoparticles might be in any of the various phases: vapor/gas, liquid or solid phase (or all three phases may be present and interact through vapor–liquid, solid–liquid, vapor–solid interfaces).

As stated above, not only the dose and elemental composition of the nanoparticles define their properties, but also such important features as their surface area, surface characteristics, shape of the particles, tendency to aggregate, surface charges, transport properties and rheological properties (viscosity, non-Newtonian behavior, flow properties, *etc.*). All these features play crucial roles in nanoparticles' distribution through the environment, ecosystem and especially, through live organisms, including the human body. These factors could contribute towards possible (genetic) toxicity.[8,14,45–47,51,61–64] Modeling the physico-chemical effects of nanomaterials could provide information about the factors influencing an aggregation mechanism, distribution in the environment (log P) and interaction pathways.

Obviously, a comprehensive investigation of the basic physico-chemical properties of nanoparticles provides key clues for the mechanisms of their unusual reactivity and mechanistic properties, biological activity, and toxicity.[11,21,25,48,63,65–67] For this purpose, the computational methods starting from quantum-mechanical *ab initio* methods to mesoscale dynamics (MesoD) techniques are very useful, since they allow one to model the structural and electronic properties of nanoparticles. Characteristics such as dipole moment, molecular volume, surface charges, energy of frontier orbitals, Fermi energy, band gap, electronic charges, molecular electrostatic potentials, transition energies, ionization potentials, electron affinity and many other physical properties could be accurately predicted using computational methods. These computed properties can then be used as descriptors for the structure–property/activity relationships analysis in the QSPR/QSAR approaches.

10.4   QSAR Methodology and Basic Principles

There is a continuing interest in the use of QSAR and other read-across approaches in risk assessment. A need exists to ensure that the domains of applicability of such approaches are adequately characterized and recognized.[18]

QSAR models bridge the intersections of chemistry, statistics and biology. The basic principle of QSAR is that the molecular structure of a chemical compound influences its physico-chemical properties and biological activities (or physical properties).[68] If the biological activity (property) and some aspects of structure can be expressed numerically, various mathematical techniques can be applied to link them.[10,19,69–74] This creates the foundations of the QSAR approaches.

The development of a QSAR model requires three components:

- A data set which provides experimental values of a biological activity (property) for a group of chemicals.

- The molecular structure and/or property data (*i.e.* descriptors, variables, or predictors) for the considered group of chemicals.
- Statistical methods, to find the relationship between these two data sets.

The challenge of QSAR approach is finding co-efficients C_0, C_1, ...C_n such that eqn (10.1) and the prediction error are minimized for a given group of n compounds.

$$\text{Biological activity(property)} = C_0 + (C_1 P_1) + \ldots + (C_n P_n) \qquad (10.1)$$

Here P_1, P_2, ...P_n represent molecular descriptors, *i.e.* numerical values that characterize the structural properties of molecules. The descriptors vary in the complexity of encoded information and in computing time.

QSAR offers numerous possibilities for predicting biological activity, toxicity and physical/mechanical properties of explored chemical entities. Many medicinal and agricultural agents have been predicted by applying a QSAR approach.[75] However, as in any methodology, there are advantages and some disadvantages of this technique.

Advantages of QSAR:

- Quantifying the relationship between structure and activity provides an understanding of the effect of structure on activity.
- By applying QSAR it is also possible to make predictions leading to the synthesis of novel analogues, with improved activity.
- The results can be used to help understand interactions between functional groups in the molecules of the greatest activity with their biological target.

Disadvantages of QSAR:

- False correlations may arise because of the quality of biological data which are subject to considerable experimental error (noisy data).
- If the training dataset is not large enough, the data collected may not reflect the complete property space. Consequently, many QSAR results cannot be used to confidently predict the most likely compounds of the best activity.
- Calculated features may not be reliable as well. This is particularly critical for 3D features because 3D structures of ligands binding to a receptor may not be available. A common approach is to use a minimized structure, but that may not represent the reality well.

As stated above, there are many successful applications of this method but one cannot expect the QSAR approach to work well all the time.

10.5 Extending the QSAR Paradigm to Nanoparticles

The previous section in this chapter discusses the principles of QSAR and some advantages of this approach for conventional compounds. Now the question is: how one can use the QSAR methodology for nanostructured compounds?

Many well-developed QSAR/QSPR methods have been proposed for conventional compounds, where compounds vary only in functional groups and molecular structure. These methods have proved their effectiveness many times in drug design based on small compounds. But nanoparticles are different from conventional chemicals: they are large, often having similar structures to each other, and they differ in sizes. To address this issue one has to consider a paradigm shift from classic QSAR to nano-QSAR methodology which takes into account more complex relationships, not only structural variances in chemical entities.

Shaping the accumulated knowledge in this regard we can depict three problems that need to be considered in order to extend the QSAR approach to nanoparticles, according to recently published review paper (ref. 66):

1. QSAR has been mainly developed for small organic compounds with diverse structure types, while nanoparticles are large and structurally limited in diversity.
2. Not enough experimental data for nanoparticles has been accumulated; there is scarce and/or inconsistent empirical data, and no systematic data yet.
3. Regular QSAR descriptors which are applicable to organic compounds are generally not applicable to nanoparticles.

What are the crucial tasks that should be accomplished in order to build predictive nano-QSAR models for nanomaterials?

First of all, proper and truthful experimental data for the series of related but different size nanoparticles have to be available for QSAR analysis. Toxicological data is only available for a few nanoparticles and exposure routes. It is most important to obtain sound toxicological data which will be used to develop QSAR models for toxicity predictions. It is also important to develop the nanomaterials' property database – collected data on experimental physico-chemical properties, biological activity and toxicity endpoints for various sizes of nanomaterials.

Secondly, the structural descriptors suitable for modeling nanoparticle reactivity have to be identified. These descriptors can be selected/adapted from available pool of descriptors designed for conventional (small) compounds or calculated by applying quantum-mechanical (molecular dynamics) methods, or selected from known experimental physical properties. A close connection and interaction between all three areas – experimental toxicology, physico-chemical nanoparticle characterization (descriptor characterization) and nano-QSAR methods – is essential for successfully understanding the mechanisms of nanomaterials' toxicity (Figure 10.2). In addition, the parameters from modeling by computational approaches (quantum chemistry methods, molecular modeling and protein–ligand docking techniques) to the chemical interactions of nanoparticles with biological systems can also be applied as descriptors.

The availability of all these key elements may give a principal prospect for building a good nano-QSAR model (Figure 10.3).

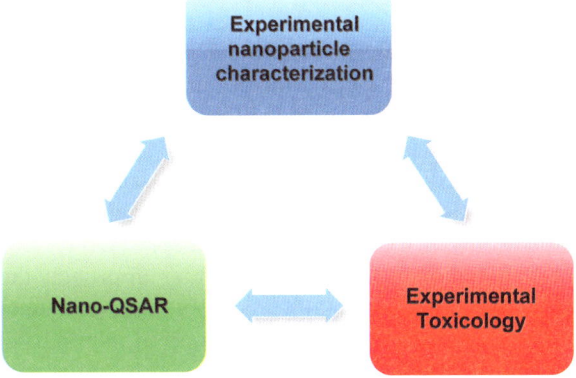

Figure 10.2 Interaction of three important areas necessary to assess the property and toxicity of nanomaterials.

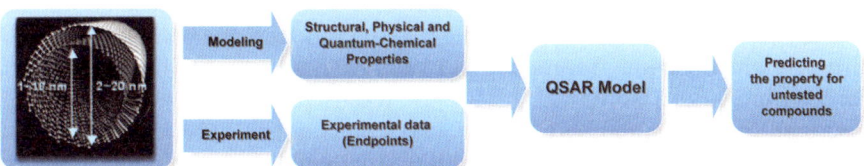

Figure 10.3 The representation of the combined use of experimental and modeling approaches necessary to develop nano-QSAR.

10.6 Nano-QSAR Modeling of Physico-chemical Properties

In order to govern the unique physico-chemical properties of nanoparticles one needs a basic knowledge as well as a means of controlling their other characteristics such as solubility, stability, reactivity, surface properties, *etc.*[11,21,25,60,63] All these properties become extremely important when nanomaterials are produced in large quantities and one has to optimize them to reduce the cost of manufacturing. Unfortunately, such an approach, although perhaps cost-efficient, could negatively affect the environment directly by applied technology or by the dissemination of produced species. To develop knowledge about possible risks before starting manufacturing one can use structure–property/activity relationship modeling where these properties are used as endpoints.

Very few studies could be found in the literature devoted to predictions of physico-chemical properties of nanomaterials by QSPR methods. However, there is a number of studies where authors undertook some efforts to correlate selected parameters of investigated nanoparticles with physico-chemical properties.

10.6.1   Solubility

One of the most important physico-chemical properties of nanomaterials is solubility. Solvent effects on the behavior of nanoparticles in the environment may be substantial.[76] Knowledge of the solubility of nanoparticles in various solvents can be also useful in manufacturing these species. Since nanoparticle systems are very large from a quantum-chemical point of view it is obvious that *ab initio* methods are often not applicable to evaluate them. Only semi-empirical, MD methods, or density functional theory (DFT)[77] approaches can be applied to calculate their properties. Another approach facilitates using stock QSAR to generate conventional descriptors that could be responsible for important nanoparticles' properties. One of group of nanoparticles where one can still use conventional descriptors is represented by fullerenes (or fullerene derivatives). The data on the solubility of the fullerene C_{60} in organic solvents can provide important information that would be of interest in chemistry and biochemistry. A quantitative structure–solubility relationships approach for this characteristic is not only able to predict numerical data but also allows the evaluation of the solubility of fullerene C_{60} as a function of a complex physico-chemical phenomenon. Several studies were devoted to solubility predictions for C_{60} in various solvents.[48,65,78–85] In recent review (ref. 66) we briefly discussed that topic and here we expound it in detail.

One of the first fullerene (C_{60})-solubility studies was performed by Abraham *et al.*[48] by analyzing experimental solubility data of C_{60} in 20 organic solvents. They used a total of 41 log P (octanol–water partition coefficient) and log L (Ostwald solubility coefficient) values which were processed to obtain the Abraham solvation descriptors[86–88] for C_{60}. The Abraham solvation descriptor showed a good correlation with C_{60} solubility in investigated solvents. The performed analysis indicated that a water solubility of 10^{-18} mol dm^{-3} is the optimum value. The authors also showed that solubilities can be also transformed into air–solvent partitions, L, using optimum values of the saturated vapor concentration, 10^{-24} mol dm^{-3}, and the air–water partition coefficient, 10^{-6}. It indicates that C_{60} is dipolar/polarizable and has no hydrogen bond acidity, but is a slightly stronger hydrogen bond base than compounds such as acetophenone. The explored descriptors indicated that C_{60} behaves more like a polyalkene than an aromatic system. Interestingly, it was shown that C_{60} is very lipophilic, and the distribution between blood and brain largely favors the brain.

The next C_{60} solubility prediction study was performed by Sivaraman *et al.*[81] The authors examined the solubility of C_{60} in 75 organic solvents, using their own experimental data reported previously. Topological, indicator indices, and polarizability parameters computed from the refractive index were used to form the regression models. In this study a number of regression models (about 36), which applied different solvent datasets split by type of solvent were generated. The different models were suggested for individual data sets such as alkanes, alkyl halides, alcohols, cycloalkanes, alkylbenzenes, and aryl halides. Naturally enough, splitting the total pool of data to small specific datasets provides a good predictive ability for such selected groups. The characteristics obtained are better than those for the models developed including the combined groups. The inclusion of an indicator parameter which represents a combination of atom contributions and contributions of substituents' position in benzenes improved the predictive ability significantly. The maximum number of descriptors in this study used in one model is 45 (29 in training and 16 in the test set); the model has 5 descriptors and the correlation coefficient $r = 0.965$.

Another C_{60} solubility study was performed by Danauskas and Jurs.[78] These authors used a slightly larger dataset than in the previous study of C_{60} solubility. They included data on solubility in 96 solvents collected from various sources. Models predicting fullerene solubility at 298 K were developed using multiple linear regression and feed-forward computational neural networks (CNN). The set of solvents spans C_{60} solubilities ranging from -3.00 to 2.12 log, where solubility $= (1 \times 10^4)$ (mole fraction of C_{60} in saturated solution). For each solvent the authors calculated four types of molecular structure descriptors: topological, electronic, geometric, and geometric/electronic hybrids. The best linear and CNN models were discussed in their paper. The authors admitted that two types of linear models failed to give a good prediction for an external prediction set and only the CNN model satisfied the requirements. The 76-compound training set for the CNN model had a root mean-square (rms) error of 0.255 log solubility units, while the 10-compound cross-validation set had an rms error of 0.253. The 10-compound external prediction set had an rms error of 0.346 log solubility units. The CNN model consisted of 9 descriptors of various types, from topological (number of fifth-order path, molecular connectivity, distance edge *etc.*) to quantum-chemical (hardness, polarizability).

The same year another study was published on fullerene C_{60} solubility where the authors used a linear solvation energy relationship (LSER) approach.[89] According to this approach, the solvation energy, *i.e.*, the interaction Gibbs energy of the solute with the solvent, depends in a linear manner on a small number of independently determinable physico-chemical effects. The authors built a model (10.2) for 120 compounds out of 145 (leaving 25 outliers with deviations $> 2\sigma$), and correlation coefficient $r^2 = 0.9909$ (*i.e.*, 99% of the data variance "explained"), where $\sigma = 0.418$ and $F_{4,116} = 3268$.

$$\log x_2 = (-0.1294 \pm 0.0036)E_T(30)_1 + (0.2197 \pm 0.0099)R_1$$
$$+ (-0.0573 \pm 0.0027)V_1 + (0.758 \pm 0.183)\beta_1 \tag{10.2}$$

where x_2 represents the measured or hypothetical solubility (x_2'), $E_T(30)_1$ is the Dimroth–Reichardt "general polarity" parameter, R_1 is molar refraction $R_1 = V_1 f(n)_1$, V_1 is the molar volume (at 298 K; $V_1(303 \text{ K}) = V_1(298 \text{ K})(1 + 5\alpha_{P1})$, where α_{P1} is the isobaric expansivity, β_1 is the electron pair donacity.

According to this study the "ideal" solubility of the fullerene was estimated, and the "non-ideal" part of the Gibbs energy of solution was then related to solute–solvent interactions.

In 2004 Hansen and Smith[90] attempted to use Hansen solubility parameters (HSPs) for the prediction of fullerene C_{60} solubility in 89 organic solvents. HSPs are widely used to correlate and predict the behavior of solvents.[91] The authors decided that it was therefore logical to use this approach to correlate the solubility of C_{60}. For that purpose the solubility properties can be conveniently visualized using HSP with a three-dimensional coordinate system with axes δ_D, δ_P and δ_H. A computer program then locates the "sphere" in HSP space which encompasses the good solvents and excludes the bad ones with a minimum of error. The HSP δ_D, δ_P and δ_H values for C_{60} have been found to be 19.7, 2.9, and 2.7, respectively. These values were assigned using solubility data for 87 liquids and by considering the 15 solvents characterized by log (solubility) > 3.0 as being good solvents. These parameters are indicative of a largely non-polar solid, and have been used in a systematic search for additional predicted good solvents for C_{60}. This study could be considered as a classification study which just divided all solvents into the categories of "good" and "bad".

In 2005 Huang published another model based on C_{60} solubility data of only 50 solvents.[92] He provided a detailed review of all previous models published and suggested his own model based on Flory–Huggins interaction parameters, χ, which were calculated and analyzed through the model of Thomas, Eckert *et al.*[93] The author found that a fullerene had dispersion and polar components near those that characterize aromatics, a small acidity, and a zero basicity component. The induction parameter was near unity. The author suggested that the model is able to correlate solubility data ranging from alkanes to alcohols.

Starting from 2007 our group published a number of papers devoted to C_{60} solubility in organic solvents.[80,83–85] The discussion below covers only general information about approaches applied and the statistical results obtained; the details are given in the published papers.

The group, in collaboration with various research teams, performed a number of studies related to the structure–solubility relationship for the fullerene in different solvents[80,83–85,94] and for carbon nanotubes in water.[95] A number of models published used the simplified molecular input line entry system (SMILES) as a basis for descriptors calculated for C_{60} [83,84,94]. The authors assumed, taking into account an increased number of internet databases which apply SMILES for the further analysis of physico-chemical parameters and biological activity, that there is a need to develop SMILES-based predictive models.[20,96–99] The SMILES-based optimal descriptors in QSPR modeling of the fullerene C_{60} solubility serve as a function of the solvents' molecular structure represented by the SMILES notation. The number of solvents used for modeling ranges from 36 (ref. 94) to 122.[83,84] Optimal descriptors in these studies have been defined as shown in eqn (10.3).

$$DCW(SMILES) = \sum_{k=1}^{n} CW(SF_k), \tag{10.3}$$

where SF_k is a fragment of SMILES, and $CW(SF_k)$ represents the correlation weights of the SF_k. Generally, an application of the least squares method yields eqn (10.4).

$$\log S = C_0 + C_1\, DCW(SMILES). \tag{10.4}$$

Then the predictive potential of the model calculated with eqn (10.4) can be estimated using data on organic solvents of the external test set. In the study,[83] the obtained models showed a relatively good performance. The final model and statistical characteristics of QSPR for solubility, log S, (S is expressed as a molar fraction) in ref. 94 are as shown in the model eqn (10.5) and Figure 10.4.

$$\log S = -6.6196(\pm 0.0154) + 0.5014(\pm 0.0026)DCW(SMILES)$$

$$n = 92,\ r^2 = 0.8612,\ q^2 = 0.8537,\ s = 0.401,\ F = 558 (\text{training set}), \tag{10.5}$$

$$n = 30,\ r^2 = 0.8908,\ r^2_{pred} = 0.8748,\ s = 0.435,\ F = 228 (\text{test set})$$

In the next study of C_{60} solubility,[84,100] the slightly improved method of optimal descriptors was applied to get better correlations. This method takes into account combinations of different SMILES attributes. Here the optimal descriptors are defined as shown in eqn (10.6).

$$DCW = CW(Nb) \cdot CW(Ndb) \cdot \prod CW(ss_k) \tag{10.6}$$

where Nb is a number of brackets in given SMILES, these are indicated below as '(000', '(001', *etc*..; Ndb is a number of the double covalent bonds indicated in SMILES by '=', these are indicated below as the '=000', '=001', *etc*..; ss_k represents two SMILES consequent elements in the SMILES strings; and $CW(SA_k)$ is the correlation weight of SMILES attribute of the SA_k, ($SA_k = Nb$

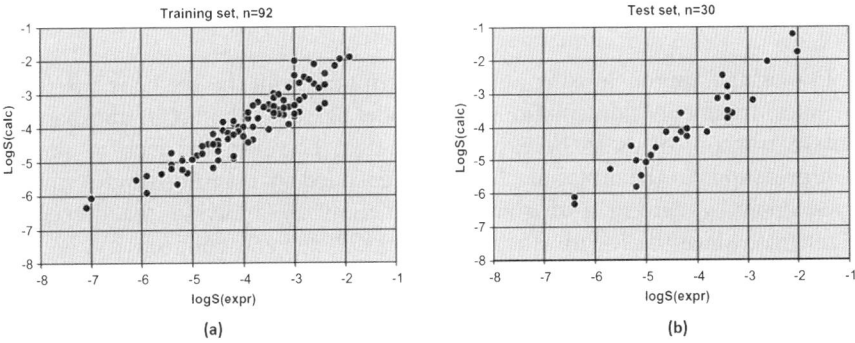

(a)   (b)

Figure 10.4   Plot of experimental *versus* calculated solubility (log S) of fullerene C_{60} for the training set (a) and for the test set (b) according to eqn (10.5). Reproduced with permission from ref. 83. Copyright Elsevier, 2007.

or Ndb or ss_k). $CW(ss_k)$ is a correlation weight of two components' SMILES fragment. In the reviewed study the three splits of data into training and test sets have been examined to estimate the ability of the applied approach to give a robust prediction. The following model and statistical data are obtained: see eqn (10.7).

$$\log S = -111.0059(\pm 0.3578) + 104.4283(\pm 0.3482)DCW$$

$$n=92,\ r^2=0.9372,\ q^2=0.9339,\ s=0.270,\ F=1342\ (\text{training set}) \qquad (10.7)$$

$$n=28,\ r^2=0.9151,\ r^2_{pred}=0.9032,\ s=0.334,\ F=280\ (\text{test set})$$

In another study devoted to C_{60} solubility prediction the optimal descriptors calculated with the International Chemical Identifier (InChI)[101] were used to construct a one-variable model.[85] The following formula has been used to calculate the correlation weights using an InChI structure representation: see eqn (10.8).

$$DCW(InChI) = W(I_k) \qquad (10.8)$$

where the I_k is an InChI attribute and $W(I_k)$ is the correlation weight for the I_k.

Here, the highest value of correlation weight $W(I_k)$ obtained for the training set is used to calculate the model of fullerene C_{60} solubility: see eqn (10.9).

$$\log S = C_0 + C_1\ DCW(InChI) \qquad (10.9)$$

The final model obtained by eqn (10.9) in the study could be expressed by model (10.10).

$$\log S = -7.9824(\pm 0.1397) + 0.3250(\pm 0.0010)DCW(InChI)$$

$$n=92,\ r^2=0.9447,\ q^2=0.9418,\ s=0.253,\ SDEP=0.258,\ F=1538\ (\text{training set}) \quad (10.10)$$

$$n=30,\ r^2=0.9398,\ r^2_{pred}=0.9315,\ s=0.348,\ F=437\ (\text{test set})$$

In fact, the approach using optimal descriptors from the InChI is based on chemical fragments' contributions. The model based on the InChI approach is not so transparent as to be easily interpreted; however, the lack of simple interpretations of the molecular fragments encoded by InChI attributes can be overridden by getting a reasonable prediction by applying the described algorithm. Since the InChI attributes (chemical element, connectivity, bonds, charges, *etc.*) have a clear genesis it is possible to extract robust heuristic information from eqn (10.10), which probably is not less important than the groups' contributions. For instance, the InChI-based optimal descriptors may give some clues for the mechanistic interpretations of fullerene C_{60} solubility.[85] Finally, it can be noted that the described InChI-based model is better than models that are based on SMILES notations.[83,84,94]

The next model for C_{60} solubility published recently[80] is rather different from all the previous ones and is based on quantum-chemical and topological descriptors. This study aimed to find a simple, transparent relationship and computationally fast approach, possibly mechanistically interpretable, to predict the solubility of C_{60} in various organic solvents. The applied approach evaluates the predictive potential of the structure-based topological descriptors and quantum-chemical parameters obtained by high level *ab initio* calculations in QSPR modeling of the fullerene C_{60} solubility in organic solvents. One of the initial sets of utilized descriptors was applied and this represents the set of constitutional, topological, and molecular descriptors that were calculated by the *DRAGON* software.[102] Another set represents the quantum-chemical descriptors that have been calculated using Gaussian 03 software[103] by DFT methodology. Five models were obtained and discussed, from a one-variable to a five-variable model. The four-variable model (10.11) showed the best statistical performance (Figure 10.5).

$$\log S = 0.532\ TI2 + 0.698\ X1Sol + 15.694\ FDI - 0.103\ H - 052 - 21.218$$

$$r^2 = 0.861,\ q^2 = 0.841,\ F = 134.80,\ s = 0.411 \tag{10.11}$$

$$r_{test}^2 = 0.903,\ F = 259.56,\ s = 0.355$$

The model (10.11) includes the topological descriptor *X1Sol*, which represents the solvation connectivity index (χ_1) that encodes the solvation property of the compound [104]. Briefly, this molecular descriptor is defined in order to model solvation entropy and dispersion interactions in solution. The *TI2* descriptor represents a topological descriptor, second Mohar index. The Mohar index is derived from a Laplacian matrix,[104] which is a distance matrix. Another important descriptor is the *FDI* descriptor, which is a geometrical descriptor – a folding degree index. The values of this descriptor are defined in a range $0 \leq FDI$

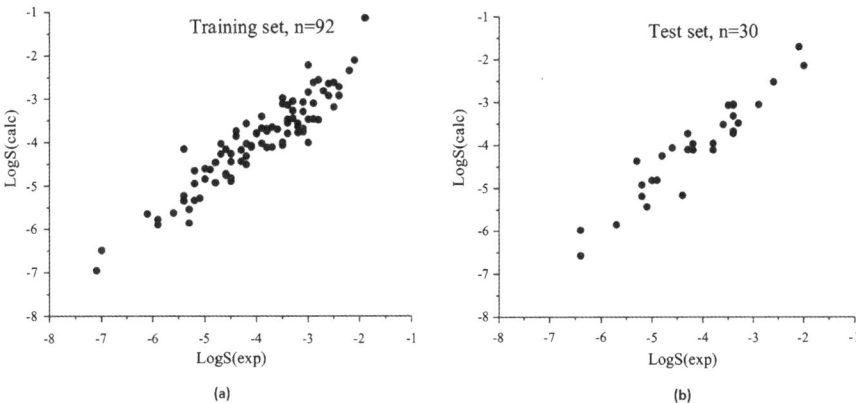

Figure 10.5 Plot of experimental *versus* calculated solubility (log S) of fullerene C_{60} for the training set (a) and for the test set (b) according to the model (10.11). Reproduced with permission from ref. 80. Copyright Springer, 2011.

≤ 1. The *FDI* descriptor converges to one for linear molecules (of infinite length) and decreases in accord with the folding degree of the molecule. The *H-052* descriptor represents the atom-centered fragments, describing H (hydrogen) attached to $C(sp^3)$ with 1X (heteroatom) attached to the next C.[104] This study demonstrates that an application of the GA-MLRA (genetic algorithm combined with multiple linear regression analysis) technique in combination with quantum-chemical and topological descriptors yields reliable models. The models are quite simple, interpretable, transparent, and statistically comparable to the previously published results. The model is equivalent to the model published in ref. 79 and the model suggested in the other publications.[83–85,94]

There have been several other publications regarding the solubility of C_{60} published recently,[100,105,106] but most of them use the same methodology that was discussed above.

Another important study which is worth mentioning here is a prediction of the solubility in water and octanol–water partition coefficients for carbon nanotubes (CNTs) based on chiral vectors.[95] Molecules of CNTs contain hundreds or even thousands of atoms. Because of the large number of atoms in structure the application of the "classic" descriptors becomes quite problematic. So, there was a need to find a simple parameter for CNT structure–solubility purposes. The authors used a known parameter for the CNT structure – a chiral vector (n, m).[107,108] The components of the chiral vector contain information about rolling up the graphite layer in the CNT. It is also known that certain physico-chemical behaviors of CNTs are correlated with the numerical values of chiral vector components. For example, $m - n = 3k$ (k is an integer) is known to be a necessary criterion for conductivity in CNTs.[49] In the reviewed study,[95] the multiple linear regression analysis was used to find correlations between chiral vectors of CNTs and solubility (log S), and also for correlations between the chiral vector and the octanol–water partition coefficient (log P). Two-variable models for the water solubility and octanol–water partition coefficient have been found. Indeed, the obtained models (eqn (10.12) and (10.13)) showed a remarkable performance.

$$\log S = -5.1041 - 3.5075n - 3.5941m$$

$$r^2 = 0.9999, \ s = 0.0534, \ F = 126.611 \ \text{(training set)} \qquad (10.12)$$

$$r^2 = 0.9999, \ s = 0.0933, \ F = 67.456 \ \text{(test set)}$$

$$\log P = -3.9193 + 3.7703n - 3.6001m$$

$$r^2 = 0.9991, \ s = 0.364, \ F = 2.927 \ \text{(training set)} \qquad (10.13)$$

$$r^2 = 0.9996, \ s = 0.287, \ F = 5.928 \ \text{(test set)}$$

Moreover, the obtained models are simple and transparent. It was concluded that components of the chiral vectors of CNTs can be used as structural descriptors in QSPR analysis aimed to predict the water solubility and octanol–water partition coefficient of carbon nanotubes.

10.6.2 Elasticity (Young's Modulus)

The mechanical (technological) properties of nanomaterials are also among the vital characteristics available for predictions. The application of QSPR to predicting nanomaterial properties provides an example where an attempt to estimate the ability of a QSPR approach to evaluate the mechanical characteristics of nanomaterials has been successful. In the discussed study, a SMILES-like approach was used to predict the elasticity (Young's modulus) of selected nanomaterials.[109] Toropov and Leszczynski attempted to use technological characteristics to encode nanomaterials, *i.e.* the SMILES-like nomenclature for a given nanomaterial used to incorporate data on atom composition and the technological conditions of its synthesis. These data were used as a basis for calculating optimal descriptors. Interestingly, the nomenclature used in this study does not represent standard SMILES, since the function of the nomenclature used is limited to encoding the available information on the nanomaterials' genesis as commercial products. The applied SMILES characteristics reflect detailed (2D, 3D, and even quantum-chemical) information on molecular architecture.

The descriptor used for modeling Young's modulus (YM) has been defined in the same way as a regular optimal descriptor, eqn (10.14).

$$DCW = \prod_{k=1}^{N} CW(I_k) \tag{10.14}$$

where I_k is the component information on the nanostructure (*e.g.*, Al, N, BULK, *etc.*); $CW(I_k)$ is the correlation weight of the component I_k; and N represents the total number of these components in the given nanostructure. The sequence of components applied to a given nanomaterial such as its code and descriptor calculated with eqn (10.14) provides a mathematical function of the code. Using the Monte Carlo method one can calculate the values of the $CW(I_k)$ which yield correlation coefficients between YM and the DCW that are as large as possible for the training set. Then, based on the $CW(I_k)$ values, the authors computed a YM by a least-squares method model (10.15).

$$YM = C_0 + C_1\, DCW \tag{10.15}$$

The developed model has the following representation and statistical data: model (10.16).

$$YM = -3720.235 + 3945.175\, DCW$$

$$r^2 = 0.9757,\ s = 18.25,\ F = 761 \tag{10.16}$$

$$r^2 = 0.8952,\ r^2_{pred} = 0.8880,\ s = 34.69,\ F = 51$$

Interestingly, the information concerning nanomaterials in view of the data used as descriptors (see the table in ref. 109) corresponds to instructions on how to carry out the synthesis of a given substance at manufacturing process. Accordingly, the authors stated that the suggested approach can be used as a tool for estimation of the YM value for nanomaterials. This also pertains to nanostructures that can be produced under technological conditions which have not been applied before. In other words, the suggested approach can help to save time and material resources in designing nanomaterials with the required knowledge about values of YM. It is worth noting that the accuracy of the prediction can be increased by increasing the number of nanomaterials used in the training set. Finally, it can be noted that 29 totally different nanomaterials characterized by experimental studies were utilized to build a predictive model for a module on elasticity.

Another study which attempts to predict the Young's modulus of nanomaterials applies a finite-element simulation technique for estimating the mechanical properties of single-walled carbon nanotubes (SWCNTs). This investigation yields relationship between the SWCNT size and YM.[110] A finite-element (FE) simulation technique for SWCNTs has been developed in this work, and the authors claim that the simulation can be easily performed by commercial code *ANSYS*. The key modeling concept of the study is that simulated molecular bonds are presented as beam elements. They proposed and verified a simplifying method to model the non-linear nature of the covalent bond between two carbon atoms in the nanotube wall. A 3D FE model that is able to assess the mechanical properties of SWCNTs is proposed. The 3D FE model is developed using *ANSYS* commercial FE code. For the modeling of the C–C bonds, a 3D elastic BEAM4 element is used (Figure 10.6). The specific element is represented by a uni-axial element with tension, compression, torsion and bending capabilities. The element has six degrees of freedom at each node: translations in the nodal x, y and z directions and rotations about the nodal x-, y- and z-axes. The element is defined by two

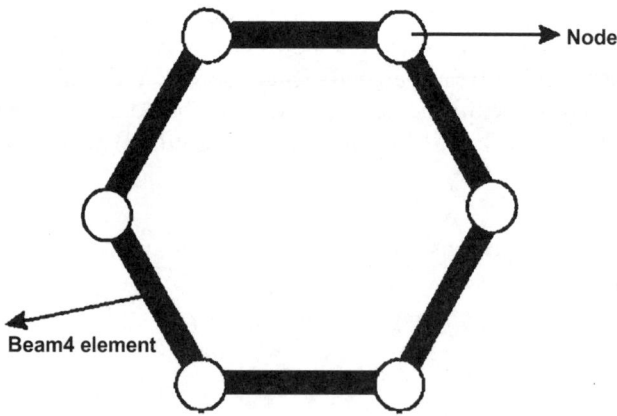

Figure 10.6   Finite-element modeling concept of the hexagonal structure of a CNT. Reproduced from open-access ref. 110. Copyright Asian Network for Scientific Information, 2011.

or three nodes as well as its cross-sectional area, two moments of inertia, two dimensions and the material properties.

Furthermore, the effect of nanotube diameter and structure is also studied in this work. Six zigzag type single-walled carbon nanotubes of different sizes are simulated and their axial Young's modulus calculated are listed in Table 10.1 and depicted in Figure 10.7. The authors stated that this method can significantly save the modeling and computing efforts when finite element analysis is performed. Numerical results for axial Young's modulus are presented to illustrate the accuracy of the established finite element models. In addition, they also investigated the relations between these mechanical properties and the nanotube size to achieve a better understanding of mechanical properties' variation among nanotubes.

The authors believe that the revealed relationship together with the outstanding advantage of the presented modeling concept can be easily

Table 10.1 Axial Young's modulus of single-walled carbon nanotubes.

Tube type	Axial Young's modulus (TPa)	Nanotube diameter (nm)
Zigzag (6,0)	1.19500	0.469732
Zigzag (8,0)	1.21700	0.626310
Zigzag (10,0)	1.22700	0.782887
Zigzag (12,0)	1.22898	0.939464
Zigzag (14,0)	1.23160	1.096042
Zigzag (22,0)	1.23280	1.722351

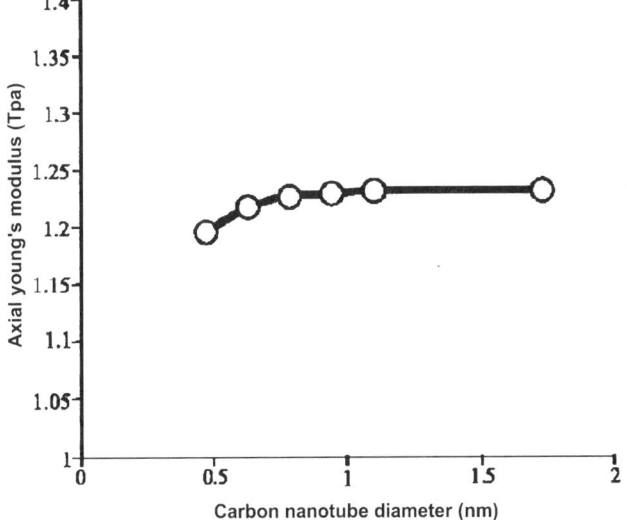

Figure 10.7 The relation between the axial Young's modulus and diameter of carbon nanotubes. Reproduced from open-access ref. 110. Copyright Asian Network for Scientific Information, 2011.

extended to multi-walled CNTs (MWCNTs) with a higher number of layers. They assume that this method will be an effective and convenient tool in predicting and studying the mechanical behavior of MWCNTs.

Another study where scientists offered a method for predicting the Young's modulus of nanowires from first-principles calculations was published recently.[111] Using the concept of surface stress, the authors developed a model that is able to predict the YM of nanowires as a function of nanowire diameters from the calculated properties of their surface and bulk materials. Both the equilibrium strain effect and the surface stress effect were taken into consideration to account for the geometric size influence on the elastic properties of nanowires. Wang and Li successfully combined first-principles DFT calculations of material properties with linear elasticity theory of clamped-end three-point bending. The calculated properties include YM, Poisson's ratio, surface energy, and surface stress. Taking those theoretical data as inputs, they predicted the Young's moduli of Ag, Au, and ZnO nanowires as functions of nanowire diameter (D).

To validate the developed model and approach, the authors compared the theoretical results with the experimental data. It is found that the model

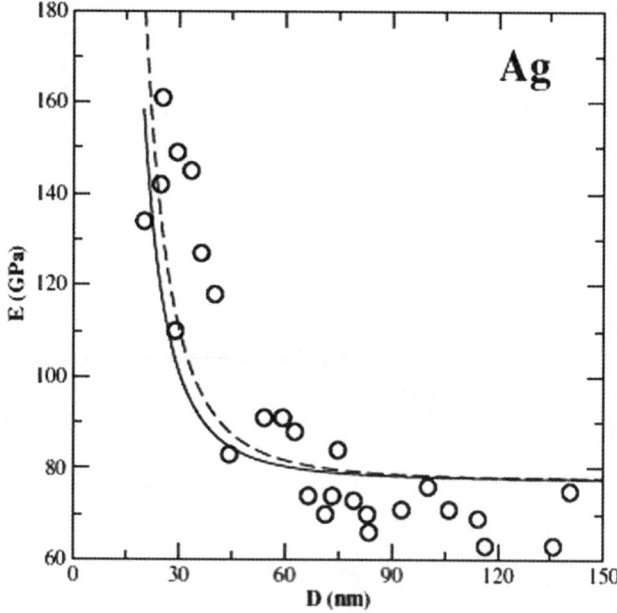

Figure 10.8   Size dependency of the Young's modulus of Ag nanowires enclosed by (111) surfaces. The solid line and dashed line show the model predictions using the surface properties of (111)/(112) and (111)/(110), respectively. For comparison, the experimental data from ref. 112 are plotted as circles. Reproduced with permission from ref. 111. Copyright American Institute of Physics, 2008.

predictions for Ag and Au nanowires agree excellently with those experimental data. For ZnO nanowires with $D > 20$ nm, the predictions also agree qualitatively with a series of experimental results. Wang and Li believe that the agreement between the model and experimental data is remarkable since no empirical parameters were introduced in the theoretical approach. The obtained model reveals two major effects that the surface exerts on the elastic deformation process of nanomaterials. First, the surface may have different minimum-energy lattice parameters from the bulk lattice parameters of the material and thus result in an equilibrium strain in the nanowire core region. Second, the surface may have tensile or compressive surface stress that is the energy required to elastically deform the surface. A tensile surface stress would lead to an increase in YM with decreasing size of nanowires, while a compressive surface stress would lead to a decrease in YM with decreasing size of nanowires. The model indicates that the tensile positive surface stress is the reason for Ag and Au nanowires to exert an enhanced YM when reducing nanowire diameters (Figures 10.8 and 10.9). The model also points out that the compressive negative surface stress in ZnO nanowires with $D > 20$ nm would lead to a softened YM when reducing nanowire diameters (Figure 10.10). For

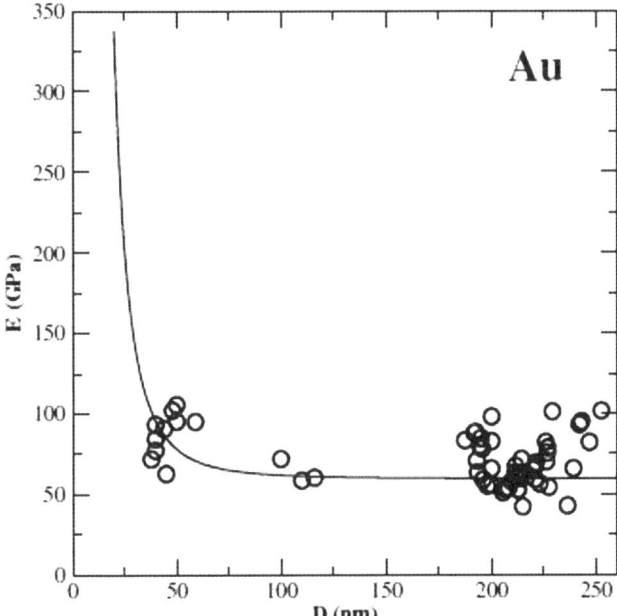

Figure 10.9   Size dependency of Young's modulus of Au nanowires enclosed by (111) surfaces. The two dashed lines (which overlap with each other) show the model predictions using the surface properties of (111)/(112) and (111)/(110). For comparison, the experimental data from ref. 113 are plotted as circles. Reproduced with permission from ref. 111. Copyright American Institute of Physics, 2008.

both Ag and Au nanowires, the theoretical predictions agree well with the experimental data published. For ZnO nanowires, the predictions are qualitatively consistent with some of the experimental data for ZnO nanostructures. Consequently, the authors found that surface stress plays a very important role in determining the YM of nanowires. Since surface stress governs the Young's modulus of nanomaterials, it is reasonable to expect that the elastic properties of nanomaterials could be engineered by altering the surface stress through rational control of the adsorptions, charges, structure, and impurities in the surfaces.

In conclusion, the physico-chemical properties of nanomaterials can be modeled and accurately predicted. Various approaches and descriptors can be used, both adapted from conventional descriptors or based on quantum-mechanical calculations. However, the applicable approach and nature of descriptors that need to be calculated depend on the type of nanomaterial studied. For example, for the fullerene-like compounds the conventional descriptors can be adapted; the only problem is that not all commercial programs generating descriptors are able to process large molecules like fullerenes and calculate descriptors for them. For nanotubes, nanowires and metal oxide nanomaterials, only specific theoretical descriptors can be used, as well as those calculated by the application of quantum-mechanical methods.

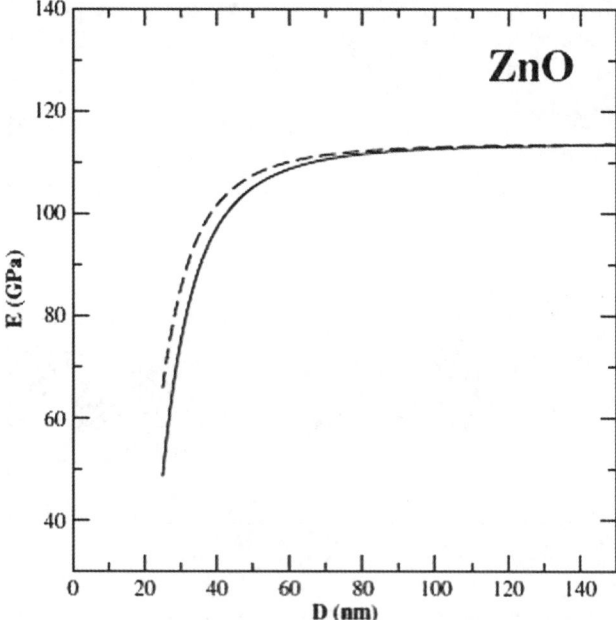

Figure 10.10   Predicted size dependency of Young's modulus of ZnO nanowires enclosed by (1010) surfaces (solid line) or (1120) surfaces (dashed line). Reproduced with permission from ref. 111. Copyright American Institute of Physics, 2008.

10.7   Nanoparticle Toxicity: Concerns and Challenges

The number of nanotechnology products in development and on the market is on the rise. Because of widespread commercial applications, this technology presents unique exposure and risk assessment challenges. For example, both the exposure and the environmental fate of various metal- and carbon-based nanoparticles need to be assessed without delay. Since nanomaterials are included in consumer products and food, a better understanding of the risks associated with nanotechnology products remains a vital issue. The use of nanotechnology in medicine (*e.g.*, in drugs and medical devices) requires a better assessment of the associated potential safety risks.[18] Many uncertainties persist as to whether the unique properties of engineered nanomaterials pose occupational health risks. These uncertainties arise because of gaps in knowledge about the potential routes of exposure, the movement of nanomaterials once they enter the body, and the interaction of the materials with the body's biological systems (Progress Toward Safe Nanotechnology in the Workplace: A Report from the NIOSH Nanotechnology Research Center, Centers for Disease Control and Prevention National Institute for Occupational Safety and Health, 2007).

There are still many unanswered questions regarding the toxicity of nanoparticles:

- Is particle surface area a more appropriate measure of exposure than particle mass?
- Can nanoparticles penetrate into various organs once they are in the bloodstream?
- By which mechanisms do nanoparticles generate reactive oxygen species?
- Do nanoparticles cause adverse health effects in workers if they penetrate the skin?
- How do the shape, durability, and chemical composition of nanoparticles affect their biological activity?
- Do nanoparticles that are bound together split into smaller, possibly more potent structures in biological fluids?
- Are *in vitro* assays predictive of *in vivo* responses to nanoparticles?
- What is the role of surface area in biological activity?

All these questions have to be answered in order to acquire an understanding of nanomaterial toxicity mechanisms and to be able to perform a required risk assessment.

10.8   Nano-QSAR and Prediction of Toxicity

The biological activities, such as pharmacological activities and toxicity, are vital characteristics that are connected with the peculiar physico-chemical properties of nanomaterials. It is known now that particle size, hydrophobicity, and protein identity all contribute to nanoparticle–protein association.[43]

For instance, fullerenes are capable of specific interactions with proteins as evidenced by the production of fullerene-specific antibodies.[114] Fullerenes have been identified as inhibitors of cysteine and serine proteinases[115] as well as the protease specific for the human immunodeficiency virus (HIV)[116,117] and are known to interact directly with the virus. For example, fullerene C_{60} derivatives can interact with the active site of HIV-1 protease (HIV-1 PR). The protease specific to HIV-1 has been shown to be a viable target for antiviral therapy.[117,118] The active site of this enzyme can be roughly described as an open-ended cylinder which is lined almost exclusively by hydrophobic amino acids. Since the C_{60} molecule has approximately the same radius as the cylinder that defines the active site of the HIV-1 PR and since C_{60} (and its derivatives) is primarily hydrophobic, an opportunity therefore exists for a strong hydrophobic interaction between the C_{60} as well as its derivatives and active site surfaces. This interaction makes the C_{60} derivatives efficient inhibitors of the HIV-1 PR.

The modeling and predictions of these properties (endpoints) are crucial to understand the mechanisms of actions of nanoparticles, decrease animal use for toxicity experiments, and reduce the costs of experiments. Last but not least, it allows for a quick prediction of the property of interest for new, untested nanoparticles. Therefore, the next important step in nanomaterial investigation and analysis is to find a reliable relationship between structural and/or physical features of nanoparticles and their exhibited biological activity or toxicity. In this review we discuss only one of the many biological properties – toxicity – since this is the most important issue that needs to be cleared for risk assessment of engineered nanomaterials. However, this is also the most comprehensive task, and for this type of study the experimental toxicity data obtained for the considered nanoparticles should be used.

Oberdörster *et al.* in their comprehensive article[119] compiled the list of possible physico-chemical properties to which toxicity of nanoparticles can be related to. The list includes:

- size
- size distribution
- agglomeration state
- shape
- porosity
- surface area
- chemical composition
- structure-dependent electronic configuration
- surface chemistry
- surface charge
- crystal structure

Apparently, toxicity is one of the vital characteristics of nanoparticles. Unfortunately, there is still only limited information about the experimentally measured toxic effects of nanoparticles. In addition, such studies are still not systematic. In fact, the toxicity properties of nanomaterials have been investigated by various groups of researchers, but considerable uncertainties in nanoparticles' mechanism of toxicity still exist. Many studies showed a much higher toxicity displayed by some of the nanoparticles when compared to bulk size particles.[22,33,120–122] For example, it is well known that the metal oxide nanoparticles possess a higher toxicity than bulk size counterparts of the same chemical composition.[10,19] Some isolated and limited toxicity data were published in the last few years,[22–40,44,51,120,121,123–127] and due to increasing interest in such data more similar studies can be expected in the near future. Obviously, this scarce toxicity information makes the study of structure–toxicity relationships of nanoparticles difficult.

Since there had not been systematic data on nanomaterials' toxicity for a long time, there are only few original papers published where authors attempted to find correlations between toxicity and nanoparticles' properties and developed predictive models.[128–130] One of the first computational studies focusing on investigation of nanotoxicity is work performed by Liu and Hopfinger,[128] which was also briefly discussed in ref. 66. The authors have used molecular dynamics simulation (MDS) of the interaction of a carbon nanotube with a fully hydrated dimyristoylphosphatidylcholine (DMPC) lipid bilayer in its physiologically relevant liquid-crystalline phase to probe the interactions between membrane penetrants and phospholipid monolayer assemblies and to construct quantitative structure–activity relationship models of membrane interaction (MI-QSAR). Four possible sources of cellular toxicity, due to the insertion of a carbon nanotube into a DMPC membrane bilayer were explored by the authors using the MI-QSAR methodology. Comparisons of (i) the structural organization of the membrane bilayer, (ii) dynamical features of the membrane bilayer, and (iii) transport of small polar molecules across the membrane bilayer were carried out with, and without, a carbon nanotube inserted into the bilayer. A fourth step of this study was performed by the same authors to determine how the transport of solvated ions through the inserted nanotube might alter the structure of the membrane bilayer. The rigidity of the inserted carbon nanotube actually leads to an open, nearly unoccupied cylindrical ring around the nanotube, which can be seen in Figure 10.11a. This packing of DMPC molecules about the nanotube in the bilayer, characterized by the open cylindrical ring, is in contrast to the packing of the DMPC molecules about one another in the bilayer in the absence of the nanotube, as shown in Figure 10.11b.

This study reveals that two substantial changes in the bilayer occur due to insertion of the carbon nanotube. First, it was stated that there is an alteration in the packing of the DMPC bilayer molecules which extends at least 18 Å from the nanotube and includes the creation of a relatively open, unoccupied cylindrical ring of 2–4 Å thickness directly around the nanotube. Second, the same bilayer structure, which undergoes the change in structural organization, also becomes much more rigid than in case when the nanotube is not inserted. Solvated calcium ions are predicted to preferentially transport through the

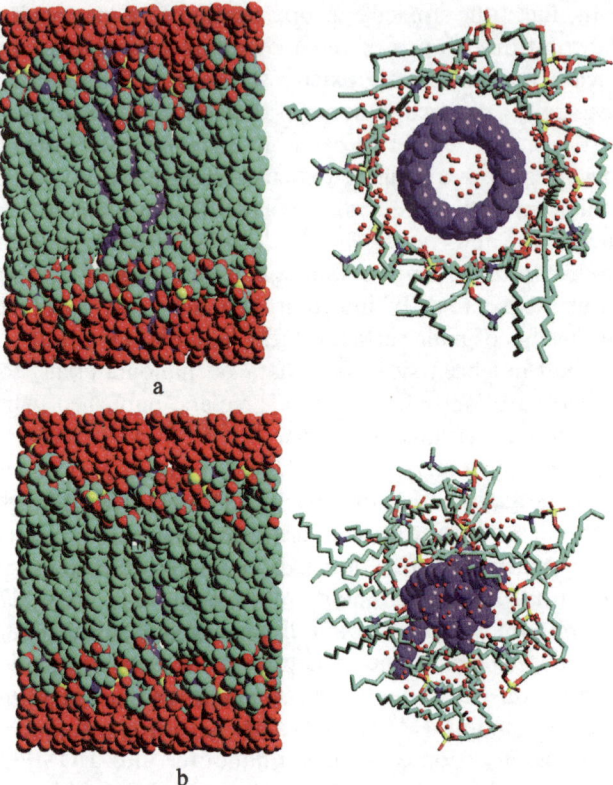

Figure 10.11   (a) Side view of the nanotube in the DMPC bilayer with waters on the top and bottom and all hydrogen atoms suppressed in the rendering. The nanotube is shown in violet. To the right, the top view of the nanotube in the DMPC bilayer with only the first layer of DMPC molecules around the nanotube shown. There is an obvious cylindrical hole formed around the nanotube. (b) Side view of center DMPC of the DMPC bilayer with waters on the top and bottom. All hydrogen atoms have been suppressed. The center DMPC is shown in violet. To the right, the top view of the center DMPC of the DMPC bilayer with only the first adjacent DMPC packing layer shown. There is no cylindrical hole around the center DMPC molecule. Reproduced with permission from ref. 128. Copyright American Chemical Society, 2008.

inserted nanotube as compared to hydrated sodium ions, but the solvated calcium ion also produces an alteration in the local bilayer structure as it passes through the nanotube. The authors find out that the total diffusion coefficient of ethanol through the membrane bilayer increases by about 35% in the presence of the inserted nanotube. Urea and caffeine also exhibit increases in their diffusion coefficients for transport through the bilayer, due to the inserted nanotube, but these increases are smaller than that of ethanol. It was stated that each of the three penetrants also diffuses more directly through the membrane bilayer in the presence of the nanotube, especially caffeine and urea.

The work reported in this paper represents the first attempt to identify possible sources of toxicity from a nanomaterial, namely a carbon nanotube inserted into a membrane bilayer. These possible sources of toxicity are all structural and/or dynamic properties of a biological system, which are computed to be significantly altered due to the presence of the nanotube. Thus, the authors proposed an indirect and non-validated scheme to permit the identification and prioritization of molecular properties which should be considered as potentially leading to toxicity from a nanomaterial.

The above-discussed work is quite interesting and gives helpful ideas for possible sources of nanoparticle toxicity; however, an absence of experimental data to compare leaves the obtained results unconfirmed and not really QSAR-like. In this regard, the second nano-QSAR study which was published two years later is fortunately based on experimental data, where authors studied manufactured nanoparticles (MNPs).[129] The authors have considered two representative sets of MNPs using *in vitro* cell-based assays: (i) 51 various MNPs with diverse metal cores (taken from ref. 131) and (ii) 109 MNPs with similar core, but different surface modifiers (taken from ref. 132). Both of these studies can be regarded as separate QSAR investigations, as it was stated in our review (ref. 66).

In case study (i), the structure of MNPs was characterized by their experimental properties treated as molecular descriptors. Conversely, the authors pointed out that case study (ii) could be considered as a conventional QSAR investigation since 109 MNPs with the same metal core (analogous to s common chemical scaffold for organic molecules) were characterized by conventional chemical descriptors of surface-modifying organic molecules. In case study (i) five types of MNPs were represented in the data set: each MNP had a metal core, that is, Fe_2O_3-predominant, Fe_3O_4-predominant, CdSe, or Fe(III); an organic coat, either acidic, basic, amphiphilic, or lipophilic; and various surface modifiers for some of the MNPs. To further demonstrate the overall feasibility of quantitative nanostructure–activity relationship (QNAR) modeling, the authors used experimentally measured physical parameters (descriptors) of MNPs to build binary classification models (*i.e.*, models capable of assigning MNPs to one of two distinct classes defined by their biological activity). Four such structural descriptors were available for 44 of the 51 MNPs: nanoparticle size, ranging from 20 to 74 nm, R1 and R2 relaxivities representing their magnetic properties, and ζ-potential representing the intensity of charge on their surface. On the other hand, the entire biological activity profile included 64 features, that is, a total number of all possible combinations of four doses, four cell lines, and four assays. In result, the built classification support vector machine (SVM) models had relatively high external prediction accuracies of 56–88% for the five independent external validation sets, with the mean external accuracy as high as 73%.

In case study (ii), where all MNPs included in the second set possessed exactly the same metal core, the structure of small organic molecule conjugated to the MNP surface was the only difference from one MNP to another.

Therefore, the authors treated this set as set of conventional organic compounds and calculated 150 MOE descriptors for all 109 organic compounds. The obtained results showed that prediction accuracies expressed as coefficients of correlation R_{abs}^2 ranged from 0.65 to 0.80 for external sets. The authors tried to interpret the descriptors in the final model and they noticed that several descriptors such as SlogP_VSA1, SlogP_VSA2, and SlogP_VSA5 represent different aspects of van der Waals surface area's contribution to compound's lipophilicity; another relatively frequent descriptor was "b_double" (representing the number of double bonds in a molecule). In the publication the authors stated: "these findings imply that the cellular behavior of a nanoparticle library based on a common core can be predicted using QNAR analysis of the surface-modifying ligands, and thus that rational design of organic compounds attached to the surface of MNPs is possible using QNAR models and descriptor analysis".

Interestingly, in the above-described study the QNAR models were built using machine learning approaches such as SVM-based classification and *k*-nearest neighbors (kNN)-based regression. The results suggest that the QNAR models can be employed for: (i) predicting biological activity profiles of novel nanomaterials, and (ii) prioritizing the design and manufacturing of nanomaterials toward better and safer products.

Recently, another nanoparticle-toxicity modeling work has been published. Puzyn *et al.*[130] in an interdisciplinary group (experimental biology, cheminformatics and computational chemists) conducted an experimental toxicity evaluation of 17 metal oxide nanoparticles and developed a nanoparticle-toxicity model, nano-QSAR, to find some important mechanistic correlations and predict toxicity for new, untested metal oxide nanoparticles. This work differs from all previous nano-QSAR works because (i) the group gathered their own experimental data; (ii) the group considered commonly applied nanoparticles – engineered metal oxide nanoparticles; and (iii) the group calculated only theoretical descriptors and developed a transparent model that can be easily tested by a reader.

The interdisciplinary team explored the *in vitro* toxicity of investigated nanoparticles on pathogen bacteria cultures (*Escherichia coli*). The studied metal oxides include: ZnO, TiO_2, SnO_2, CuO, La_2O_3, Fe_2O_3, Al_2O_3, Y_2O_3, Bi_2O_3, ZrO_2, SiO_2, In_2O_3, V_2O_3, CoO, NiO, Cr_2O_3, and Sb_2O_3 , with sizes ranging from 15 to 90 nm. The cytotoxicity of the nanoparticles was expressed in terms of the logarithmic values of molar $1/EC_{50}$ (the effective concentration of a given oxide that reduces bacterial viability by 50%). The 13 nanoparticles, including those for which toxicity data were taken from the authors' previous paper[133] (ZnO, CuO, Al_2O_3, Fe_2O_3, SnO_2, TiO_2) and those tested in Batch I (V_2O_3, Y_2O_3, Bi_2O_3, In_2O_3, Sb_2O_3, SiO_2, ZrO_2), were split into two sets, the training set (T) and validation set (V1) (Table 1 of ref. 133), ensuring that the points from V1 were evenly distributed within the range of toxicity of the training set compounds (T). Three compounds tested in Batch II (CoO, NiO, Cr_2O_3) and La_2O_3 were also then included in a validation set (V2). Based on this split dataset the authors were able to develop a nano-QSAR model, based on quantum-mechanically calculated descriptors. This model and experimental data were starting points for the hypothesis on the most probable mechanism for the cytotoxicity of these nanoparticles. Utilizing the toxicity data and structural descriptors, a simple but statistically significant ($F = 45.4$, $P = 0.0001$) model was developed. The obtained nano-QSAR equation (10.17) was based on only one descriptor that allows successfully predict the cytotoxicity of the considered metal oxide nanoparticles.

$$\log(1/EC_{50}) = 2.59 - 0.50 \ H_{Me+} \tag{10.17}$$

It is important to point out that the descriptor ΔH_{Me+}, which was selected as the best one, represents the enthalpy of formation of a gaseous cation having the same oxidation state as that in the metal oxide structure: see eqn (10.18).

$$Me(s) \rightarrow Me^{n+}(g) + n \ \bar{e} \Delta H_{Me+} \tag{10.18}$$

The completed list of the calculated molecular descriptors and details of the QSAR modeling procedure, including splitting for a training and validation set, data pre-processing, the method of modeling, internal validation, measuring goodness-of-fit and robustness, external validation of predictive ability and applicability domain can be found in the published paper.[130] The obtained results indicated that ΔH_{Me+} can be utilized as an efficient descriptor of the chemical stability (or reactivity) of metal oxides and, therefore, their cytotoxicity in *E. coli in vitro* tests. They noted that during the development of the model, various parameters were tested that were, in some cases, more interpretative than ΔH_{Me+}. For example, the tests included ΔH_L (lattice enthalpy), which describes the dissolution of nanoparticles without oxidation or reduction of the cation, and the electronic properties (energies of HOMO and LUMO) of the oxides, which describe their redox properties. In all these tests, the correlation between the tested descriptors and cytotoxicity was unsatisfactory. One important point is that the model was developed on the assumption that all metal oxide nanoparticles within the size range 15–90 nm (as in experiments) from a quantum-mechanical point of view do not possess drastic changes in properties within one type of nanoparticle, for example in electronic properties. Puzyn et al. stated that large, size-dependent change of some electronic properties (*i.e.*, ionization potential and electron affinity) for metal oxide particles occur below the size of about 5 nm. Thus, the property value variation with the increasing size of the nanomaterial does not occur beyond so-called a saturation point – starting from which the property value does not change. Therefore, the authors assumed that the physico-chemical property variations between 15 and 90 nm are negligible in the considered case.

A plot of experimentally determined *versus* predicted log values of $1/EC_{50}$ is presented in Figure 10.12. The straight green line represents a perfect agreement between experimental and calculated values. The model shows an

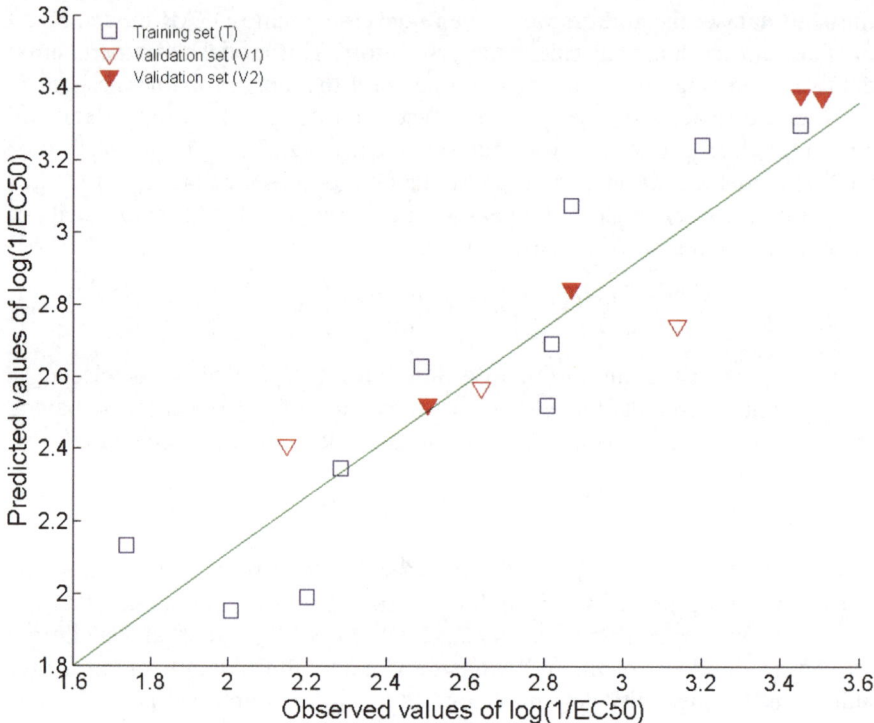

Figure 10.12 Plot of experimentally determined (observed) *versus* predicted log values of $1/EC_{50}$. Squares represent values predicted for the metal oxides from the training set; triangles represent data calculated for metal oxides from the validation sets. Reproduced from ref. 130. Copyright Nature Publishing Group, 2011.

excellent agreement between the observed toxicity values and those predicted by the nano-QSAR model, for the metal oxides from the training set (squares) and those from the validation sets (triangles). The following statistical results were stated: squared regression coefficient, $r^2 = 0.85$; cross-validated regression coefficient, $q_{CV}^2 = 0.77$; externally validated regression coefficient $q_{ext}^2 = 0.83$; root-mean-square error of calibration (RMSEC) = 0.20, of cross-validation (RMSECV) = 0.24, and of external prediction (RMSEP) = 0.19.

Based on a relatively large number of metal oxide nanoparticles (the largest ever tested in one laboratory) it can be concluded that the discussed study combines efficient experimental testing and reliable computational modeling methodologies to develop a mechanism of the cytotoxicity of metal oxide nanoparticles on *E. coli*. This combined study resulted in developed of an interpretative nano-QSAR model (see eqn (10.17)) that reliably predicts the toxicity of considered metal oxides and provides the foundations for the theoretical evaluation of the toxicity of untested nanomaterials, particularly metal oxide nanoparticles. Importantly, this study resulted in the hypothesis that mechanistically explains differences in toxicity between individual nanoparticle oxides. It is quite obvious that such a study can now trigger many new investigations of the toxicity and biological activity of various nanoparticles.

There is another recently published paper where the authors offer a hypothetical model for predicting the toxicity of high aspect ratio nanoparticles (HARNs).[134] In fact, quite interestingly, they did not develop a mathematical model but rather supply a list of rules useful for occupational and environmental health protection. Certainly, this model might be valuable to predict a potential occupational toxicity of HARNs, especially when coupled with the development of a computational tool based on the model, as was stated in the paper.

To conclude, several nano-QSAR models have been recently offered by different scientific groups. Each model has a number of advantages, as well as some negative aspects regarding a possible prediction ability, experimentally-confirmed models availability and accessibility for possible public use. As stated before, it is obvious that these studies can now trigger many new investigations (both experimental and computational) which would assist predictions of the toxicity and biological activity of various nanoparticles.

10.9 Applications of Nano-QSAR for Biological Activities

A few nano-QSAR studies have been performed recently where a SMILES-based optimal descriptors approach was utilized regarding the anti-HIV activity of fullerene derivatives.[135–137] These studies showed great potential for predicting the pharmacological activity of nanoparticles. However, the pharmacological applications of nanomaterials are currently in the early stages and the experimental data for pharmacological activity is even more scarce than those for toxicity. Therefore, the discussion of nano-QSAR applications is limited here only to physico-chemical properties and toxicity predictions. Nevertheless, we strongly believe that with the fast progress in applications of various nanomaterials their biological activities will also receive due attention in the near future.

10.10 Conclusions

Nanomaterials are becoming an important component of modern life and have been the subject of an increasing number of various investigations involving their physico-chemical properties and toxicity. Characterization of the risks posed by nanomaterials is extraordinarily complex because these materials can have a wide range of sizes, shapes, chemical compositions and surface modifications, all of which may affect toxicity.[138] There is an urgent need for a testing strategy that can rapidly and efficiently provide a screening approach for evaluating the potential hazard of nanomaterials. Predictive toxicity models could form an integral component of such an approach by evaluating which nanomaterials, as a result of their physico-chemical properties, possess potentially hazardous characteristics. However, experimental data and theoretical modeling results for physico-chemical and biological activity of nanomaterials are still very scarce. Therefore, the development and application of QSAR predictions for the nanomaterials is a very complex task, because of the "non-classical" structure of nanomaterials and relatively large size of nanospecies from computational point of view. This chapter concludes that an application of the QSAR methods for nanomaterials is nevertheless possible and is vital for modeling and predicting their various properties and activities, including toxicity. Recent advances in this field are revealed, for example how the physico-chemical properties (such as solubility in various organic solvents, solubility in water, Young modulus and other properties) can be predicted for nanomaterials. A few current modeling studies carried out to predict the toxicity of nanoparticles are also discussed to stress that toxicity also can be modeled and predicted for the series of nanoparticles with various cores and organic covers and, for the series of metal oxide nanoparticles, by applying the quantum-chemical methods in combination with the nano-QSAR models.

Nanotechnology and nanotoxicology are relatively new fields. In the case of nanotoxicology, the concepts of dosimetry, dose metrics, exposure assessment and risk characterization are still being assessed.[11] Therefore, the research related to nanoparticles' property and toxicity assessment by computational methods is only in its very early stage. We believe that many cutting-edge developments in predicting nanomaterial properties can be expected soon, perhaps leading towards generating a reliable nano-QSAR methodology within the next few years.

Acknowledgements

The authors are thankful for the support of the NSF-CREST Interdisciplinary Nanotoxicity Center NSF-CREST – Grant # HRD-0833178; NSF-EPSCoR Award #: 362492-190200-01\NSFEPS-0903787. They also thank the Department of Defense through the U. S. Army Engineer Research and Development Center, Vicksburg, MS, for generous support through the following contract: High Performance Computational Design of Novel Materials (HPCDNM) – Contract #W912HZ-06-C-0057.

This work was supported by the Polish National Science Center (grant no. 557-8180-1360-11).

A.G. thanks the European Social Fund, the State Budget and the Pomorskie Voivodeship Budget according to the Operational Programme Human Capital, Priority VIII, Action 8.2, Under-action 8.2.2: 'Regional Innovation Strategy' for granting her with a fellowship in frame of the project "InnoDoktorant — Scholarships for PhD students, IVth edition".

References

1. G. Oberdörster, V. Stone and K. Donaldson, *Nanotoxicology*, 2007, **1**, 2–25.
2. R. Shenhar and V. M. Rotello, *Acc. Chem. Res.*, 2003, **36**, 549–561.

3. A. S. Edelstein and R. C. Cammarata, *Nanomaterials: Synthesis, Properties and Applications*, CRC Press, 1998.
4. *Metal Nanoparticles: Synthesis, Characterization, and Applications*, ed. D. L. Feldheim and C. A. Foss, CRC Press, New York, 2001.
5. K. J. Klabunde, *Nanoscale Materials in Chemistry*, John Wiley & Sons, Inc., New York, 2001.
6. G. Oberdörster, E. Oberdörster and J. Oberdörster, *Environ. Health Perspect.*, 2005, **113**, 823–839.
7. D. Srivastava, M. Menon and K. Cho, *Comput. Sci. Eng.*, 2001, **3**, 42–55.
8. M. N. Moore, *Environ. Int.*, 2006, **32**, 967–976.
9. M. J. Pitkethly, *Materials Today*, 2004, **7**, 20–29.
10. T. Puzyn, D. Leszczynska and J. Leszczynski, *Small*, 2009, **5**, 2494–2509.
11. P. Rivera Gil, G. Oberdörster, A. Elder, V. Puntes and W. J. Parak, *ACS Nano*, 2010, **4**, 5527–5531.
12. M. Auffan, J. Rose, J. Y. Bottero, G. V. Lowry, J. P. Jolivet and M. R. Wiesner, *Nat. Nanotechnol.*, 2009, **4**, 634–641.
13. R. D. Handy and B. J. Shaw, *Health Risk Soc.*, 2007, **9**, 125–144.
14. G. Jiang, Z. Shen, J. Niu, L. Zhuang and T. He, *Prog. Chem.*, 2011, **23**, 1769–1781.
15. J. Kreuter, *Int. J. Pharm.*, 2007, **331**, 1–10.
16. E. K. Rushton, G. Oberdörster and J. N. Finkelstein, Materials Research Society Fall Meeting, Boston, MA, 2006.
17. J. S. Tsuji, A. D. Maynard, P. C. Howard, J. T. James, C.-W. Lam, D. B. Warheit and A. B. Santamaria, *Toxicol. Sci.*, 2006, **89**, 42–50.
18. M. S. Bonnefoi, S. E. Belanger, D. J. Devlin, N. G. Doerrer, M. R. Embry, S. Fukushima, E. S. Harpur, R. N. Hines, M. P. Holsapple, J. H. Kim, J. S. MacDonald, R. O'Lone, S. D. Pettit, J. L. Stevens, A. S. Takei, S. S. Tinkle and J. W. Van Der Laan, *Crit. Rev. Toxicol.*, 2010, **40**, 893–911.
19. T. Puzyn, A. Gajewicz, D. Leszczynska and J. Leszczynski, in *Recent Advances in QSAR Studies: Methods and Applications*, ed. T. Puzyn and J. Leszczynski, M.T. Springer, London, New York, 2010, **8**, 383–409.
20. A. A. Toropov and E. Benfenati, *Curr. Drug Discovery Technol.*, 2007, **4**, 77–116.
21. A. E. Nel, L. Mädler, D. Velegol, T. Xia, E. M. V. Hoek, P. Somasundaran, F. Klaessig, V. Castranova and M. Thompson, *Nat. Mater.*, 2009, **8**, 543–557.
22. V. Aruoja, H. C. Dubourguier, K. Kasemets and A. Kahru, *Sci. Total Environ.*, 2009, **407**, 1461–1468.
23. P. V. Asharani, Y. Lian Wu, Z. Gong and S. Valiyaveettil, *Nanotechnology*, 2008, **19**, 255102.
24. P. V. Asharani, Y. Lian Wu, Z. Gong and S. Valiyaveettil, *Nanotoxicology*, 2011, **5**, 43–54.
25. W. Bai, Z. Zhang, W. Tian, X. He, Y. Ma, Y. Zhao and Z. Chai, *J. Nanopart. Res.*, 2010, **12**, 1645–1654.

26. H.-W. Chen, S.-F. Su, C.-T. Chien, W.-H. Lin, S.-L. Yu, C.-C. Chou, J. J. W. Chen and P.-C. Yang, *FASEB J.*, 2006, **20**, 2393–2395.

27. W. S. Cho, M. Cho, J. Jeong, M. Choi, H. Y. Cho, B. S. Han, S. H. Kim, H. O. Kim, Y. T. Lim and B. H. Chung, *Toxicol. Appl. Pharmacol.*, 2009, **236**, 16–24.

28. R. Duffin, L. Tran, D. Brown, V. Stone and K. Donaldson, *Inhalation Toxicol.*, 2007, **19**, 849–856.

29. G. Federici, B. J. Shaw and R. D. Handy, *Aquat. Toxicol.*, 2007, **84**, 415–430.

30. N. M. Franklin, N. J. Rogers, S. C. Apte, G. E. Batley, G. E. Gadd and P. S. Casey, *Environ. Sci. Technol.*, 2007, **41**, 8484–8490.

31. C. M. Goodman, C. D. McCusker, T. Yilmaz and V. M. Rotello, *Bioconjugate Chem.*, 2004, **15**, 897–900.

32. S. M. Hussain, K. L. Hess, J. M. Gearhart, K. T. Geiss and J. J. Schlager, *Toxicol. In Vitro*, 2005, **19**, 975–983.

33. K. Kasemets, A. Ivask, H. C. Dubourguier and A. Kahru, *Toxicol. In Vitro*, 2009, **23**, 1116–1122.

34. M. Li, L. Zhu and D. Lin, *Environ. Sci. Technol.*, 2011, **45**, 1977–1983.

35. T. Li, B. Albee, M. Alemayehu, R. Diaz, L. Ingham, S. Kamal, M. Rodriguez and S. Whaley Bishnoi, *Anal. Bioanal. Chem.*, 2010, **398**, 689–700.

36. M. Mortimer, K. Kasemets and A. Kahru, *Toxicology*, 2010, **269**, 182–189.

37. J. Wu, W. Liu, C. Xue, S. Zhou, F. Lan, L. Bi, H. Xu, X. Yang and F. D. Zeng, *Toxicol. Lett.*, 2009, **191**, 1–8.

38. X. Y. Yang, R. E. Edelmann and J. T. Oris, *Aquat. Toxicol.*, 2010, **100**, 202–210.

39. Y. Zhang, W. Chen, J. Zhang, J. Liu, G. Chen and C. Pope, *J. Nanosci. Nanotechnol.*, 2007, **7**, 497–503.

40. L. Braydich-Stolle, S. Hussain, J. J. Schlager and M.-C. Hofmann, *Toxicol. Sci.*, 2005, **88**, 412–419.

41. O. R. Moss and V. A. Wong, *Inhalation Toxicol.*, 2006, **18**, 711–716.

42. G. Oberdörster, *Environ. Health Perspect.*, 2012, 120.

43. T. Cedervall, I. Lynch, S. Lindman, T. Berggård, E. Thulin, H. Nilsson, K. A. Dawson and S. Linse, *Proc. Natl. Acad. Sci. U. S. A.*, 2007, **104**, 2050–2055.

44. J.-K. Lee and M. H. Cho, *Toxicol. Sci.*, 2006, **89**, 338–347.

45. O. Choi and Z. Hu, *Environ. Sci. Technol.*, 2008, **42**, 4583–4588.

46. N. Lubick, *Environ. Sci. Technol.*, 2008, **42**, 8617.

47. A. M. Schrand, M. F. Rahman, S. M. Hussain, J. J. Schlager, D. A. Smith and A. F. Syed, *Wiley Interdiscip. Rev.: Nanomed. Nanobiotechnol.*, 2010, **2**, 544–568.

48. M. H. Abraham, C. E. Green and W. E. Acree, Jr., *Perkin 2*, 2000, 281–286.

49. J. L. Ormsby and B. T. King, *J. Org. Chem.*, 2004, **69**, 4287–4291.

50. A. A. Toropov, D. Leszczynska and J. Leszczynski, *Mater. Lett.*, 2007, **61**, 4777–4780.

51. N. M. Schaeublin, L. K. Braydich-Stolle, A. M. Schrand, J. M. Miller, J. Hutchison, J. J. Schlager and S. M. Hussain, *Nanoscale*, 2011, **3**, 410–420.

52. J. Jiang, G. Oberdörster and P. Biswas, *J. Nanopart. Res.*, 2009, **11**, 77–89.
53. K. W. Powers, M. Palazuelos, B. M. Moudgil and S. M. Roberts, *Nanotoxicology*, 2007, **1**, 42–51.
54. A. S. Barnard, *J. Mater. Chem.*, 2006, **16**, 813–815.
55. S. J. L. Billinge and I. Levin, *Science*, 2007, **316**, 561–565.
56. S. T. Bromley, I. D. P. R. Moreira, K. M. Neyman and F. Illas, *Chem. Soc. Rev.*, 2009, **38**, 2657–2670.
57. T. Çağin, J. Che, Y. Qi, Y. Zhou, E. Demiralp, G. Gao and W. A. Goddard Iii, *J. Nanopart. Res.*, 1999, **1**, 51–69.
58. P. Heino, *Microsyst. Technol.*, 2009, **15**, 75–81.
59. G. C. Schatz, *Proc. Natl. Acad. Sci. U. S. A.*, 2007, **104**, 6885–6892.
60. A. S. Barnard, *Rep. Prog. Phys.*, 2010, 73.
61. G. Bystrzejewska-Piotrowska, J. Golimowski and P. L. Urban, *Waste Manage.*, 2009, **29**, 2587–2595.
62. A. M. El Badawy, R. G. Silva, B. Morris, K. G. Scheckel, M. T. Suidan and T. M. Tolaymat, *Environ. Sci. Technol.*, 2011, **45**, 283–287.
63. B. Fubini, M. Ghiazza and I. Fenoglio, *Nanotoxicology*, 2010, **4**, 347–363.
64. S. Gill, R. Lobenberg, T. Ku, S. Azarmi, W. Roa and E. J. Prenner, *J. Biomed. Nanotechnol.*, 2007, **3**, 107–119.
65. R. Flunt, R. Makitra, O. Makohon and J. Pyrih, *Org. React. (Tartu)*, 1997, **31**, 21–26.
66. A. Gajewicz, B. Rasulev, T. C. Dinadayalane, P. Urbaszek, T. Puzyn, D. Leszczynska and J. Leszczynski, Advancing risk assessment of engineered nanomaterials: Application of computational approaches, *Adv. Drug Deliver Rev.*, 2012, 64, DOI:10.1016/j.addr.2012.05.014
67. C. Medina, M. J. Santos-Martinez, A. Radomski, O. I. Corrigan and M. W. Radomski, *Br. J. Pharmacol.*, 2007, **150**, 552–558.
68. P. Gramatica, *Chem. Today*, 1991, 18–24.
69. D. K. Agrafiotis, W. Cedeño and V. S. Lobanov, *J. Chem. Inf. Comput. Sci.*, 2002, **42**, 903–911.
70. L. Eriksson, E. Johansson, F. Lindgren, M. Sjöström and S. Wold, *J. Comput.-Aided Mol. Des.*, 2002, **16**, 711–726.
71. D. M. Hawkins, S. C. Basak and X. Shi, *J. Chem. Inf. Comput. Sci.*, 2001, **41**, 663–670.
72. H. Kubinyi, *Drug Discov. Today*, 1997, **2**, 538–546.
73. M. Pérez González, C. Terán, L. Saíaz-Urra and M. Teijeira, *Curr. Top. Med. Chem.*, 2008, **8**, 1606–1627.
74. R. Perkins, H. Fang, W. Tong and W. J. Welsh, *Environ. Toxicol. Chem.*, 2003, **22**, 1666–1679.
75. D. B. Boyd, in Reviews in Computational Chemistry, K. B. Lipkowitz and D. B. Boyd, Eds., VCH Publishers, New York, 1990, **1**, 355–371.
76. J. A. Schwarz and C. I. Contescu, *Surfaces of Nanoparticles and Porous Materials*, CRC Press, New York, 1999.
77. P. Hohenberg, and W. Kohn, *Phys. Rev.*, 1964, **136**, 864–871.

78. S. M. Danauskas and P. C. Jurs, *J. Chem. Inf. Comput. Sci.*, 2001, **41**, 419–424.

79. H. Liu, X. Yao, R. Zhang, M. Liu, Z. Hu and B. Fan, *J. Phys. Chem. B*, 2005, **109**, 20565–20571.

80. T. Petrova, B. F. Rasulev, A. A. Toropov, D. Leszczynska and J. Leszczynski, *J. Nanopart. Res.*, 2011, **13**(8), 3235–3247.

81. N. Sivaraman, T. G. Srinivasan, P. R. V. Rao and R. Natarajan, *J. Chem. Inf. Comput. Sci.*, 2001, **41**, 1067–1074.

82. A. L. Smith, L. Y. Wilson and G. R. Famini, *Proc. Electrochem. Soc.*, 1996, **96–10**, 53–62.

83. A. A. Toropov, B. F. Rasulev, D. Leszczynska and J. Leszczynski, *Chem. Phys. Lett.*, 2007, **444**, 209–214.

84. A. A. Toropov, B. F. Rasulev, D. Leszczynska and J. Leszczynski, *Chem. Phys. Lett.*, 2008, **457**, 332–336.

85. A. A. Toropov, A. P. Toropova, E. Benfenati, D. Leszczynska and J. Leszczynski, *J. Math. Chem.*, 2009, **46**, 1232–1251.

86. M. H. Abraham, C. M. Du, J. W. Grate, R. A. McGill and W. J. Shuely, *J. Chem. Soc., Chem. Commun.*, 1993, 1863–1864.

87. F. H. Quina, E. O. Alonso and J. P. S. Farah, *J. Phys. Chem.*, 1995, **99**, 11708–11714.

88. Y. H. Zhao, J. Le, M. H. Abraham, A. Hersey, P. J. Eddershaw, C. N. Luscombe, D. Boutina, G. Beck, B. Sherborne, I. Cooper and J. A. Platts, *J. Pharm. Sci.*, 2001, **90**, 749–784.

89. Y. Marcus, A. L. Smith, M. V. Korobov, A. L. Mirakyan, N. V. Avramenko and E. B. Stukalin, *J. Phys. Chem. B*, 2001, **105**, 2499–2506.

90. C. H. Hansen and A. L. Smith, *Carbon*, 2004, **42**, 1591–1597.

91. C. M. Hansen, *Prog. Org. Coat.*, 2004, **51**, 77–84.

92. J.-C. Huang, *Fluid Phase Equilib.*, 2005, **237**, 186–192.

93. E. R. Thomas, B. A. Newman, T. C. Long, D. A. Wood and C. A. Eckert, *J. Chem. Eng. Data*, 1982, **27**, 399–405.

94. A. A. Toropov, D. Leszczynska and J. Leszczynski, *Chem. Phys. Lett.*, 2007, **441**, 119–122.

95. A. A. Toropov, D. Leszczynska and J. Leszczynski, *Comput. Biol. Chem.*, 2007, **31**, 127–128.

96. D. Vidal, M. Thormann and M. Pons, *J. Chem. Inf. Model.*, 2005, **45**, 386–393.

97. A. Toropov, K. Nesmerak, I. Raska, K. Waisser and K. Palat, *Comput. Biol. Chem.*, 2006, **30**, 434–437.

98. A. A. Toropov and E. Benfenati, *Comput. Biol. Chem.*, 2007, **31**, 57–60.

99. A. A. Toropov, A. P. Toropova, D. V. Mukhamedzhanova, I. Gutman and A. Chemistry, *Indian J. Chem., Sect. A: Inorg., Bio-inorg., Phys., Theor. Anal. Chem.*, 2005, **44**, 1545–1552.

100. A. P. Toropova, A. A. Toropov, E. Benfenati, G. Gini, D. Leszczynska and J. Leszczynski, *Mol. Diversity*, 2011, **15**, 249–256.

101. Unofficial InChI FAQ, http://wwmm.ch.cam.ac.uk/inchifaq.

102. Talete srl, DRAGON (Software for Molecular Descriptor Calculation) Version 3.0 – 2003 – http://www.talete.mi.it/.
103. M. J. Frisch, G. W. Trucks, H. B. Schlegel, G. E. Scuseria, M. A. Robb, J. R. Cheeseman, J. A. Montgomery, Jr., T. Vreven, K. N. Kudin, J. C. Burant, J. M. Millam, S. S. Iyengar, J. Tomasi, V. Barone, B. Mennucci, M. Cossi, G. Scalmani, N. Rega, G. A. Petersson, H. Nakatsuji, M. Hada, M. Ehara, K. Toyota, R. Fukuda, J. Hasegawa, M. Ishida, T. Nakajima, Y. Honda, O. Kitao, H. Nakai, M. Klene, X. Li, J. E. Knox, H. P. Hratchian, J. B. Cross, V. Bakken, C. Adamo, J. Jaramillo, R. Gomperts, R. E. Stratmann, O. Yazyev, A. J. Austin, R. Cammi, C. Pomelli, J. W. Ochterski, P. Y. Ayala, K. Morokuma, G. A. Voth, P. Salvador, J. J. Dannenberg, V. G. Zakrzewski, S. Dapprich, A. D. Daniels, M. C. Strain, O. Farkas, D. K. Malick, A. D. Rabuck, K. Raghavachari, J. B. Foresman, J. V. Ortiz, Q. Cui, A. G. Baboul, S. Clifford, J. Cioslowski, B. B. Stefanov, G. Liu, A. Liashenko, P. Piskorz, I. Komaromi, R. L. Martin, D. J. Fox, T. Keith, M. A. Al-Laham, C. Y. Peng, A. Nanayakkara, M. Challacombe, P. M. W. Gill, B. Johnson, W. Chen, M. W. Wong, C. Gonzalez, and J. A. Pople, *Gaussian 03, Revision E.01*, Gaussian, Inc., Wallingford CT, 2004.
104. R. Todeschini and V. Consonni, *Handbook of Molecular Descriptors*, Wiley-VCH, Weinheim, 2000.
105. M. Goodarzi, P. R. Duchowicz, M. P. Freitas and F. M. Fernández, *Fluid Phase Equilib.*, 2010, **293**, 130–136.
106. E. Pourbasheer, S. Riahi, M. R. Ganjali and P. Norouzi, *Fullerenes, Nanotubes, Carbon Nanostruct.*, 2011, **19**, 585–598.
107. F. Torrens, *Nanotechnology*, 2005, **16**, S181–S189.
108. F. Torrens, *Mol. Simul.*, 2005, **31**, 107–114.
109. A. A. Toropov and J. Leszczynski, *Chem. Phys. Lett.*, 2006, **433**, 125–129.
110. E. Mohammaadpour, M. Awang and M. Z. Abdullah, *J. Appl. Sci.*, 2011, **11**, 1653–1657.
111. G. Wang and X. Li, *J. Appl. Phys.*, 2008, 104.
112. G. Y. Jing, H. L. Duan, X. M. Sun, Z. S. Zhang, J. Xu, Y. D. Li, J. X. Wang and D. P. Yu, *Phys. Rev. B*, 2006, 73.
113. B. Wu, A. Heidelberg and J. J. Boland, *Nat. Mater.*, 2005, **4**, 525–529.
114. B. X. Chen, S. R. Wilson, M. Das, D. J. Coughlin and B. F. Erlanger, *Proc. Natl. Acad. Sci. U. S. A.*, 1998, **95**, 10809–10813.
115. H. Tokuyama, S. Yamago, E. Nakamura, T. Shiraki and Y. Suguira, *J. Am. Chem. Soc.*, 1993, **115**, 7918–7919.
116. R. Sijbesma, G. Srdanov, F. Wudl, J. A. Castoro, C. Wilkens, S. H. Friedman, D. L. Decamp and G. L. Kenyon, *J. Am. Chem. Soc.*, 1993, **115**, 6510–6514.
117. S. H. Friedman, P. S. Ganapathi, Y. Rubin and G. L. Kenyon, *J. Med. Chem.*, 1998, **41**, 2424–2429.
118. G. L. Marcorin, T. Da Ros, S. Castellano, G. Stefancich, I. Bonin, S. Miertus and M. Prato, *Org. Lett.*, 2000, **2**, 3955–3958.

119. G. Oberdörster, A. Maynard, K. Donaldson, V. Castranova, J. Fitzpatrick, K. Ausman, J. Carter, B. Karn, W. Kreyling, D. Lai, S. Olin, N. Monteiro-Riviere, D. Warheit and H. Yang, *Part. Fibre Toxicol.*, 2005, 2.

120. M. Heinlaan, A. Ivask, I. Blinova, H. C. Dubourguier and A. Kahru, *Chemosphere*, 2008, **71**, 1308–1316.

121. H. Wang, R. L. Wick and B. Xing, *Environ. Pollut.*, 2009, **157**, 1171–1177.

122. T. Xia, M. Kovochich, J. Brant, M. Hotze, J. Sempf, T. Oberley, C. Sioutas, J. I. Yeh, M. R. Wiesner and A. E. Nel, *Nano Lett.*, 2006, **6**, 1794–1807.

123. I. Blinova, A. Ivask, M. Heinlaan, M. Mortimer and A. Kahru, *Environ. Pollut.*, 2010, **158**, 41–47.

124. E. Navarro, F. Piccapietra, B. Wagner, F. Marconi, R. Kaegi, N. Odzak, L. Sigg and R. Behra, *Environ. Sci. Technol.*, 2008, **42**, 8959–8964.

125. K. M. Reddy, K. Feris, J. Bell, D. G. Wingett, C. Hanley and A. Punnoose, *Appl. Phys. Lett.*, 2007, 90.

126. F. Rispoli, A. Angelov, D. Badia, A. Kumar, S. Seal and V. Shah, *J. Hazard. Mater.*, 2010, **180**, 212–216.

127. S. M. Cook, W. G. Aker, B. F. Rasulev, H. M. Hwang, J. Leszczynski, J. J. Jenkins and V. Shockley, *J. Hazard. Mater.*, 2010, **176**, 367–373.

128. J. Liu and A. J. Hopfinger, *Chem. Res. Toxicol.*, 2008, **21**, 459–466.

129. D. Fourches, D. Pu, C. Tassa, R. Weissleder, S. Y. Shaw, R. J. Mumper and A. Tropsha, *ACS Nano*, 2010, **4**, 5703–5712.

130. T. Puzyn, B. Rasulev, A. Gajewicz, X. Hu, T. P. Dasari, A. Michalkova, H.-M. Hwang, A. Toropov, D. Leszczynska and J. Leszczynski, *Nat. Nanotechnol.*, 2011, **6**, 175–178.

131. S. Y. Shaw, E. C. Westly, M. J. Pittet, A. Subramanian, S. L. Schreiber and R. Weissleder, *Proc. Natl. Acad. Sci. U. S. A.*, 2008, **105**, 7387–7392.

132. R. Weissleder, K. Kelly, E. Y Sun, T. Shtatland and L. Josephson, *Nat. Biotechnol.*, 2005, **23**, 1418–1423.

133. X. Hu, S. Cook, P. Wang and H. M. Hwang, *Sci. Total Environ.*, 2009, **407**, 3070–3072.

134. C. L. Tran, R. Tantra, K. Donaldson, V. Stone, S. M. Hankin, B. Ross, R. J. Aitken and A. D. Jones, *J. Nanopart. Res.*, 2011, 1–16.

135. A. A. Toropov, A. P. Toropova, E. Benfenati, D. Leszczynska and J. Leszczynski, *Eur. J. Med. Chem.*, 2010, **45**, 1387–1394.

136. A. A. Toropov, A. P. Toropova, E. Benfenati, D. Leszczynska and J. Leszczynski, *J. Comput. Chem.*, 2010, **31**, 381–392.

137. A. P. Toropova, A. A. Toropov, E. Benfenati, D. Leszczynska and J. Leszczynski, *J. Math. Chem.*, 2010, **48**, 959–987.

138. K. A. Clark, R. H. White and E. K. Silbergeld, *Regul. Toxicol. Pharmacol.*, 2011, **59**, 361–363.

CHAPTER 11

Development and Evaluation of Structure–Reactivity Models for Predicting the In Vitro Oxidative Stress of Metal Oxide Nanoparticles

ENRICO BURELLO* AND ANDREW WORTH

Systems Toxicology Unit, Institute for Health and Consumer Protection, Joint Research Centre, European Commission, Ispra, Varese, Italy
*E-mail: enricoburello@yahoo.com

11.1 Introduction

Under physiological conditions, cells maintain a reduced intracellular state. This process is the result of a balance between the levels of oxidized and reduced species present in the cell. Oxidizing substances can create an imbalance in this state, for example by depleting electrons from aqueous redox species or by acting like catalysts. The overall result is the decrease of antioxidants and/or an increase in the production of reactive oxygen species (ROS), a cellular condition which eventually evolves into inflammatory and cytotoxic responses. ROS include hydrogen peroxide and superoxide ions; antioxidants are molecules such as vitamin C and glutathione which scavenge the unwanted oxidants.

RSC Nanoscience & Nanotechnology No. 25
Towards Efficient Designing of Safe Nanomaterials: Innovative Merge of Computational Approaches and Experimental Techniques
Edited by Tomasz Puzyn and Jerzy Leszczynski
© The Royal Society of Chemistry 2012
Published by the Royal Society of Chemistry, www.rsc.org

The dynamic equilibrium between antioxidant defense mechanisms which act to restore redox equilibrium and cellular responses which can lead to adverse outcomes is encapsulated in the "hierarchical oxidative stress paradigm",[1] which posits that ROS production leads to incremental cellular responses that can be classified as antioxidant defense, pro-inflammatory effects, and cytotoxicity (Figure 11.1). This paradigm is supported by studies on the adverse health effects of ambient ultrafine particles. In such studies, excellent correlation coefficients have been established between the capacity of ultrafine particles to generate ROS abiotically and their ability to induce oxidative stress responses (*e.g.* heme oxygenase 1 expression) in epithelial cells and macrophages.[2]

At a lower oxidative stress level (tier 1, Figure 11.1) the nanoparticles induce cytoprotective responses, *e.g.* through the activation of the homeostatic antioxidant defense pathway, inducing the expression of several antioxidant and phase II metabolizing enzymes (*e.g.*, heme oxygenase 1 and glutathione-*S*-transferase). If this level of protection fails, the oxidative stress (tier 2) will lead to cytokine and chemokine expression through transcriptional activation of the gene promoters by redox-sensitive mitogen-activated protein (MAP) kinase and the transcription factor nuclear factor κ-B (NF-κB) signaling cascades. Further escalation (tier 3) will trigger disturbance of the mitochondrial function, resulting in cytotoxicity (apoptosis and necrosis). These pro-oxidative effects depend on nanomaterial properties such as electronically reactive surface states, the presence of transition metals or redox cycling organic chemical impurities (*e.g.* carbon nanotubes), photoactivation leading to the generation of electron/hole pairs (*e.g.* TiO_2), dissolution of the particle surface, and shedding of toxic metal ions (*e.g.* ZnO and chalcogenides).

The majority of the studies collected from the literature do not assess the nanomaterials tested according to the hierarchic oxidative stress model; only a few of the most well-known and used materials such as TiO_2, ZnO and some

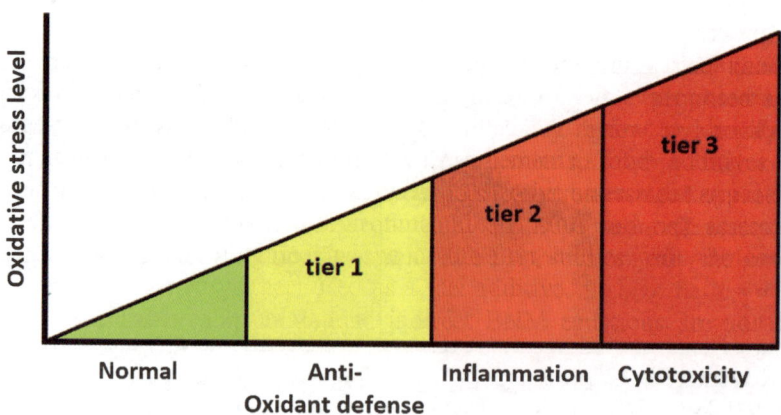

Figure 11.1   Hierarchic oxidative stress model used to develop cellular screening assays for nanomaterials. (Adapted with permission from ref. 1. Copyright 2009 American Chemical Society.)

iron oxides have been tested with specific experiments which are based on this approach. Other studies have been performed in cell-free experiments to focus on the generation of oxidative stress by metal oxide nanoparticles in controlled conditions.

Here we use literature *in vitro* data to describe the application of our model, based on an electron-transfer mechanism, for the prediction of ROS generation by oxide nanoparticles.[3] The model predictions are confronted with numerous literature studies that employ *in vitro* assays to test the cellular toxicity of the following nanoparticles: Al_2O_3, CuO, CeO_2, Fe_2O_3, Fe_3O_4, NiO, SiO_2, TiO_2 and ZnO. The calculated properties of oxide nanomaterials, which reflect their reactivity, can be used to guide the experimental design of *in vitro* toxicological and acellular tests as well as to interpret their outcomes.

11.2 Mechanism of Electron Transfer

Metal oxide nanomaterials have redox properties which can affect the redox balance of cells and cause oxidative stress. Based on this assumption, we have developed a theoretical model that can describe the generation of oxidative stress by oxide nanoparticles.

The model uses reactivity descriptors to calculate the energy structure of nanoparticles and predicts their oxidative stress potential by comparing their conduction and valence band energy levels with relevant redox potentials of biological reactions occurring inside cells. If these two energy levels are comparable, then electrons can be transferred and the oxide nanoparticle, which acts as a catalyst or an electron donor/acceptor, is responsible for disrupting the cellular redox state (Figure 11.2).

The potential transfer of electrons between the metal oxide and an aqueous species is evaluated by comparing the energy structure of the solid and the adsorbed reactants. Relevant energy levels for the metal oxide are the top of the valence band, E_V, and the bottom of the conduction band, E_C, while the energy level of adsorbates undergoing an electron transfer can be approximated by the standard redox potential, $E°$. The relative position of E_V or E_C with respect to $E°$ dictates whether an electron transfer between the oxide and an adsorbate is feasible: redox couples with $E°$ above the conduction band can transfer electrons to the conduction band, whereas redox couples in the band gap can accept conduction band electrons or (photo-induced) valence band holes. The redox couples below the valence band can only be reduced by valence band electrons. The energy of the valence band edge is a measure of the ionization potential of a material. The lowest unoccupied electronic level in most semiconductors coincides with the bottom of the conduction band, and the band edge energy E_C is a measure of the electron affinity of the oxide. The Fermi level or energy, E_F, represents the chemical potential of electrons in a semiconductor. In essence, the Fermi level is the absolute electronegativity, χ, of a pristine semiconductor, a value which corresponds to the energy halfway between the conduction and valence band edges. Aqueous redox species exchanging electrons with an oxide semiconductor can either accept or donate

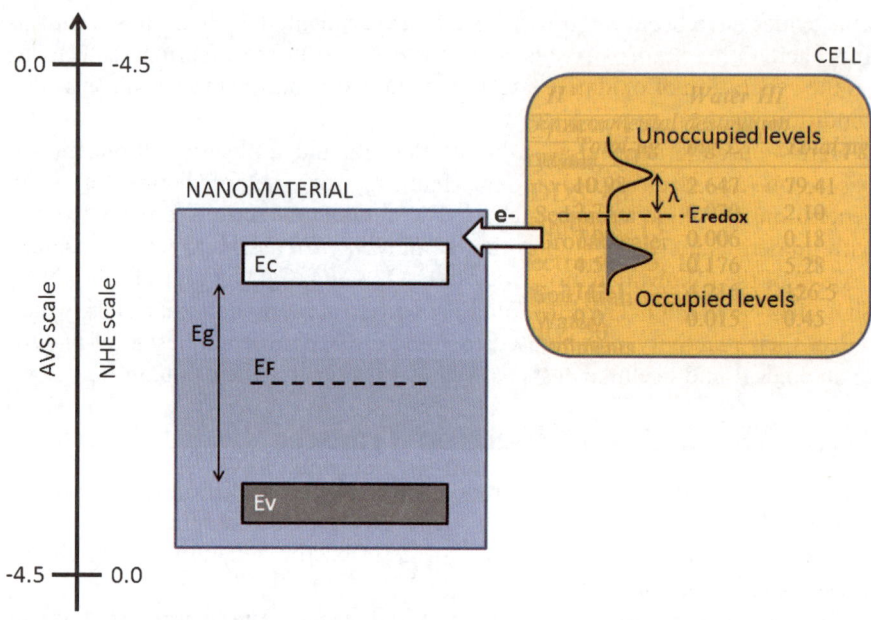

Figure 11.2 Schematic representation of electron transfer mechanism in the generation of oxidative stress by oxide nanoparticles. The positions of the energy levels are represented in the absolute vacuum scale (AVS) and in the normal hydrogen electrode (NHE) scales (E_g is the band gap of a nanomaterial; E_C and E_V are the conduction and valence band energy levels, respectively; E_F is the Fermi energy; E_{redox} is the redox potential of a biological reaction; λ is the reorganization energy; and e^- is an electron transferred from the cell's material to the metal oxide).

electrons. Upon electron transfer from or to an aqueous species, the electronic structure of the species changes. Upon the acceptance of an electron, a previously unoccupied electronic level becomes occupied, whereas upon electron donation an electron is removed from an occupied level. For an electron acceptor the energy of the lowest unoccupied level is relevant, whereas for an electron donor it is the highest occupied energy level. Furthermore, because of the polar nature of water molecules, H_2O dipoles in the solvation shell of a redox species will re-orientate when there is a change in the charge of the redox species. This re-orientation will result in an additional energy gain or loss when an electron is transferred from or to an aqueous redox species. The free energy change associated with this re-orientation process is known as reorganization energy (λ). Moreover, thermal fluctuation of the solvation structure causes a corresponding thermal distribution of the energy levels of both the lowest unoccupied and highest occupied energy levels. While these energy distributions are difficult to quantify, the redox potential of a redox couple undergoing a one-electron transition, which lies between the maxima of these distributions, is used. The Fermi level of electrons of a redox couple is equivalent to the redox potential of aqueous redox couples on the absolute energy scale.

Following Auffan *et al.*,[4] the standard redox potentials of couples active in biological media range from -4.12 eV to -4.84 eV relative to the AVS; other redox potentials calculated with respect to the NHE can be converted using: $E_{(AVS)} = -E_{(NHE)} - 4.5$. Qualitatively, by comparing this range of biological redox couples with the energy band structure of 71 among the most used oxides (Table 11.1) we can see that as the conduction band drops below -4.12 eV, there is a possibility for electron transfer from the cell's material to the nanoparticles.

11.3 Energy Band Structure Calculation of Metal Oxides

The energy structure of metal oxides is calculated using chemical reactivity concepts developed within the density functional theory framework.[5] In this context, any chemical system is characterized by an electronic chemical potential (μ) and an absolute chemical hardness (η). These quantities are reactivity descriptors of isolated chemical species that can be used to represent the electron transfer process of combined systems. The chemical potential characterizes the escaping tendency of the electron cloud from the equilibrium state, and is equivalent to the negative of the electronegativity:

$$\mu = \left(\frac{\partial E}{\partial N}\right)_Z = -\chi \tag{11.1}$$

where E is the ground-state electronic energy, N is the number of electrons and Z is the nuclear charge.

The second derivative of E defines the absolute chemical hardness, a quantity that characterizes the resistance towards charge transfer:

$$\eta = \frac{1}{2}\left(\frac{\partial \mu}{\partial N}\right)_Z \tag{11.2}$$

The chemical hardness corresponds to half the band gap of a chemical compound, *i.e.* the difference between the HOMO (highest occupied molecular orbital) and the LUMO (lowest unoccupied molecular orbital) orbital energies in a simple orbital theory.

Using the finite difference approximation, these reactivity descriptors are computed from the vertical ionization potential (*IP*) and electron affinity (*EA*):

$$\mu = -\frac{IP + EA}{2} \tag{11.3}$$

$$\eta = \frac{IP - EA}{2} \tag{11.4}$$

For an undoped intrinsic oxide, the chemical hardness corresponds to half the energy gap, E_g, between the bottom of the conduction band (the vacant or only partially occupied set of many closely spaced electronic levels) and the top of the valence band (the highest energy continuum of energy levels in a solid that is fully occupied by electrons at 0 K), while electronegativity corresponds to the Fermi level, E_F, (the energy level at which the probability of occupation by an electron is 1/2). For an intrinsic semiconductor, the Fermi level lies at the mid-point of the band gap:

$$\eta = 0.5E_g \tag{11.5}$$

$$\chi = E_F \tag{11.6}$$

In this study, band gap values were calculated from the standard enthalpy of formation ($E_{\Delta H^\circ}$) of oxides.[6,7] Enthalpies were first converted into eV units using eqn (11.7), in which (ΔH°_f) is the standard enthalpy of formation, N_A is the Avogadro number and n_e is the number of electrons involved in the formation reaction:

$$E_{\Delta H^\circ} = -\frac{2\Delta H^\circ_f 2.612 \times 10^{19}}{N_A n_e} \tag{11.7}$$

The optical band gap (E_g) was then calculated using eqn (11.8):

$$E_g = Ae^{\left(0.34E_{\Delta H^\circ}\right)} \tag{11.8}$$

The pre-exponential term A is a property of the cation and it generally corresponds to a value of 1 for d-block elements, 0.8 for s-block elements, 1.35 for p-block elements and 0.5 for f-block elements. The predicted band gap values were then compared to experimental values obtained from literature; the correlation between observed and predicted values was fairly good ($R^2 = 0.84$), as shown in Figure 11.3:

The Fermi energy, or electronegativity (χ) was calculated using a set of empirical equations:[8]

$$\chi_{cation}(P.u.) = 0.274z - 0.15zr - 0.01r + \alpha + 1 \tag{11.9}$$

$$\chi_{cation}(eV) = \frac{(\chi_{cation}(P.u.) + 0.206)}{0.336} \tag{11.10}$$

$$\chi_{oxide}(eV) = 0.45\chi_{cation}(eV) + 3.36 \tag{11.11}$$

where *P.u.* denotes Pauling units, z is the formal charge of the cation, r is the Shannon ionic radius and α is a correcting term specific for each cation.

The bottom of the conduction band (E_C) and the top of the valence band (E_V) were obtained by adding and subtracting half the band gap energy value to the Fermi energy, respectively:

$$E_C = -\chi_{oxide} + 0.5E_g \tag{11.12}$$

$$E_V = -\chi_{oxide} - 0.5E_g \tag{11.13}$$

Table 11.1   Conduction and valence band energies of 71 metal oxides. Materials capable of direct ROS production are highlighted in bold.

No.	Metal oxide	E_C	E_V
1	Y_2O_3	−0.58	−9.96
2	Lu_2O_3	−0.82	−9.91
3	MgO	−1.32	−8.83
4	Al_2O_3	−1.40	−9.92
5	SiO_2	−1.45	−11.50
6	Li_2O	−1.48	−7.89
7	CaO	−1.58	−8.23
8	BaO	−1.76	−7.70
9	TiO	−1.81	−8.06
10	BeO	−1.95	−9.06
11	HfO_2	−1.95	−9.47
12	SrO	−2.07	−7.55
13	Ti_2O_3	−2.44	−8.42
14	Sc_2O_3	−2.70	−8.07
15	K_2O	−2.83	−6.11
16	VO	−2.91	−7.49
17	La_2O_3	−2.98	−7.42
18	Na_2O	−3.02	−6.16
19	Cs_2O	−3.04	−5.81
20	ZrO_2	−3.06	−8.34
21	Er_2O_3	−3.07	−7.72
22	NbO	−3.10	−7.28
23	Ho_2O_3	−3.15	−7.70
24	Tb_2O_3	−3.15	−7.62
25	Dy_2O_3	−3.19	−7.65
26	Rb_2O	−3.19	−5.64
27	Ce_2O_3	−3.22	−7.34
28	Gd_2O_3	−3.23	−7.47
29	Nd_2O_3	−3.26	−7.44
30	Yb_2O_3	−3.27	−7.49
31	CeO_2	−3.37	−7.37
32	MnO	−3.41	−7.26
33	GeO_2	−3.44	−8.86
34	Ga_2O_3	−3.48	−7.80
35	GeO	−3.49	−6.89
36	Eu_2O_3	−3.54	−7.02
37	V_2O_3	−3.59	−7.78
38	PbO	−3.63	−6.43
39	NiO	−3.63	−7.47
40	Tl_2O	−3.65	−6.18
41	ZnO	−3.77	−7.24
42	NbO_2	−4.00	−8.07
43	SnO_2	−4.03	−7.74
44	CdO	−4.05	−6.31
45	Cr_2O_3	−4.07	−7.58
46	In_2O_3	−4.08	−6.89
47	**CoO**	**−4.20**	**−6.90**
48	**TiO_2**	**−4.27**	**−7.49**

Table 11.1 (*Continued*)

No.	Metal oxide	E_C	E_V
49	FeO	−4.28	−6.76
50	Fe_2O_3	−4.29	−7.66
51	Mn_2O_3	−4.36	−7.44
52	PbO_2	−4.36	−7.30
53	Ta_2O_5	−4.46	−8.61
54	Ag_2O	−4.52	−5.64
55	Cu_2O	−4.53	−6.53
56	Tl_2O_3	−4.66	−6.26
57	CuO	−4.69	−6.73
58	MoO_2	−4.77	−7.59
59	WO_2	−4.78	−7.60
60	Nb_2O_5	−4.79	−8.15
61	HgO	−4.81	−5.81
62	MnO_2	−4.84	−7.89
63	CrO_2	−4.86	−7.73
64	Ni_2O_3	−5.15	−6.93
65	MoO_3	−5.15	−8.95
66	V_2O_5	−5.20	−8.00
67	As_2O_5	−5.40	−7.99
68	Sb_2O_5	−5.66	−7.08
69	WO_3	−5.68	−8.42
70	Fe_3O_4	−5.73	−5.83
71	CrO_3	−6.55	−8.80

According to the Nernstian relation, band edges shift to higher or lower energy levels following a linear relation with respect to the solution's pH.[9] Therefore the following energy shift was added to the band energies:

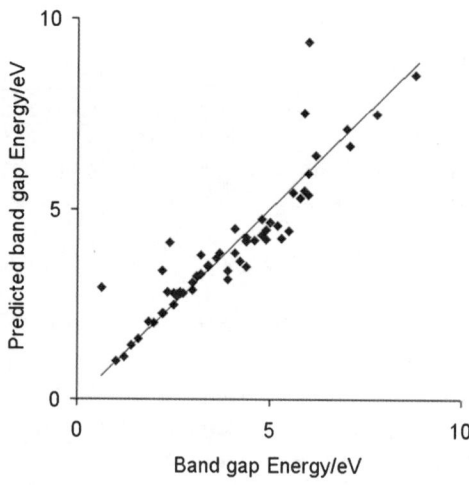

Figure 11.3 Observed *vs.* predicted band gap energy values of oxides ($R^2 = 0.84$).

$$E_{shift} = 0.059(PZZP - pH) \qquad (11.14)$$

where *PZZP* is the point of zero zeta-potential of the bulk oxide and pH was taken as 7. Because not all *PZZP* values were available for all the oxides included in this study, we estimated these values using an empirical equation that correlates the *PZZP* points with the electronegativity of the oxides' cations (see Figure 11.6):

$$PZZP = -8.1387\chi_{cation}(P.u.) + 18.6850 \qquad (11.15)$$

Thus eqn (11.12) and (11.13) become (11.16) and (11.17), respectively:

$$E_C = -\chi_{oxide} + 0.5E_g + 0.059(PZZP - pH) \qquad (11.16)$$

$$E_V = -\chi_{oxide} - 0.5E_g + 0.059(PZZP - pH) \qquad (11.17)$$

The calculation method here developed represents a simplified model of metal oxide reactivity in a number of respects. The calculations pertain to bulk materials and can therefore be applied to particles whose diameter is larger than 20–30 nm.[4] For smaller nanoparticles or particles with surface states within the band gap, the band structure changes. In such cases, the model should ideally be updated with more accurate energy state values, which could be calculated by using *ab initio* density functional theory (DFT) methods.

In addition, to determine the reactivity of a metal oxide nanoparticle one should know the type and extent of surface planes that the material exposes to the environment. This comprises knowledge of the type and density of surface defect sites, which exhibit different reactivities from the bulk material. The presence on the surface of under-coordinated cations and anions with distorted geometry due to surface relaxations or reconstructions will lead to different adsorption and reactivity profiles.

Thus, an extension of the theoretical approach developed here would require a large amount of information, which is only available at present for a few well-known nanomaterials (*e.g.* TiO_2).

On the other hand, the distribution of energy states in a redox couple becomes more complicated if the redox couple undergoes a multi-electron transfer, because each one-electron step has a population of two energy states (occupied and unoccupied) associated with it. Instead of the standard redox potential of the overall reaction, the standard redox potential of each one-electron step should be considered for a multi-electron transfer reaction. These one-electron redox potentials are mostly unavailable because the intermediate products are often unstable. However, using $E°$ for the overall reaction leads to a misleading comparison of energy levels between the oxide and the aqueous species; this arises from the fact that $E°$ for the overall reaction does not lie midway between the maxima of the occupied and unoccupied energy levels.

11.4   Comparison of Model Predictions with Literature Data

In this section we describe the application of the oxidative stress model to a number of metal oxide nanoparticles and compare the predictions with literature *in vitro* studies.

11.4.1   Titania (Rutile and Anatase)

Titanium dioxide is the most widely studied oxide nanomaterial. Numerous studies have found that oxidative stress is a prominent feature of the cellular response to TiO_2 nanoparticles, including evidence of increased ROS production, depletion of cellular antioxidants, increase in oxidative products (such as lipid peroxidation) or evidence that toxicity is diminished on pretreatment with antioxidants.[10] Moreover, most studies confirm the ability of titania nanoparticles to induce oxidative stress when illuminated with ultraviolet (UV) light. The irradiation of TiO_2 results in the promotion of an electron from the valence to the conduction band, with the concomitant generation of a hole in the valence band (reaction (11.18) below): the electron in the conduction band is the reducing species whereas the valence band hole can accept an electron and is therefore the oxidizing agent.

The electronic structure of TiO_2 has been calculated using a wide variety of theoretical approaches and there is wide agreement that the surface electronic structure is not too different from that of the bulk. In our model, the conduction and the valence band edge positions for titania are calculated at -4.3 and -7.5 eV *vs.* AVS, respectively. The components of the electron–hole pair, when transferred across the interface, are capable of reducing and oxidizing an adsorbate, forming a singly oxidized electron donor and a singly reduced electron acceptor. The ability of a semiconductor to undergo photo-induced electron transfer to adsorbed species on its surface is governed by the band energy positions of the semiconductor and the redox potentials of the adsorbate. The relevant potential level of the acceptor species is thermodynamically required to be below the conduction band potential of the semiconductor. The potential level of the donor needs to be above the valence band position of the semiconductor in order to donate an electron to the vacant hole. Electrons promoted in the conduction band can reduce dioxygen (reaction (11.19); experimentally, the conduction band of TiO_2 is almost isoenergetic with the reduction potential of dioxygen) to form superoxide ions which are thought to abstract hydrogen atoms from various biological substrates; similarly, the hole in the valence band can yield a hydroxyl radical (OH$^\bullet$; reaction (11.20)).

$$TiO_2 + h\nu \rightarrow TiO_2\left(e^-/h^+\right) \tag{11.18}$$

$$TiO_2(e^-) + O_{2\,ads} \rightarrow TiO_2 + O_2^{\bullet-} \tag{11.19}$$

$$TiO_2(h^+) + OH^-_{ads} \rightarrow TiO_2 + OH^{\bullet}{}_{ads} \tag{11.20}$$

$$TiO_2(h^+) + H_2O_{ads} \rightarrow TiO_2 + OH^{\bullet}{}_{ads} + H^+ \tag{11.21}$$

All these processes result in radicals which can undergo further reactions. It is generally held that hole capture is directly through OH^- and not water first (reaction (11.21)): although the $1b_1$ orbital of water lies above the 1π level of OH^- (so one expects water molecules to be better at capturing a hole than OH^-), the radical-cation of water may be neutralized before decomposing into an $OH^{\bullet}$ radical. In addition, it is mostly assumed that the surface is hydroxylated and therefore the hole is directly transferred to OH^-.

From these considerations, we propose the scheme depicted in Figure 11.4 to explain the formation of the electron–hole pair and the generation of ROS by titania nanoparticles.

In vitro tests described in the literature report on the use of two main crystal forms of TiO_2, termed rutile and anatase. The anatase phase, which is greater for small (<15 nm) TiO_2 nanoparticles, is more effective in the production of hydroxyl radicals and the subsequent decomposition of organic compounds than the rutile phase. Consistent with this, anatase has been demonstrated to be the most toxic form of titania, having a greater capacity to generate ROS on exposure to light. Although the difference in band gap between anatase (3.2 eV) and rutile (3.0 eV) is in the range of 0.2 eV and the Fermi energy difference is 0.1

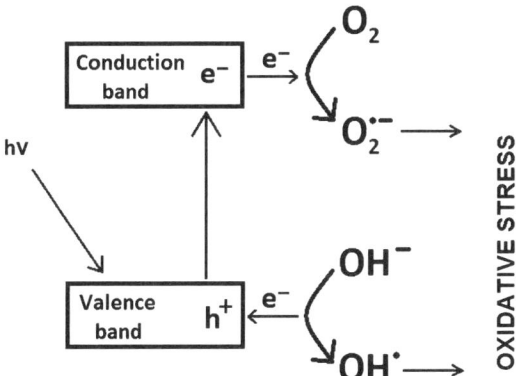

Figure 11.4 Mechanism of hydroxyl radical and superoxide generation by photo-activated titania nanoparticles. The electron transfer from the semiconductor to molecular oxygen is potentially favorable, since the conduction band of the oxide has a higher energy value than the experimental redox potential for the generation of superoxide species, which lies at about –4.3 eV. The reduction potential for $OH^{\bullet}$ has been assigned a value of +2.74 V.[11] It is therefore thermodynamically favorable for the hole site formed in the valence band to oxidize water to form the hydroxyl radicals.

eV, anatase results in better radical production. The higher activity of anatase with respect to that of rutile can be explained by the higher aptitude of anatase (i) to photoadsorb oxygen, in O_2^- and O^- forms; (ii) to photodesorb it; and (iii) to have a lower relative electron–hole recombination rate. Most commercial titania powders are a mixture of rutile and anatase (e.g. the most often used Degussa P25 contains approximately 80–90% of anatase and the rest is rutile). These mixtures are optimal for certain photocatalytic reactions and non-photo-induced catalysis. Moreover, there is growing evidence that anatase is more active than rutile for O_2 photo-oxidation, but not necessarily for all photocatalytic processes.

In principle, the reactivity of titania nanoparticles depends on the type and extent of surface planes exposed to the environment, i.e. on the surface density and coordination number of Ti and O atoms on the surface (which represent acidic and basic sites, respectively), the defects due to Ti vacancies and surface reduction/oxidation, and the degree of hydroxylation. Other factors that can play an important role in reactivity are the surface relaxations, the ionic strength and the type of electrolyte present in solution, the irradiation of the material (which can modify the hydrophilicity/hydrophobicity of the surface) and the presence of doping atoms. For rutile, for example, it is important to consider the planes exposed by the crystal surface, because each plane has a different reactivity and isoelectric point, as determined experimentally. These differences have a strong impact on the surface potential and colloidal interactions of this material.

The model calculates the electronegativity values (i.e. the Fermi level) from bulk titania, that is for Ti^{4+} ions with coordination number 6 and radius of approximately 0.6 nm. However, the crystal structure of rutile, based on the Wulff construction[†] and calculated surface energies, indicates that the (110), (001) and (100) surfaces can be found (Figure 11.5). These planes have Ti and O surface atoms with different coordination numbers and surface densities, which leads to different isoelectric points (IEP) (Table 11.2). In Table 11.2, the electronegativity values in column 5 were calculated from the cations formal charge and radius. The data in columns 1, 2 and 6 were taken from Diebold;[12] and the data in columns 3 and 4 from Bullard.[13]

Atomic force microscopy (AFM) measurements have been used to calculate different IEP points for each plane of rutile[11] (Table 11.2). Differences of IEP between the 110 and 100 planes can be ascribed to differences in the coordination number of titania atoms on the surface: the lower the coordination number, the lower the IEP, the stronger the acidity and the higher the electronegativity. The (001) surface, however, does not fit this description: despite having the lowest coordination number for its cationic positions, strong relaxations and facile reconstruction energetics substantially increase the average coordination value thereby changing the overall isoelectric point. Moreover, the exclusive presence of two-fold coordinated oxygen sites

---

[†] The Wulff construction is a method for determining the equilibrium shape of a crystal based on energy minimization arguments which are used to show that certain crystal planes are preferred over others, giving the crystal its shape.

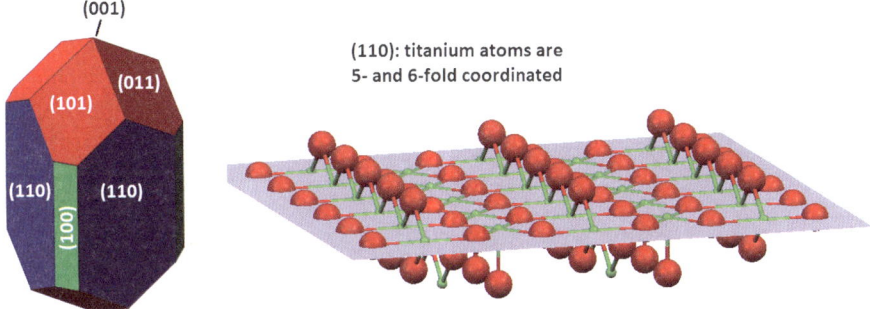

Figure 11.5   Left: equilibrium shape of a macroscopic TiO$_2$ crystal using the Wulff
construction. Right: (110) surface of rutile with 5-fold and 6-fold
coordinated Ti atoms and 2-fold and 3-fold coordinated O atoms. The
surface is cut in such a manner that the same number of Ti→O and
O→Ti bonds are broken, and the surface is autocompensated.

on the ideal (001) surface, in contrast with a mixture of 2-fold and 3-fold sites
on the (110) and (100) surfaces, suggests that stronger and more numerous
Lewis base sites exist on the (001) surface. More precisely, the isoelectric point
of rutile can be better explained by taking into account both the
electronegativity of Ti atoms and their surface density. This assumption
reflects that the isoelectric point depends on both the extent of surface acid
sites and their intrinsic acidity.

Since the habit of the rutile structure is dominated by the presence of (110)
surfaces (56% of the crystal surface area), the measured IEP values for rutile
particles are very close to the range determined for this plane (5.5–4.8). In
contrast, the (100) surface, whose IEP was measured to be substantially lower,

Table 11.2   Coordination numbers, cation density, isoelectric point, electronegativity and surface formation energies of Ti and O atoms at
different rutile and anatase surfaces.

	Ti CN[a]	O CN	Cation density/nm^2	IEP[b]	X$_{cation}$/eV, (CN)	Surface formation energies/J m^{-2}
Rutile						
110	6/5	3/2	6.0	5.5–4.8	1.591, (6) and 1.649, (5)	0.31
100	5	3/2	7.4	3.7–3.2	1.649, (5)	N/a
001	5/4	2	4.8	5.8–5.5	1.649, (5) and 1.703 (4)	N/a
Anatase						
101	6/5	3/2	N/a	N/a	1.591, (6) and 1.649, (5)	0.44
001	5	3/2	N/a	N/a	1.649, (5)	0.90
Bulk	6	3	N/a	5.7^c	1.591 (6)	N/a

[a]Coordination number. [b]Isoelectric point. ^{c}IEP $= -8.1387X + 18.6850$, $X =$ electronegativity of
cation in Pauling units.

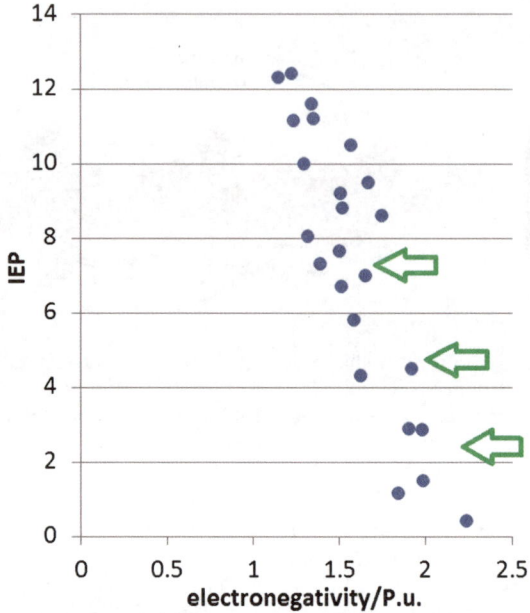

Figure 11.6 Experimental IEP vs. cation electronegativity for 24 metal oxides. Oxides with IEP < 8 are shifted compared to those with higher IEP. This shift probably reflects that surface acidities are underestimated when IEPs are calculated using electronegativity values for cations with coordination number of the bulk material (e.g. 6 for titania) instead of using the surface coordination.

constitutes a relatively small percentage of surface area, as indicated by the Wulff construction. Our calculation of IEP for bulk titania is 5.7. This value is calculated using a regression equation where the electronegativity of the cation (in Pauling units, P.u.) is correlated to the experimental IEP values obtained from the literature.[14] A diagram showing the IEP values *vs.* the electronegativity is given in Figure 11.6: it can be seen that oxides with IEP < 8 are shifted compared to those with high IEP > 8. This shift probably reflects that surface acidities are underestimated when calculating electronegativity with a coordination number of 6 (the coordination of bulk titanium); the basic character of oxides is mainly related to the oxygens on the surface (which, in turn, are influenced by the surrounding cations).

In summary, these findings suggest that electronegativity can be used as a descriptor for characterizing the reactivity and IEP values of metal oxide nanoparticles. Electronegativity is related to the chemical potential of the material.

11.4.2   Magnetite and Maghemite

Iron oxide nanoparticles used for cytotoxicity tests are mainly in the form of magnetite (Fe_3O_4) and maghemite (Fe_2O_3). Most of the studies considered here use naked maghemite nanoparticles whereas an equal number of studies involve both coated and uncoated magnetite particles. In general, the toxicity of maghemite nanoparticles depends on the cell type used. Toxicity has been reported for MSTO cells, rat colon cells, and lung epithelial cells.[15] In contrast, no toxicity was observed in 3T3 and human aortic endothelial cells.[16] One study reports the production of IL-6 and IL-8,[17] which indicates a tier 2 oxidative stress level in the hierarchical model.

Studies on coated magnetite nanoparticles found no variation in cell viability for rat brain endothelial cells,[18] human fibroblasts[19] and rat macrophages;[20] naked magnetite particles, instead, are toxic and reduce cell viability in a number of different cell lines comprising human fibroblasts,[21] rat macrophage,[20] liver cells[22] and human monocyte macrophages.[23]

Toxicological tests employing particles composed of different oxides but doped with iron atoms also showed a higher cytotoxicity when compared to undoped particles.[24]

These studies show that the toxicity of an iron oxide nanoparticle is reduced when the surface is coated. Although naked nanoparticles adsorb biological material such as proteins and electrolytes that can mask the surface reactivity, they are still capable of producing radicals, probably as a result of partial surface coverage and transient interactions with macromolecules.

By comparing in our model the range of redox potentials of biological reactions occurring inside a cell with the band energy structure of maghemite and magnetite, it can be seen that the conduction bands of Fe_2O_3 (-4.3 eV) and Fe_3O_4 (-5.7 eV) are within and below that range, respectively, indicating that both oxides can extract electrons from the cell's biomaterial and are therefore potentially capable of inducing oxidative stress.

Most often the toxicity ascribed to iron oxide nanoparticles is explained in terms of Fenton reactions (reaction (11.22)). Literature shows examples where both Fe^{2+} and Fe^{3+} cations are involved in the production of radicals, with Fe^{2+} being the most reactive. It is well known that reducing metal complexes such as iron(II) can promote the Fenton reaction by reacting with the hydrogen peroxide in the cell and yielding hydroxyl radicals:

$$H_2O_2 + Fe(II) \rightarrow OH^\bullet + OH^- + Fe(III) \qquad (11.22)$$

In the case of nanoparticles, in order to donate an electron the valence band of the iron oxide must lie above the redox potential for the reduction of water peroxide. The redox potential for the decomposition of H_2O_2 into hydroxyl ions and hydroxyl radicals is measured around -4.9 eV *vs.* AVS ($+0.38$ eV *vs.* NHE in Table 11.3). The model we developed calculates the position of the magnetite valence band around -5.8 eV, a value that is far from the redox potential for the H_2O_2 reduction. A work function value of -5.20 ± 0.15 eV, however, is measured for the $Fe_3O_4(100)$ surface,[25] which is comparable with the redox potential for the decomposition of hydrogen peroxide. Maghemite has an even lower valence band (around -7.7 eV), which reflects its lower degree of reactivity compared to magnetite.

This one-electron reductive cleavage can be extended to alkyl hydroperoxides too, yielding alkoxyl radicals, RO', whose oxidizing power, although smaller than that of OH', is still considerable:[26]

$$ROOH + Fe(II) \rightarrow RO^{\bullet} + OH^{-} + Fe(III) \qquad (11.23)$$

The reverse one-electron oxidative cleavage of hydroperoxides to peroxyl radicals (ROO') with ferric iron ions may also occur, but this reaction is slower because most ferric complexes do not possess enough oxidizing power to promote it:[26]

$$ROOH + Fe(III) \rightarrow ROO^{\bullet} + H^{+} + Fe(II) \qquad (11.24)$$

When this mechanism is applied to nanoparticles, it turns out that only the valence band of magnetite (-5.8 eV) is above the redox potential for the generation of RO' radicals (-6.5 eV), and therefore Fe_3O_4 should be able to produce radicals from organic hydroperoxides such as *t*-butyl-hydroperoxide.

The debate as to whether it is the particle that is involved in the Fenton reaction rather than dissolved ions was recently addressed by Voinov *et al.*,[27] who conducted an acellular study on magnetite and maghemite nanoparticles. By performing a series of spin-trapping electron paramagnetic resonance (EPR) experiments, the authors showed that the surface of unprotected γ-Fe_2O_3 nanoparticles mediates the production of highly reactive hydroxyl radicals (OH') under conditions of the biologically relevant superoxide-driven Fenton reaction, also known as the Haber–Weiss cycle. The following catalytic cycle (reactions (11.25)–(11.27)) was proposed for maghemite nanoparticles:

$$O_2^{\bullet-} + Fe^{3+} \rightarrow O_2 + Fe^{2+} \qquad (11.25)$$

Table 11.3 Standard redox potentials of oxygen species *vs.* NHE, at pH $= 7$ (taken from ref. 11).

Redox couples	E°/V
$O_2/O_2^{\bullet-}$	-0.33
$O_2^{\bullet-}/H_2O_2$	$+0.89$
O_2/H_2O_2	$+0.28$
$H_2O_2/OH^{\bullet} + H_2O$	$+0.38$
$OH^{\bullet} + H_2O/H_2O$	$+2.32$
H_2O_2/H_2O	$+1.35$
O_2/H_2O	$+0.82$
$ROO^{\bullet}/ROOH$	$+1.0$
$ROOH/RO^{\bullet} + H_2O$	$+2.0$
$RO^{\bullet} + H_2O/ROH$	$+1.6$
$ROOH/ROH$	$+1.8$

$$2O_2^{\bullet-} + 2H^+ \rightarrow H_2O_2 + O_2 \tag{11.26}$$

$$H_2O_2 + Fe^{2+} \rightarrow OH^\bullet + OH^- + Fe^{3+} \tag{11.27}$$

A test conducted with iron ions leached from the nanoparticles confirmed that the observed catalytic activity of the γ-Fe_2O_3 and Fe_3O_4 nanoparticles is attributed primarily to reactions at the surface of the nanoparticle, rather than being caused by the dissolved metal ions released by the nanoparticles as previously thought. In general the solubility of Fe(III) is low, while Fe(II) oxides are sparingly soluble. This means that except at extreme pH values, these compounds maintain a very low level of total [Fe] in solution. In the pH range 4–10 and in the absence of complexing or reducing agents the concentration of iron atoms in solution is less than 10^{-6} M.[28] For iron oxide nanoparticles, which have a high surface area and relatively high surface free energies, particle size will have a marked effect on the solubility of these compounds. Different authors[29] have derived slightly different sets of equations relating K_{SO} (the solubility product) of some iron oxides, in particular goethite and hematite, to the particle size. These equations should enable an estimate of the rise in solubility as particle size drops.

11.4.3 Zinc Oxide

Literature studies report high levels of toxicity for zinc oxide towards a number of cell types, including MSTO and 3T3 cells,[15] Chinese hamster ovary cells,[30] human aortic endothelium cells,[16] neuro 2A cells,[31] rat epithelial lung cells,[32] RAW 264.7[18] and BEAS-2B cells.[33]

According to the hierarchical oxidative stress hypothesis, ZnO was observed to produce oxidative stress responses at all tiers in RAW 264.7 and BEAS-2B cells:[33] at the tier 1 level, ZnO could generate robust HO-1 mRNA and protein expression as well as increased Nrf2 and NQO-1 mRNA expression. At tier 2, ZnO is capable of activating pro-inflammatory signaling pathways such as the Jun kinase (JNK) and NF-κB cascades. An increase in TNF-α, IL-1 and IL-8 production was also observed. In human aortic endothelial cells an increase in the expression of ICAM, IL-8, MPE1 and mRNA content was observed. Furthermore, ZnO is capable of inducing oxidative stress at the tier 3 level, resulting in more severe damage to the cells. An increased intracellular Ca^{2+} level and mitochondrial damage was also reported.

In summary, the cell viability in all assays tested was dramatically reduced indicating that ZnO is highly toxic; this toxicity was directly related to particle dissolution (especially in acidifying endosomal compartments) and release of toxic Zn^{2+} in the cell culture medium as well as to the uptake of the particle remnants by specific endosomal compartments in the cells. In particular, the Zn^{2+} accumulation in lysosomes was associated with organellar clumping, oxidative cell injury, intracellular Ca^{2+} release, mitochondrial depolarization, cytokine release, and eventually cytotoxicity.[33]

It is worth mentioning that proteins and organic substances might increase the dissolution rates of this material through at least two mechanisms: aqueous complexation (that is, aqueous species complexing free ions released from the material's surface) and ligand enhanced dissolution (that is, adsorbed natural organic material and organic acids extracting surface metal atoms from the nanoparticle surfaces). The latter mechanism has been demonstrated for iron and aluminum oxides and oxyhydroxides, and is also likely to occur for ZnO. Moreover, it was observed that the addition of ZnO agglomerates shrink when transferred from water to tissue culture medium. This could reflect the effect of serum components on particle dispersion or dissolution. Serum proteins and lipids have been shown to improve nanoparticle dispersion and may also assist in dissolution. This size reduction of ZnO could facilitate cellular uptake and intracellular dissolution, particularly in the size-limited endosomal compartments.

According to our model, the bottom of the conduction band of zinc oxide (which is located at -3.8 eV) is above the range of biological redox potentials, whereas the valence band is more than 2 eV below that range. This suggests that ZnO particles are not directly responsible for the generation of radicals and the observed oxidative stress, and that particle solubility and ion toxicity might be involved in causing the oxidative stress. Moreover, the dissolution of this oxide increases under acidic (and basic) conditions too (see Figure 11.7), for example when nanoparticles are sequestered in lysosomes where the pH is around 4.5. In this precise situation, the nanoparticle–lysosome system behaves like a 'Trojan horse' type of carrier enabling the transport of toxic metal ions inside the cells and thereby eluding the cell's membrane, which acts as a physical barrier for most ions.

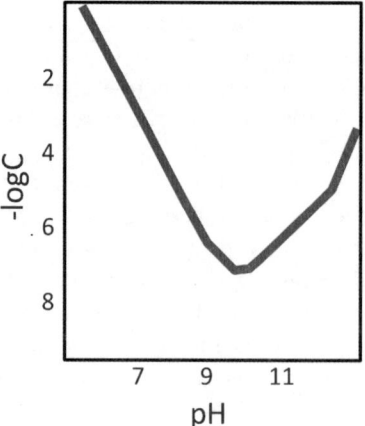

Figure 11.7   Zinc oxide solubility curve as a function of pH.

One of the most important physical properties of the solid that will affect its solubility is particle size. For crystals with diameter < 1 μm, the high surface area may increase the solubility. This occurs because the surface properties, especially the surface free energy, rather than the properties of the bulk, govern the dissolution behavior. The relationship between the change in solubility associated with a change in particle size can be obtained by dividing a bulk solid, suspended in water, into a finely divided material of molar surface area A. The free energy, ΔG, associated with this process is:

$$\Delta G = 0.66\bar{\gamma}A \tag{11.28}$$

where $\bar{\gamma}$ is the mean free surface energy (interfacial tension) of the solid/liquid interface. As $G_T^{\circ} = RT \ln K_{SO}$, the solubility product can be related to the molar surface energy by the relationship:

$$\log K_{SO(A)} = \log K_{SO(A=0)} + \frac{0.66\bar{\gamma}M\alpha}{2.3RTd\sigma} \tag{11.29}$$

where M is the molecular weight, σ the density of the solid, α is a geometric shape factor and d is the particle size. The experimental data for ZnO (and CuO as well) have been shown to agree with the theoretical predictions.[34]

11.4.4 Ceria

Toxicological studies on ceria (CeO_2) nanoparticles are contradictory. A study reported a dose- and time-dependent decrease in cell viability in human bronchoalveolar carcinoma A549 cells associated with increased levels of oxidative stress and reduced glutathione and α-tocoferol levels.[35] Other studies reported that ceria nanoparticles are not toxic towards CRL8798 breast epithelial cells and display low levels of toxicity in MCF-7 breast carcinoma cells.[36] Another study reported a small increase in IL-6 and IL-8 secretion in lung epithelial BEAS-2B cells.[17] Finally, another study[37] showed that ceria nanoparticles are relatively non-toxic to cultured cells; instead, they can act as radical scavengers and redox cycling antioxidants. For example, they are able to protect HT22 nerve cells from oxidative stress caused by exogenous glutamic acid.

Several of these studies point out that the antioxidant activity of ceria nanoparticles is caused by the redox properties of this material, which has oxygen vacancies on its surface that lead to the formation of Ce^{3+} cations. The presence of both reduced and oxidized cations (Ce^{3+}/Ce^{4+}) on the surface of the nanoparticle imparts redox properties to this oxide making it capable of catalysis.[38] X-ray photoelectron spectroscopy (XPS) and X-ray absorption near edge spectroscopy (XANES) suggest that the concentration of Ce^{3+} relative to Ce^{4+} increases as particle size decreases with a conservative [Ce^{3+}] minimum of 6% in 6 nm nanoparticles and 1% in 10 nm particles. It is also known that for crystallites with an average size of 20 nm, oxygen vacancies present in the reduced ceria undergo rapid ordering leading to the immediate formation of Ce_2O_3. The ease of the reversibility is important because it provides the structural mechanism for the rapid storage and release of oxygen that takes place in ceria in redox reactions.

An acellular study showing the antioxidant activity of CeO_2 nanoparticles[39] demonstrated that these particles act as a catalyst by mimicking superoxide dismutase (SOD) activity, with a catalytic rate constant exceeding that determined for the SOD enzyme. The following mechanism was proposed:

$$O_2^{\bullet-} + Ce^{4+} \rightarrow O_2 + Ce^{3+} \tag{11.30}$$

$$O_2^- + Ce^{3+} + 2H^+ \rightarrow H_2O_2 + Ce^{4+} \tag{11.31}$$

The presence on the surface of both reduced and oxidized cerium atoms and oxygen vacancies changes the band structure of bulk ceria nanoparticles. For the bulk material, the conduction and valence band energies are located at -3.4 and -7.4 eV, respectively. The abstraction of an electron from the superoxide anion is not favorable in this situation ($E°$ of $O_2/O_2^{\bullet-} = -0.33$ eV). The presence of defects on the surface might however create surface states in the band gap that reside approximately 3 eV above the valence band and can accept an electron for oxidizing the superoxide anion (Figure 11.8).

By comparing the range of redox potentials of biological reactions with the energy structure of bulk ceria, we find that the former are comprised within the band gap of this oxide; this suggests that for this semiconductor the electron transfer can be fully forbidden since the redox potential is within the band gap and far away from either band edges. Such electron transfer would, however, be possible if surface states located within the band gap are present in the material.

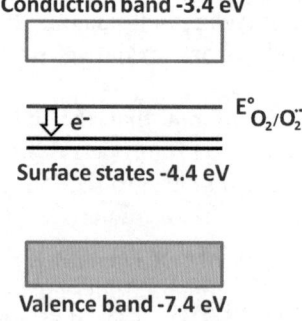

Figure 11.8   Mechanism of superoxide oxidation by surface states present in the band gap and originated at defect sites on the surface of ceria nanoparticles.

11.4.5 Copper Oxide (CuO)

Copper oxide nanoparticles, when tested, are typically found to be highly toxic, although the mechanism underlying toxicity is not yet clearly established since both particle surface reactivity and solubility have been reported as possible explanations. The conduction band of this oxide (-4.7 eV), calculated with our model, lies in the range of biological redox potentials, suggesting that the cytotoxic effects observed for this material might depend on its ability to pull electrons from the cell components. This material is potentially capable of accepting electrons from the oxidation of biological matter and subsequently transferring them to molecular acceptors forming reactive compounds, *e.g.* radicals. Tests for ROS production in the human lung epithelial cell line A549[40] concluded that the toxicity was not explained by Cu^{2+} ions released to the cell medium: cells exposed to Cu in the form of CuO showed approximately an eight times steeper linear dose–response curve when compared to Cu ions from $CuCl_2$ salt. In contrast, another study[41] reported that, in Chinese hamster oocytes (CHO) and HeLa cell models, CuO nanoparticles are toxic, and showed that the dissolution of this material was related to the observed difference in toxicity between carbon-coated CuO and pure CuO nanoparticles. Solubility measurements in ultra-pure water at pH $= 7.0$ and in acidic conditions at pH $= 5.5$, which simulate the intracellular dissolution after the uptake from lysosomes, were quite different, with the coated material being the least capable of releasing ions in solution. Control experiments using pure carbon nanoparticles as well were used to exclude significant surface effects. Reference experiments with ionic copper solutions produced a similar response of cultures when exposed to copper oxide nanoparticles or ionic copper, confirming the mechanism of toxicity by ion leaching. These observations are in line with a Trojan horse-type mechanism.

Another study on NIH3T3 and A549 cell lines[42] also examined the toxic effect of possibly released cationic ions from copper oxide by removing the nanoparticles from the cell culture media where they stayed for 48 h, and then using the media for toxicity evaluation. The authors observed little toxic effect and concluded that possibly released cationic ions contributed little to the cytotoxicity.

11.4.6 Nickel Oxide (NiO)

Nickel oxide nanoparticles were tested in human keratinocyte HaCaT cells and human lung carcinoma A549 cells.[43] This study analyzed the influence of the depletion of medium components by adsorption to metal oxide nanoparticles on cell viability and proliferation. The particles were removed from the dispersion by centrifugation, and the supernatant was applied to the cells. Both the cell viability and the proliferation of human keratinocyte HaCaT cells and human lung carcinoma A549 cells were affected by the supernatant. In particular, cell proliferation was strongly inhibited by the supernatant of TiO_2 and CeO_2 dispersions. The supernatant showed depletion of serum proteins and Ca^{2+} by adsorption to metal oxide nanoparticles (titanium oxide has a high affinity for calcium ions). When the adsorption effect was blocked by the pretreatment of particles with FBS, the inhibitory effect was lost. However, in NiO, which showed ion release, a decrease of inhibitory effect by pretreatment was not shown. The authors reported that cytotoxicity was associated with the solubility of the material and therefore cell viability was affected by released ions rather that by the particles *per se*.

The same authors[44] analyzed the cellular responses induced by NiO particles in human lung carcinoma A549 cells; the study was also paralleled by *in vivo* intratracheal instillation to measure the concentrations of lipid peroxide heme oxygenase-1, surfactant protein-D and lactate dehydrogenase in bronchoalveolar fluid. The exposure of NiO particles induced oxidative stress and activation of antioxidant systems in both cultured cells and *in vivo*. The increase of dichlorofluorescin and diphenyl-1-pyrenylphosphine (DPPP) fluorescence and total hydroxyoctadecanoic acid (tHODE) levels indicated that NiO nanoparticles induced intracellular ROS generation in the cytosol and then lipid oxidation at the cell membrane. By combining *in vivo* and *in vitro* results, the authors proposed the following mechanism: first, primary oxidative stress occurs due to the direct effect of NiO particles, release of Ni^{2+} and subsequent generation of ROS by Fenton-like reactions. The secondary oxidative stress occurred after 3 days and is caused by indirect effects: this later event is induced by secretion of pro-inflammatory cytokines and subsequent macrophage accumulation.

Our model shows that the conduction band of NiO is above the range of biological redox potentials ($E_C = -3.6$ and $E_V = -7.5$ eV). Therefore, the model predicts that this material is not capable of producing radicals and interfering with the redox equilibrium of the cell. However, *in vitro* studies have revealed that cellular uptake of NiO particles allows the release of Ni^{2+} ions within the cells; therefore the toxicity of this material is probably dependent on its intracellular solubility.

11.4.7 Silica

Our model predicts that silica energy levels are not in a favorable position for interfering with the cell's redox state. This oxide is an insulator with conduction and valence band energy levels at approximately -1.4 and -11.5 eV, respectively. Therefore, in principle, this material could be toxic either because there are additional surface states in the band gap due to the material's defects or impurities or because there is an indirect mechanism of oxidative stress involved. The literature evidence is that this material can be toxic.[15,45,46] However, one study did not detect ROS production,[24] suggesting that this oxide could elicit toxicity *via* a different mechanism that does not involve direct ROS production. The doping of silica with other transition metals results in an increase in ROS production and cell toxicity.[24]

11.4.8 Alumina

Most toxicity studies on alumina nanoparticles report that this oxide is not toxic: no measurable effects were detected in a number of cells lines including neuro 2A cells,[31] HT-22 nerve cells, mouse macrophages and lung epithelial BEAS-2B cells.[37] These tests did reveal, however, lactate dehydrogenase (LDH) leakage at high doses in neuro 2A cells and an increase in the levels of IL-6 and IL-8 in lung epithelial BEAS-2B cells.[17] In contrast, an EC_{50} of approximately 5/6 µg mL^{-1} was measured for murine alveolar macrophage RAW 264.7 cells, human alveolar macrophage THB-1 cells and human epithelial A549 cells.[47] Another study[48] reported that alumina nanoparticles are internalized in mouse skin epithelial cells and significantly increase manganese superoxide dismutase protein levels, indicating that the effect of alumina may occur, in part, *via* alteration of the cellular redox status.

According to our model an alumina particle is not a potential ROS producer since its energy band levels ($E_C = -1.4$ eV and $E_V = -9.9$ eV) are not in a favorable position to accept or release electrons from/to the cell. Discrepancies with observed toxicity may arise from different toxicity pathways which this oxide may elicit when in contact with cells.

11.5 Conclusions

The model presented in this report can qualitatively predict the oxidative stress potential of metal oxide particles towards cells: by knowing the energy structure of the oxide and the redox potential of couples involved in the oxidation/reduction reactions, it is in principle possible to predict the ability to generate ROS. In addition, in the case of titania, we show that the reactivity descriptors employed here, the electronegativity (Fermi energy) and the band gap of the bulk oxide, can be used to predict the photocatalytic activity of this material and the isoelectric point of the planes exposed by rutile crystals.

It should be noted that ROS production also depends on other properties of the materials, *e.g.* their solubility, and additional information is required to fully understand their toxicity. For a quantitative prediction of the level of toxicity, a more detailed description of the metal oxide structure is required and therefore a thorough characterization and structure calculation of the material is needed to fully understand its catalytic and redox properties. For some materials this information is available; however, these data pertain to idealized nanomaterials, whereas those used during the tests can display a large structural variability that depends on the synthesis method and environmental conditions. One important contribution for elucidating the mechanism of toxicity of these materials would be the use of acellular experiments to identify possible redox cycling or catalytic mechanisms that involve specific cellular components and molecules and lead to the formation of ROS. These studies are starting to appear in the literature[27] and constitute a valuable piece of information that helps to understand the structure–toxicity relationship of nanomaterials. These experiments could be complemented with quantum mechanics calculations to describe the surface structure and reactivity of nanoparticles; both experimental and computational methods are applicable because in both cases the control of the test conditions is more accessible compared to *in vitro* model tests.

Acknowledgements

This work was supported by the NanoTEST project under contract no. HEALTH2008-201335 of the European Commission. The authors report no conflicts of interest. The authors alone are responsible for the content and writing of the paper.

References

1. H. Meng, T. Xia, S. George and A. E. Nel, A predictive toxicological paradigm for the safety assessment of nanomaterials, *ACS Nano*, 2009, **3**, 1620–1627.
2. N. Li, C. Sioutas, A. Cho, D. Schmitz, C. Misra, J. Sempf, M. Wang, T. Oberley, J. Froines and A. Nel, Ultrafine particulate pollutants induce oxidative stress and mitochondrial damage, *Environ. Health Perspect.*, 2003, **111**, 455–460.
3. E. Burello and A. Worth, A theoretical framework for predicting the oxidative stress potential of oxide nanoparticles, *Nanotoxicology*, 2011, **5**, 228–235.
4. M. Auffan, J. Rose, M. R. Wiesner and J. Y. Bottero, Chemical stability of metallic nanoparticles: a parameter controlling their potential cellular toxicity *in vitro*, *Environ. Pollut.*, 2009, **157**, 1127–1133.
5. W. Kohn, A. D. Becke and R. G. Parr, Density Functional Theory of Electronic Structure, *J. Phys. Chem.*, 1996, **100**, 12974–12980.
6. J. Portier, H. S. Hilal, I. Saadeddin, S. J. Hwang, M. A. Subramanian and G. Campet, Thermodynamic correlations and band gap calculations in metal oxides, *Prog. Solid State Chemistry*, 2004, **32**, 207–217.
7. *CRC Handbook of Chemistry and Physics*, ed. D. R. Lide, Taylor and Francis, Boca Raton, FL, Edition 86th, 2006.
8. J. Portier, G. Campet, A. Poquet, C. Marcel and M. A. Subramanian, Degenerate semiconductors in the light of electronegativity and chemical hardness, *Int. J. Inorg. Mater.*, 2001, **3**, 1039–1043.
9. M. A. Butler and D. S. Ginley, *J. Electrochem. Soc.*, 1978, **125**, 228–232.
10. A. M. Fond and G. J. Meyer, Biotoxicity of metal oxide nanoparticles, in *Nanotechnologies for the Life Sciences* (Nanomaterials – Toxicity, Health and Environmental Issues), Weinheim, Wiley-VCH, 2007, vol. 5.
11. P. M. Wood, The potential diagram for oxygen at pH 7, *Biochem. J.*, 1988, **253**, 287–289.

12. U. Diebold, The Surface Science of Titanium Dioxide, *Surf. Sci. Rep.*, 2003, **48**, 53–229.

13. J. W. Bullard, *Anisotropic and tunable characteristics of the colloidal behavior of metal oxide surfaces*, PhD Thesis, Department of Materials Science and Engineering, MIT, USA, 2006. Available at: http://hdl.handle.net/1721.1/37372

14. H. H. Kung, *Transition Metal Oxides: Surface Chemistry and Catalysis*, Elsevier Science Publishers B.V., Amsterdam, The Netherlands, 1989.

15. T. J. Brunner, P. Wick, P. Manser, P. Spohn, R. N. Grass, L. K. Limbach, A. Bruinink and W. J. Stark, In Vitro Cytotoxicity of Oxide Nanoparticles: Comparison to Asbestos, Silica, and the Effect of Particle Solubility, *Environ. Sci. Technol.*, 2006, **40**, 4374–4381.

16. A. Gojova, B. Guo, R. S. Kota, J. C. Rutledge, I. M. Kennedy and A. I. Barakat, Induction of inflammation in vascular endothelial cells by metal oxide nanoparticles: effect of particle composition, *Environ. Health Perspect.*, 2007, **115**, 403–409.

17. J. M. Veranth, E. G. Kaser, M. M. Veranth, M. Koch and G. S. Yost, Cytokine responses of human lung cells (BEAS-2B) treated with micron-sized and nanoparticles of metal oxides compared to soil dusts, *Part. Fibre Toxicol.*, 2007, **4**, 2.

18. F. Cengelli, D. Maysinger, F. Tschudi-Monnet, X. Montet, C. Corot, A. Petri-Fink, H. Hofmann and L. Juillerat-Jeanneret, Interaction of functionalized superparamagnetic iron oxide nanoparticles with brain structures, *J. Pharmacol. Exp. Ther.*, 2006, **318**, 108–116.

19. A. K. Gupta and M. Gupta, Cytotoxicity suppression and cellular uptake enhancement of surface modified magnetic nanoparticles, *Biomaterials*, 2005, **26**, 1565–1573.

20. F. Hu, K. G. Neoh, L. Cen and E. T. Kang, Cellular response to magnetic nanoparticles "PEGylated" via surface-initiated atom transfer radical polymerization, *Biomacromolecules*, 2006, **7**, 809–816.

21. A. K. Gupta and A. S. Curtis, Surface modified superparamagnetic nanoparticles for drug delivery: Interaction studies with human fibroblasts in culture, *J. Mater. Sci.: Mater. Med.*, 2004, **15**, 493–496.

22. S. M. Hussain, K. L. Hess, J. M. Gearhart, K. T. Geiss and J. J. Schlager, In vitro toxicity of nanoparticles in BRL 3A rat liver cells, *Toxicol. In Vitro*, 2005, **19**, 975–983.

23. K. Müller, J. N. Skepper, M. Posfai, R. Trivedi, S. Howarth, C. Corot, E. Lancelot, P. W. Thompson, A. P. Brown and J. H. Gillard, Effect of ultrasmall superparamagnetic iron oxide nanoparticles (Ferumoxtran-10) on human monocyte-macrophages in vitro, *Biomaterials*, 2007, **28**, 1629–1642.

24. L. K. Limbach, P. Wick, P. Manser, R. N. Grass, A. Bruinink and W. J. Stark, Exposure of engineered nanoparticles to human lung epithelial cells: influence of chemical composition and catalytic activity on oxidative stress, *Environ. Sci. Technol.*, 2007, **41**, 4158–4163.

25. M. Fonin, R. Pentcheva, Y. S. Dedkov, M. Sperlich, D. V. Vyalikh and M. Scheffler, Surface electronic structure of the $Fe_3O_4(100)$: Evidence of a half-metal to metal transition, *Phys. Rev. B*, 2005, **72**, 104436–104444.

26. J. Chaudière, Some chemical and biochemical constraints of oxidative stress in living cells, in *Free Radical Damage and its Control*, ed. C. A. Rice-Evans and R. H. Burdon, Elsevier Science Publishers B.V., Amsterdam, The Netherlands, 1994.

27. M. A. Voinov, J. O. Sosa Pagán, E. Morrison, T. I. Smirnova and A. I. Smirnov, Surface-mediated production of hydroxyl radicals as a mechanism of iron oxide nanoparticle biotoxicity, *J. Am. Chem. Soc.*, 2011, **133**, 35–41.

28. R. M. Cornell and U. Schwertmann, *The Iron Oxides: Structure, Properties, Reactions, Occurrences and Uses*, Wiley-VCH, Weinheim, 2003. ISBN: 3-527-30274-3.

29. F. Trolard and Y. Tardy, The stability of gibbsite, boehmite, aluminous goethites and aluminous hematites in bauxites, ferricretes and laterites as a function of water activity, temperature and particle size, *Geochim. Cosmochim. Acta*, 1987, **51**, 945–957.

30. E. K. Dufour, T. Kumaravel, G. J. Nohynek, D. Kirkland and H. Toutain, Clastogenicity, photo-clastogenicity or pseudo-photo-clastogenicity: Genotoxic effects of zinc oxide in the dark, in pre-irradiated or simultaneously irradiated Chinese hamster ovary cells, *Mutat. Res., Genet. Toxicol. Environ. Mutagen.*, 2006, **607**, 215–224.

31. H. A. Jeng and J. Swanson, Toxicity of metal oxide nanoparticles in mammalian cells, *J. Environ. Sci. Health, Part A: Toxic/Hazard. Subst. Environ. Eng.*, 2006, **41**, 2699–2711.

32. C. M. Sayes, R. Wahi, P. A. Kurian, Y. Liu, J. L. West, K. D. Ausman, D. B. Warheit, and V. L. Colvin, Correlating nanoscale titania structure with toxicity: a cytotoxicity and inflammatory response study with human dermal fibroblasts and human lung epithelial cells, *Toxicol. Sci.*, 2006, **92**, 174–185.

33. T. Xia, M. Kovochich, M. Liong, L. Madler, B. Gilbert, H. Shi, J. I. Yeh, J. I. Zink and A. E. Nel, Comparison of the mechanism of toxicity of Zinc oxide and Cerium oxide nanoparticles based on dissolution and oxidative stress properties, *ACS Nano*, 2008, **2**, 2121–2134.

34. P. W. Schindler, Heterogeneous equilibria involving oxides, hydroxides, carbonates and hydroxide carbonates, in *Equilibrium Concepts in Natural Water Systems*, ed. R. Gould, *Advances in Chemistry*, American Chemical Society, Washington DC, 1967, vol. 67, pp. 196–221.

35. W. Lin, Y. W. Huang, X. D. Zhou and Y. Ma, Toxicity of cerium oxide nanoparticles in human lung cancer cells, *Int. J. Toxicol.*, 2006, **25**, 451–457.

36. R. W. Tarnuzzer, J. Colon, S. Patil and S. Seal, Vacancy engineered ceria nanostructures for protection from radiation-induced cellular damage, *Nano Lett.*, 2005, **5**, 2573–2577.

37. D. Schubert, R. Dargusch, J. Raitano and S. W. Chan, Cerium and yttrium oxide nanoparticles are neuroprotective, *Biochem. Biophys. Res. Commun.*, 2006, **342**, 86–91.

38. R. Wang, P. A. Crozier and R. Sharma, Structural Transformation in Ceria Nanoparticles during Redox Processes, *J. Phys. Chem. C*, 2009, **113**, 5700–5704.

39. C. Korsvik, S. Patil, S. Seal and W. T. Self, Superoxide dismutase mimetic properties exhibited by vacancy engineered ceria nanoparticles, *Chem. Commun.*, 2007, 1056–1058.

40. H. L. Karlsson, P. Cronholm, J. Gustafsson and L. Moller, Copper oxide nanoparticles are highly toxic: a comparison between metal oxide nanoparticles and carbon nanotubes, *Chem. Res. Toxicol.*, 2008, **21**, 1726–1732.

41. A. M. Studer, L. K. Limbach, L. Van Duc, F. Krumeich, E. K. Athanassiou, L. C. Gerber, H. Moch and W. J. Stark, Nanoparticle cytotoxicity depends on intracellular solubility: Comparison of stabilized copper metal and degradable copper oxide nanoparticles, *Toxicol. Lett.*, 2010, **197**,169–174.

42. M. Xu, D. Fujita, S. Kajiwara, T. Minowa, X. Li, T. Takemura, H. Iwai and N. Hanagata, Contribution of physicochemical characteristics of nano-oxides to cytotoxicity, *Biomaterials*, 2010, **31**, 8022–8031.

43. M. Horie, K. Nishio, K. Fujita, S. Endoh, A. Miyauchi, Y. Saito, H. Iwahashi, K. Yamamoto, H. Murayama, H. Nakano, N. Nanashima, E. Niki and Y. Yoshida, Protein adsorption of ultrafine metal oxide and its influence on cytotoxicity toward cultured cells, *Chem. Res. Toxicol.*, 2009, **16**, 543–553.

44. M. Horie, H. Fukui, K. Nishio, S. Endoh, H. Kato, K. Fujita, A. Miyauchi, A. Nakamura, M. Shichiri, N. Ishida, S. Kinugasa, Y. Morimoto, E. Niki, Y. Yoshida and H. Iwahashi, Evaluation of acute oxidative stress induced by NiO nanoparticles in vivo and in vitro, *J. Occup. Health*, 2011, **53**, 64–74.

45. M. Chen and A. von Mikecz, Formation of nucleoplasmic protein aggregates impairs nuclear function in response to SiO_2 nanoparticles, *Exp. Cell Res.*, 2005, **305**, 51–62.

46. K. Peters, R. E. Unger, C. J. Kirkpatrick, A. M. Gatti and E. Monari, Effects of nano-scaled particles on endothelial cell function in vitro: studies on viability, proliferation and inflammation, *J. Mater. Sci.: Mater. Med.*, 2004, **15**, 321–325.

47. K. Soto, K. M. Garza and L. E. Murr, Cytotoxic effects of aggregated nanomaterials, *Acta Biomater.*, 2007, **3**, 351–358.

48. S. Dey, V. Bakthavatchalu, M. T. Tseng, P. Wu, R. L. Florence, E. A. Grulke, R. A. Yokel, S. K. Dhar, H. S. Yang, Y. Chen and D. K. St Clair, Interactions between SIRT1 and AP-1 reveal a mechanistic insight into the growth promoting properties of alumina (Al_2O_3) nanoparticles in mouse skin epithelial cells, *Carcinogenesis*, 2008, **29**, 1920–1929.

CHAPTER 12

Modeling the Environmental Release and Exposure of Engineered Nanomaterials

FADRI GOTTSCHALK AND BERND NOWACK*

EMPA – Swiss Federal Laboratories for Materials Science and Technology, Technology and Society Laboratory, St. Gallen, Switzerland
*E-mail: nowack@empa.ch

12.1 Introduction

Engineered nanomaterials (ENMs) are used in a broad spectrum of consumer products. The particle size at the nano-range gives ENMs particular physicochemical properties that may lead to exceptional performance regarding parameters such as *e.g.* surface reactivity, conductivity, porosity, solubility, crystallinity, optical sensitivity, bio-persistence, *etc.*[1–5] Apart from a large range of potential benefits to the human health and the environment, the use of materials with such particular properties could also cause dangerous interactions with human and environmental systems.[6–13] The research community is faced with a risky lack of adequate methodologies to investigate these potentially harmful environmental effects of materials at such a small scale.[14] Consequently, there is a distinct knowledge gap concerning the risks caused by the environmental release of and exposure to such materials.[6–9,11,12,15–24] There is no doubt that current and future ENM production quantities release them into the environment.[15,17,25–28]

RSC Nanoscience & Nanotechnology No. 25
Towards Efficient Designing of Safe Nanomaterials: Innovative Merge of Computational Approaches and Experimental Techniques
Edited by Tomasz Puzyn and Jerzy Leszczynski
© The Royal Society of Chemistry 2012
Published by the Royal Society of Chemistry, www.rsc.org

One of the critical points of environmental release and exposure of such material is that the emitted nanosized material may show – compared to its bulk counterpart – a new and mostly unknown behavior, for example, dissolution and agglomeration increased surface reactivity, altered magnetism or optical characteristics and an increased strength and flexibility, electrical conductivity or absorption.[29] ENMs in commercial products do not represent a uniform group of materials, but are produced in many different sizes and forms.[4] ENMs may also be coated *e.g.* with polymers, polyelectrolyte or surfactants that can influence the water solubility or biocompatibility of the ENMs.[15,30]

The potential for negative environmental effects of ENMs is very widely discussed.[23,31–33] Nano-specific regulations that cover such effects, however, are not yet available.[34,35] Normally environmental regulations are focused on conventional chemicals and do not address the aforementioned particular colloidal physicochemical characteristics.[29] REACH[36] is focused on the effects and exposure of chemicals in general, without treating separately the particular physicochemical characteristics of nanosized materials. However, environmental benefits due to ENM applications *e.g.* in pollution prevention and remediation or green manufacturing are expected as well (see *e.g.* ref. 37 and 38). These new material characteristics and functionalities are also expected to reduce *e.g.* energy consumption, waste production and the use of other hazardous chemicals.[9,39]

Frequently used nanomaterials are *e.g.* TiO_2, Ag, ZnO, carbon nanotubes (CNTs) and fullerenes. Probably the first study concretely addressing the environmental release of ENMs showed that CNTs could be released into the environment from lithium-ion batteries and textiles during different life cycle stages of these products.[40] CNTs are also incorporated into other products such as composite materials, electronics and medical applications since they are extremely strong and lightweight and may show different electrical properties.[41,42] Fullerenes, another carbon-based form of ENM, are *e.g.* used in cosmetics, food supplements, electronics and fuel cells.[42,43] The release caused by the use of nano-TiO_2 as an active ingredient in sunscreens has also been discussed.[44] Different technical uses are evaluated for nano-Ag:[45,46] due to its antimicrobial properties nano-Ag is used in cosmetics, textiles, paints, cleaning agents and sprays[47,48] and also in wound dressings and detergents.[49] The US Environmental Protection Agency (EPA) aims to regulate the use of nanosized particles with antimicrobial properties (*e.g.* nano-Ag), as is done for other antimicrobials or pesticides that fall under the Federal Insecticide, Fungicide & Rodenticide Act (FIFRA).[50] Since nano-ZnO (and nano-TiO_2) filter UV light of a broad spectrum, these compounds are often applied *e.g.* as photo catalysts[51] and, as mentioned above, in sunscreen products.[52] Nano-TiO_2 may *e.g.* also be used for the decontamination of water, soil and air.[53]

At the beginning of the research on environmental release and effects of ENMs, the focus was on a precautionary approach[8] that addressed hypothetical questions such as whether ENMs could reach the environment and what their interaction with the environment would be after such a release. Initially the presence of ENMs in the environment had to be assumed by expecting environmental release due to an increasing production and application of all types of ENM.[54] This changed when Kaegi *et al.*[55] detected nano-TiO$_2$ in a small stream emitted from painted house facades. Others then detected nano-Ag released from textiles during washing.[56,57]

In this chapter we present firstly the current state of knowledge on the environmental release of and exposure to ENMs by reviewing the available methodologies and models used and developed to quantify such release and exposure. Secondly, the adequacy of the REACH regulations for dealing with ENM release and exposure in the environment will be examined. Furthermore, the available modeled results in the scientific literature will be reviewed and a comparison to first available measurements will be conducted to discuss to what extent the modeled output validates analytical work and *vice versa*. Finally, we will give an outlook as to what one should expect from future modeling and analytical work on ENM release and exposure.

12.2   Environmental Release and Exposure in REACH

According to the REACH guidance,[58] the release and exposure estimation for chemicals is usually conducted based on the fugacity concept by using multimedia fate models. The model of the European Union System for the Evaluation of Substances (EUSES) is a well-established software that has been created in order to perform environmental exposure (also including worker and consumer exposure/risk), effects and risk assessments.[59] EUSES is a transparent and user-friendly quantitative assessment tool for estimating potential risks caused by chemicals to human beings and the environment. Other prominent models are mass-balance multi-compartment approaches[60–65] such as *e.g.* the Berkeley–Trent (BETR)[66] and the climate zone model for chemicals (CliMoChem).[67] These models represent box models, covering several environmental compartments (Figure 12.1), which may be considered homogeneous and well mixed.

Emitted chemicals are distributed between the different compartments according to the substance characteristics and environmental properties. REACH distinguishes among different forms of release and transfer between compartments for potential contaminants:[58]

- direct and indirect emission (*via* sewage treatment plants (STPs)) to air, water, industrial and agricultural soil;
- biotic and abiotic degradation;
- diffusive transfers between two compartments such as gas absorption and volatilization depending on the concentration in both compartments;
- advective one-way transport dynamics such as deposition, run-off and erosion.

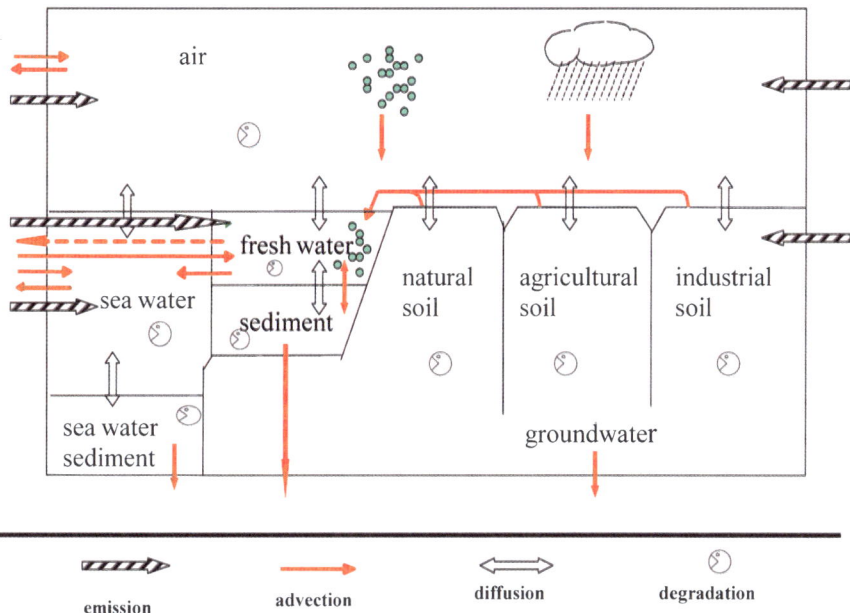

Figure 12.1   Regional emission and distribution routes. Reprinted from ref. 58. Reproduction authorized based on the source acknowledgement.

There are several limitations that have to be considered in such a regional release/exposure methodology:[58] (i) the model outputs with a regional resolution only reflect averages for entire regional compartments that are assumed to be well and homogeneously mixed. Such a risk assessment from a regional perspective may ignore a local release that might be much higher. (ii) The challenge in such modeling is to quantify realistically the main "release" parameters such as partitioning coefficients and degradation rates for a potential contaminant. Two approaches are recommended: (a) modeling of release/exposure based on "agreed" standardized regional parameters for a standard European region; and (b) modeling of regional release/exposure on the basis of "country specific" parameter values when sufficient specific information is at hand, *e.g.* from the most relevant emission sites. About two dozen standard parameter values ranging from geometrical and systemic information of the compartments (water depth, residence time of air and water *etc.*), intermedia mass transfer coefficients, precipitation data, and average connection percentages to STPs are defined. The guidance on the REACH implementation[58] distinguishes three release scenarios: a wide, dispersive and an industrial setting scenario on a local resolution and a regional case for all chemical uses. Wide dispersive scenarios cover release from consumption processes, professional, and service life chemical uses whereas the industrial setting reflects a release from single industrial sources.

A wide dispersive release stems from the use of substances and substances in mixtures by private consumers and non-industrial companies. These kinds of emissions are assumed to reflect constant and continuous material transfer that may be averaged over a one year time period. Direct emissions to the atmosphere and soils are not considered; emissions to the water body are modeled as emission *via* a 10 000 inhabitants standard STP. The total chemical tonnage used is derived for such a standard town from the total registrant's tonnage at the EU level. Assessment factors are considered to account for temporal or geographical variation in the chemical use and release scenarios.

Industrial setting scenarios: discharge to water occurs *via* municipal sewage treatment plants (STPs) and industrial wastewater treatment plants (WWTP). Emissions to the atmosphere are considered to occur *via* water treatment processes in STPs and emissions to soils from the application of STP sludge to agricultural soil and from atmospheric deposition of released material. Default daily emission rates for manufacturing (process where the substance is produced), formulation (blending and mixing materials) and industrial end uses (*e.g.* using such a substance as a processing aid or incorporating it into a matrix) are quantified. Release factors are used to describe the fraction (% or kg kg^{-1}) emitted to a particular environment. To determine such factors, environmental release categories (ERCs) are provided. Conservative default release factors for these ERCs expressed in % are listed in Appendix R.16-1 of chapter 16 of ref. 58.

In the *regional resolution* widely dispersed release and industrial release are lumped together to quantify total emissions to a water body, soil and air. Standard modeling assumes 80% (EU average) of the wastewater is cleaned in STPs. Emissions are considered to occur continuously during the total life cycle of products containing a particular compound. If regional emissions are associated with specific-use scenarios, the tonnage at regional level for each use and the release factors are equal to the parameters used at local scale.

The regional emission for each environmental compartment is obtained by summing the emissions over all life cycle stages of a particular substance. If the activities related to a particular life cycle stage can be considered to occur within a region under investigation, as is normally the case for manufacturing, formulation and industrial uses, the whole registered tonnage at the EU level is

Table 12.1 Direct emissions to environmental compartments of the different scenarios.[a] Taken from ref. 58. Reproduction authorized based on the source acknowledgement.

Release compartment scenario	Air	Water – via STP	Water – direct	Soil
Industrial setting – local scale	Y	Y	N	N
Wide dispersive uses – local scale	N	Y	N	N
Industrial settings – total regional				
Wide dispersive uses – total regional	Y	Y (80%)	Y (20%)	Y

[a]Y: release pattern taken into account in the scenario; N: release pattern not considered in the scenario.

accounted for. Otherwise only a fraction of the registered tonnage is attributed to a particular region, the remaining part being attributed to the continental scale.

12.3   Environmental Release and Exposure Assessment for ENMs

Although the current database on environmental release and exposure for ENMs is very thin, several state-of-the-art reviews are already available. Contributions may be found about ENM release, fate and/or implications in the environment.[4,28,68–74] Other publications[12,75] have reviewed potential regulations for ENMs and discussed how far ENMs have to be considered as a new material that cannot be regulated based on conventional regulation approaches for its macroscaled counterparts. Grieger *et al.*[76] screened the current literature and analyzed the level of knowledge about environmental, health and safety effects (EHS) for ENMs; knowledge gaps were found everywhere: *e.g.* lack of environmental release and fate information and human and environmental toxicity, of standardized methods and reference material to assess potential negative effects, and of life cycle and commercial aspects for nanoproducts. Finally, reviews about potential environmental exposure associated with nanoremediation are available as well.[77,78]

12.3.1   Early Qualitative Release/Exposure Analysis

What makes nanomaterials different from the bulk material/chemicals in terms of exposure potential had been recognized very early: Morgan[79] used expert elicitation – the most prominent research approach at the beginning of the environmental impact assessment for ENMs – to identify the set of particle-related properties potentially determining ENM exposure (and toxicity) (see Figure 12.2). Kandlikar *et al.*[80] proposed investigations on the degree of consensus (disagreement) between experts that estimated the importance of the crucial variables within the exposure and risk paradigm by assigning subjective degrees of belief (probability values) to the potentially key variables such as particle surface characteristics, particle size distribution, particle shape and agglomeration behavior; release metrics: mass, surface area, particle number; exposure routes: *via* inhalation, dermal exposure, *via* ingestion *etc.*; environmental fate and translocation processes and dose–response mechanisms.

Furthermore, researchers soon recognized that using a life cycle perspective and continuously examining new nanoproducts would be critical to investigate environmental release and exposure.[17,69,81–83] Robichaud *et al.*[84] performed a qualitative risk assessment for the industrial production of several ENMs (CNTs, fullerenes (C_{60}), quantum dots, aluminum oxide nanoparticles, and nano-TiO_2). By means of an industrial insurance database and from an industrial insurer's perspective, the authors were able to benchmark the risks of fabrication processes of one ENM against other ENMs and against other

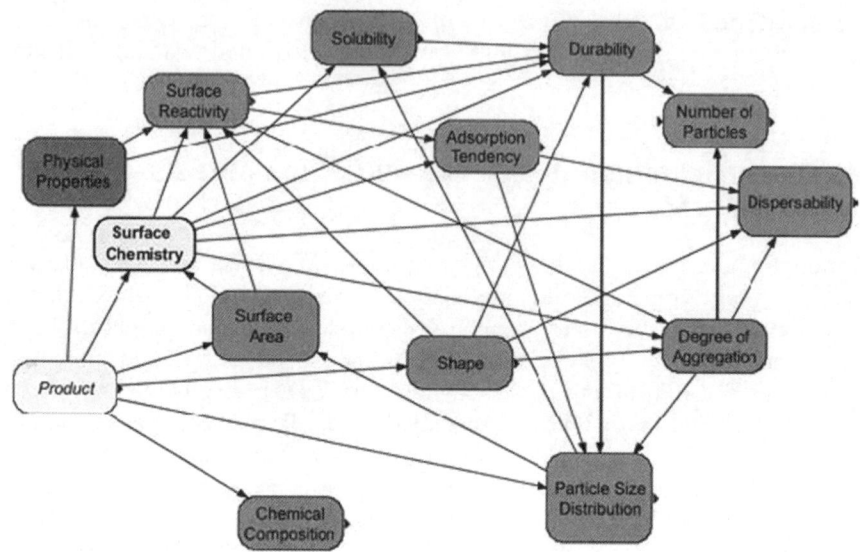

Figure 12.2   Particle-related material properties determining exposure and toxicity. Reprinted with permission from ref. 79. Copyright (2005), Society for Risk Analysis.

non-nanobased materials. Exposure quantities and physicochemical material characteristics were used to qualitatively rank risk based on ENM carcinogenicity, flammability, volatility, persistence (bioaccumulation) and toxicity. Risk protocols were developed that ranked three risk categories (incident risk, normal operations risk, and latent risk) by assessing the relative contamination up to a 100 point risk scale (100 means maximum risk).

Davis and Thomas[83] provided a basic concept named comprehensive environmental assessment (CEA) developed to prioritize research efforts for engineered nanomaterials and products containing such materials (Figure 12.3). The recent EPA-reviews[44,85] about nano-TiO$_2$ and nano-Ag have followed this CEA approach. Michelson and Rejeski[86] proposed a gap analysis approach based on a pre-market regulatory review to identify "at risk" product areas (cosmetics, dietary supplements, food additives and consumer products). Anticipative life-cycle perspective assessment approaches focused on past experience in environmental exposure and risk assessment were proposed and exemplified in one of the first studies with the case of MTBE (methyl-*tert*-butylether) used as a fuel additive.[82]

As seen in Table 12.2, unintended indirect ENM emissions are normally clearly localizable and occur mostly *via* technical facilities. Such emissions will occur *via* wastewater treatment and waste incineration plants and landfills. Hence, the main focus of the release research should be on the – currently only poorly understood – removal of ENMs during wastewater treatment[87] and during waste incineration as well as during leaching processes from landfills.[88] In contrast, unintended diffusive release occurs mostly *via* the application/

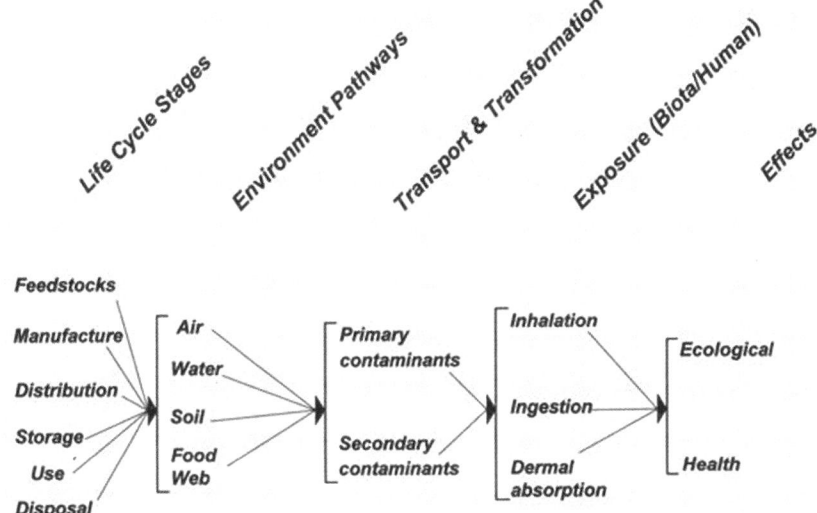

Figure 12.3 Basic structure of the comprehensive environmental assessment (CEA) approach to identifying and prioritizing research efforts for a nanoscale product. Reprinted with permission from ref. 83. Copyright (2006) American Scientific Publishers.

degradation of ENM products such as paints, cosmetics, cleaning agents, surface coatings *etc.* Consequently and as emphasized in Nowack *et al.*[89] minimizing such ENM release requires measures to be taken on the products themselves: ensure a persistent ENM binding in the matrix and minimizing degradation/abrasion of the matrix *e.g.* when exposed to rain water or UV light. However, for other uses ENM emissions cannot be avoided, *e.g.* ENMs containing cosmetics/sunscreens washed off during swimming.

Finally, unintended emissions cannot be excluded either during ENM production and manufacturing processes (ENM incorporation into products) as well as from accidents or any kind of site damage cases that could lead to extremely high water, air and soil exposure at the concerned locations. Nowack *et al.*[89] describe events such as *e.g.* incautiously handled nano-powders emitted to the air at open windows and release from wastewater treated internally in the factory and then directly discharged into natural waters. Measurements to examine worker exposure performed at ENM production sites later confirmed such release events.[90–94]

For studying the likelihood of release and exposure in the environment it is also crucial to assess in what forms ENMs are emitted.[28] Koehler and Som[95] distinguished among free nanosized particles, agglomerated and aggregated, or nanomaterial integrated in larger emitted particles. An exemplary case is discussed[28] for nano-TiO_2 emissions from exterior paints[55] where the emitted ENMs can either be emitted as free particles (*e.g.* dry dust into air) or be still

Table 12.2   Environmental release scenarios for ENMs (adapted from ref. 89).

Origin	Emission path (ENM application)	Environmental destination
Intended release (diffuse sources)	Water/groundwater/soil remediation	Water
	Agrochemicals	Sediments
	Use for water treatment in sewage treatment plants	Groundwater
		Soil, air
		Water
		Sediments
Intended release (point sources)		Soil (if biosolids with ENMs removed during water treatment are applied on land)
Unintended release (diffuse sources)	Wear during use, e.g. from tires, textiles etc.	Air, soil, water
	ENMs wash off from sunscreen (in rivers, lakes etc.)	Water
	Weathering, e.g. of outside paints	Soil, water
	Use of additives in fuels	Air, soil
	ENMs as food additive	Water
	Medical use	Water
Unintended release (point source)	Leaching from landfills	Groundwater, soil
	Recycling facility (e.g. recycling of nano-coated plastic/glass/metal material, dismantling of batteries etc.)	Air
		Water
	End-of-life treatment (incineration) of nanotextiles, nanocomposites	Air

embedded in the paint matrix when ending up in the water system or soils (abrasion due to rainwater and sunlight exposure).

Summing up, all these exposure scenarios contribute to environmental exposure that in the end strongly depends on the life cycle of the ENMs and ENM-containing products considered in the scenarios.[96] Som *et al.*[97] have underlined that completely different life cycles for the same ENM may be observed: the use of nano-TiO$_2$ in sunscreen varies *e.g.* considerably from the use of this ENM in paints leading to completely different environmental release and exposure scenarios for the same ENM. Hence, ENM-specific life-cycle concepts (Table 12.3) should be combined with current toxicological and risk assessment findings to provide a better base in decision making for regulators and industry.[9] Such concepts would include a relative evaluation/ comparison of the environmentally relevant performance of nanoproducts and their conventional counterparts.

Table 12.3   Life-cycle concepts relevant for the development of safe nanoproducts. Reproduced with permission from Som *et al.*[9]

Methods provide information on:	Risk assessment				Life cycle concepts		
	Toxicology translocation	Ecotoxicology and environmental behavior	Monitoring	Exposure – effect studies, biomonitoring (incl. occupational health)[a]	Life cycle thinking in the framework of technology assessment, e.g. foresight, roadmapping etc.	Material flow analysis	LCA
Future nanoapplication and -products					×		(×)
Mode of ENMs integration into nanoproducts (e.g. amount and types of ENMs)					×		×
Release scenarios for ENMs from nanoproducts (product life cycle stage, compartment released to, quantities, "forms")		×		×	×	×	×
Exposure routes		×		×	×	×	
Quality of exposure: potential uptake paths (e.g. respiratory system, dermal, gut)	×	×			×		
Behavior of ENM in technosphere and environmental compartments		×	×	×			
Occupational health scenarios	×		×	×	×	×	
Consumer exposure scenarios	×		×	×	×	×	
Dose/response relationships for ENMs	×	×					
Bioavailability of ENMs	×	×					
Biopersistency of ENMs	×	×	×				
Degradation of ENMs in technosphere and environmental compartments under what conditions		×	×				
Relative impact of nanoproducts on human health and the environment	×	×		×			×
Environmental sustainability: opportunity for material and energy savings or risk for increased resource consumption							×

[a]Risk assessment methods, as well as life cycle methods contribute to exposure studies.

Linkov *et al.*[98] used in such a context multi-criteria decision analysis (MCDA) in order to examine three hypothetical alternative ENMs with different societal and economic characteristics and potential for environmental effects (Figure 12.4). The relative environmental effects of the hypothetical ENM applications were assessed by contrasting such effects to the societal importance and the stakeholder preference of these applications. Koehler *et al.*[40] then got down to specifics for ENMs regarding the consideration of the life-cycle principle in exposure assessment. Possible environmentally relevant emission sources for CNTs in synthetic textiles and lithium-ion secondary batteries were investigated. Their analysis considered fully the life cycle of the ENMs and the ENM-containing products. The authors showed that

People:

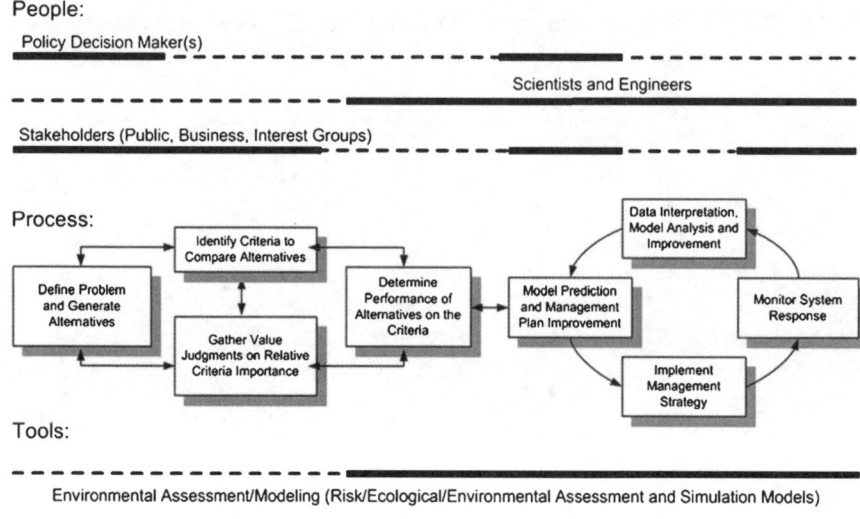

Tools:

Environmental Assessment/Modeling (Risk/Ecological/Environmental Assessment and Simulation Models)

Decision Analysis (Group Decision Making Techniques/Decision Methodologies and Software)

Figure 12.4   ENM-specific multi-criteria decision analysis (MCDA) framework: solid
lines illustrate direct involvement; dashed lines less direct involvement.
Reprinted with permission from ref. 98. Copyright (2007), Springer.

significant environmental emissions of ENM during the ENM production as
well as during the nanoproduct consumption, recycling and disposal phase
could not be excluded. This qualitative case study may be seen as a kick-start
for semi-quantitative/quantitative release and exposure research. It made clear
that the research community and society were faced with a conspicuous lack of
knowledge under which conditions and in which life-cycle phases ENMs may
be released from products.

12.3.2 Predictive Quantitative Modeling

In this section we focus on the first quantitative modeling studies conducted to
assess environmental exposure to ENMs (see Table 12.4). As mentioned
above, in standard multimedia modeling transfer kinetics for chemicals into
and among environmental compartments are usually quantified based on the
octanol–water partition coefficient (K_{ow}) and the water–gas exchange constant
(Henry constant, K_H). However, K_H cannot be used for substances as non-
volatile as most ENMs. Thus, we cannot simply consider the K_H of a bulk
substance to be predictive for a nanosized material derived from such bulk
substances in particulate form.[99] Therefore, new methods are needed to
describe ENM mass transfers.

Several modeling approaches have already been used.[100–109] All these
methods have in common the fact that they have to operate with a distinct lack
of, or at least high uncertainties in, the data that is needed to parameterize the

Study	ENM	Compartments	System/resolution	Methodology
Boxall *et al.* (2008)[100]	TiO$_2$, ZnO, CeO$_2$, Al-oxide, SiO$_2$, Au, Ag, C$_{60}$	Water, sludge, soil, air	UK/regional	Simplistic release and dilution equations.
Mueller and Nowack (2008)[103]	TiO$_2$, Ag, CNT	Water, air, soil	Switzerland/regional	Material flow scenario analysis for PEC / PNEC (predicted environmental concentrations, predicted no effect concentrations) estimations.
Blaser *et al.* (2008)[101]	Ag (silver in biocidal products containing also nanosilver)	Water	EU/local	Mass flow (release) scenario analysis combined with environmental fate modeling (sedimentation, suspension/resuspension, burial, bed load shift, water flow *etc.*)
Park *et al.* (2008)[102]	CeO$_2$	Air, soil	Generic	Release modeling for different scenarios on application of CeO$_2$ as fuel additive and based on the HIWAY2[111] model and on approaches for PM release from road transport COPERT[112] and TRENDS.[113]
Koelmans *et al.* (2009)[109]	CNT	Sediments	Generic	Material flow analysis[103] combined with mass balance modeling of sediment accumulation processes.
Gottschalk *et al.* (2009)[104]	TiO$_2$, ZnO, Ag, CNT, C$_{60}$	Water, sludge, air, sediments, soils	CH, EU, USA/local	Probabilistic material flow analysis (PMFA)[116] for PEC /PNEC estimations.
Kiser *et al.* (2009)[122]	Ti	STP effluents	USA/local	Filtration (<700 nm), sequencing batch reactor experiments, scanning electron microscopy/electron dispersive X-ray microoanalysis (SEM/EDX).
Farré *et al.* (2010)[123]	C$_{60}$ and C$_{70}$ fullerene and *N*-methylfulleropyrrolidine (C$_{60}$)	STP effluents	Spain/local	Ultrasonication extraction prior to liquid chromatography (LC) and hybrid triple quadrupole linear ion trap mass spectrometry (QqLIT-MS).
Gottschalk *et al.* (2011)[110]	TiO$_2$, ZnO, Ag	River water	Switzerland/local	PMFA[116] combined with graph theory modeling[118].
Neal *et al.* (2011)[127]	Ti	River water	UK/local	Filtration (<450 nm), inductively coupled plasma–mass and plasma–optical emission spectrometry techniques.
Mitrano *et al.* (2011)[128]	Ag	Wastewater samples	USA/local	Inductively coupled plasma–mass spectrometry in single-particle counting methods.

models. However, the methods used differ due to varying mathematical conceptualization and model geometry as well as due to different methods of model input treatment. Gottschalk and Nowack[28] distinguish mathematical methods used to cope with the distinct ENM-specific uncertainty and variability in input and output of the models that range from deterministic single-scenario modeling[100–103,106,109] to probabilistic/stochastic mass-flow computations.[104,105,110,116] Furthermore, material-flow modeling based on particle numbers taken as metric (instead of mass) has been proposed as well.[107] Other researchers have suggested adaptations of the above-mentioned release and exposure modeling guidelines R.16 (to estimate environmental concentrations of chemical substances) in REACH[58] by incorporating *e.g.* nano-specific dissolution and sedimentation coefficients.[108] However, beyond such conceptual/parametric differences these studies also vary in particular due to systemic differences in framing the ENM release.

Gottschalk and Nowack[28] distinguished between *top-down* and *bottom-up* framing methods: The top-down methods[100–102,106] focus on release from a small but relevant set of products by considering a particular market penetration of these ENM products. In the bottom-up approach[103,104,110] the models are fed with data which cover as far as possible a broad spectrum of applications of ENMs in products. Some studies also account for ENM emissions during ENM production and the manufacturing of ENM products as well as during recycling processes. Gottschalk *et al.*[110] provide the first exposure results at higher spatial and temporal resolution by modeling local predicted environmental concentrations (PECs) for ENMs along Swiss rivers.

Table 12.4 characterizes the studies that quantitatively investigated exposure concentrations in the environment. In the following we will review in chronological order the main methodological aspects and model outputs of those works. Figure 12.5 compares for different ENMs all the quantitative modeled or measured environmental concentrations available for the aquatic environment.

12.3.2.1   Top-down Studies

Boxall *et al.*[100] used simple dilution equations for ENM release and (TiO_2, ZnO, CeO_2, Al_2O_3, SiO_2, Au, Ag, C_{60}) to model concentrations in water, sludge, and soil. However, these calculations do not necessarily cover the properties of a broader life cycle of the emitted nanosized material. Emissions to water were *e.g.* considered *via* indirect entry from sewage treatment processes; to air, for example, from industrial stack and traffic sources and *via* consumption of hygiene products; and to soils *e.g.* *via* ENM use in plant protection products or remediation technologies and *via* the application of sewage sludge to soils. The problem with such modeling is that the scenarios only focus on a limited number of release events, thus reflecting only an individual non-comprehensive spectrum of ENM products and applications. It strongly depends on the accuracy of individual parameters such *e.g.* the ENM

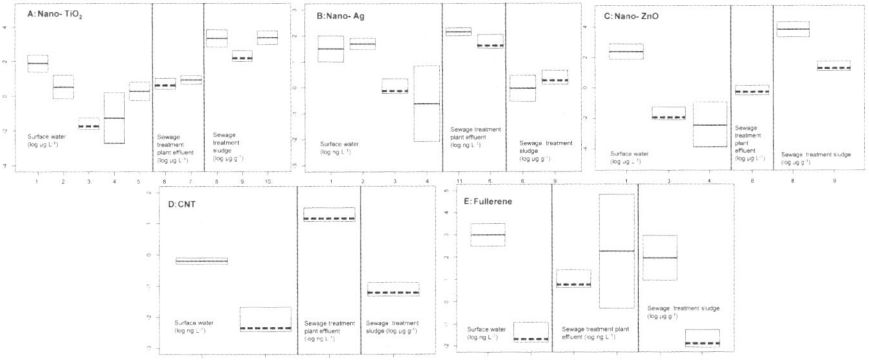

Figure 12.5   Comparison of modeled and analytical concentrations of engineered nanomaterials (ENMs). Gray boxes represent the range (and mean of the range) of measured data, white boxes represent modeled results. Sources: 1. Predicted environmental concentrations (PEC) of 10 and 100% market penetration for surface water at regional (UK) resolution, ENMs used or potentially used in cosmetics, personal care products and coatings;[100] 2. PECs of the realistic and high exposure scenarios for surface water at regional (Switzerland) resolution;[103] 3. PECs (modal value and range of the 15–85% quantiles) for surface water at regional (Europe) resolution;[104] 4. PECs for surface water at local (Switzerland) resolution;[110] 5. Concentrations of river water draining rural and urban land (UK) (filtered Ti < 0.45 μm);[127] 6. PECs (modal value and range of the 15–85% quantiles) for sewage treatment plant effluents at regional (Europe) resolution;[104] 7. Concentrations in effluents from wastewater treatment plants Arizona (USA) (filtered Ti < 0.7 μm);[122] 8. PECs of 10 and 100% market penetration for sewage treatment plant sludge (UK), ENMs used or potentially used in cosmetics, personal care products and coatings;[100] 9. PECs (modal value and range of the 15–85% quantiles) for sewage treatment plant sludge at regional (Switzerland) resolution;[104] 10. Concentrations in sewage treatment plant sludge Arizona (USA) (filtered Ti < 0.7 μm);[122] 11. Concentrations of Ag ENMs concentrations in wastewater samples in Boulder Colorado (USA) in the presence of 50 to 500ng L^{-1} dissolved Ag;[128] 12. Concentrations in effluents from wastewater treatment plants in Catalonia (Spain) of C$_{60}$ and C$_{70}$ Fullerene as well as *N*-methylfulleropyrrolidine (C$_{60}$).[123]

concentrations within a particular product, the consumption dynamics of such a product, the sludge application rates and the removal efficiency of sewage treatment processes. Hence, in some cases the results may stand or fall with the accuracy of the assumed market penetration scenarios (*e.g.* 10, 50 and 100%) for the ENM products considered. The single ENM application-based release modeling led to concentrations (when neglecting the very marginal CeO$_2$ concentrations) in water that ranged from 0.2 ng L^{-1} (nano-Al$_2$O$_3$) to 760 μg L^{-1} (ZnO). The equivalent (rounded) sludge concentrations ranged from 10 μg kg^{-1} (nano-Al$_2$O$_3$) to 21 722 mg kg^{-1} (ZnO), the same values including the same materials for soils were 10 ng kg^{-1} (nano-Al$_2$O$_3$) and 31 944 μg kg^{-1} (ZnO).

Park *et al.*[102] estimated nano-CeO_2 release from use as a diesel additive into air and soils for a street canyon and a highway. The authors considered a worst case (all the cars using CeO_2 fuel additive). ENM emission modeling conducted based on the US EPA HIWAY2[111] approach and on models for PM release calculations from road transport COPERT[112] and TRENDS[113] model was performed. The results for nano-CeO_2 air concentrations ranged from 5 to 80 ng m^{-3}. However, the authors stated that the real concentrations are expected to be marginally lower, since the model did not consider that the contaminants may be washed out and removed by fallout from the atmosphere. However, the results also represent worst cases since all vehicles on the road were assumed to be diesel powered. The modeling for soil contamination indicated that the CeO_2 concentration varied between 0.28 and 1.12 µg g^{-1} depending on the considered soil depth and the distance from the edge of the highway.

In Blaser *et al.*[101] silver emissions from biocidal and nano-functionalized plastics and textiles were quantified. Nano-Ag was modeled to cause emissions of Ag^+ into wastewater discharged (treated or untreated) into the Rhine River. This study concluded that only insignificants amounts of silver were not removed *via* sewage sludge and ended up in natural waters. Most of the material emitted ended up in sludge and was assumed to be incinerated in waste incineration plants (WIPs), spread on soils or disposed in landfills. This model produced results that strongly reflect the amount of silver removed in the treatment plants. Therefore, it will be crucial to know precisely the connection rates to and the removal efficiency of STPs for a considered region. Finally, this paper did not account for nanosized particulate emissions. An evaluation of the uncertainty and variability of the model input and output was not performed: three emission scenarios (minimal, realistic and worst case) were modeled for biocidal plastics and textiles that were predicted to consider up to 15% of the total silver emissions. The modeled water Ag concentrations along the Rhine were estimated to range between 4 ng L^{-1} (lowest value, minimum scenario) and 320 ng L^{-1} (highest value, maximum scenario). The corresponding results for sediments were 0.04 mg kg^{-1} and 14 mg kg^{-1}.

O'Brien and Cummins[106] assessed the exposure to nanosized TiO_2, Ag and CeO_2 in the atmosphere and surface waters. An exposure ranking approach based on ENM release from exterior paints, food packaging and fuel additives was used. Such a semi-quantitative three-level assessment method[114,115] links different material properties and the related release processes as well as the environmental fate of a potential contaminant to produce possible environmental exposure scenarios resulting in exposure (and risk) rankings of varying concerns. Environmental exposure concentrations taken from the literature for some specific ENM applications were assessed against elemental environmental concentrations, provisional regulatory and ecotoxicological limits. Since such results are ranked on a relative scale, no mass-based exposure metrics could be derived from this study. However, the considered use scenario for nano-TiO_2 in exterior paints (assuming 5–10% market penetration) was predicted to result in a relevant increase on current environmental surface water concentrations reaching dimensions above current regulatory/toxicological limits. Nano-Ag in food packaging, assuming again 5–10% market penetration, was said to result in a moderate increase of current environmental exposure, whereas the increase of nano-Ag water concentrations resulting from emission *via* applications in plastics and textiles were said to be significant. The same was said for the use of nano-CeO_2 as a diesel additive and 5–10 % market penetration that resulted in a significant increase of environmental exposure in the atmosphere and water by a subsequent deposition of such ENMs onto waters.

12.3.2.2 Bottom-up Studies

Mueller and Nowack[103] marked the beginning of a fully life-cycle-based quantitative modeling of ENM emissions and exposure in the environment. The environmental release and exposure for Switzerland were modeled. Release events ranging from ENM abrasion during washing and consumption/ use (*e.g.* of textiles, coatings, cosmetics *etc.*), and water treatment in STPs, to product disposal processes were considered. Two release scenarios were accounted for: a realistic and a high release scenario: The first one covered the most realistic input data received at that time. The worst case one was based on higher estimations of ENM release and production/application amounts. The calculations included the first estimation of the worldwide ENM production volumes, the ENM use amounts in different product categories and the estimation of emission factors from these products. The realistic and high emission PECs ranged for nano-TiO_2 from 0.7 to 16 µg L^{-1}, the corresponding results for nano-Ag ranged from 0.03 to 0.08 µg L^{-1}, whereas those for CNTs were about 100 times smaller than the latter values. The equivalent PECs for the soil compartment ranged from 0.4 to 4.8 µg kg^{-1}, 0.02 to 0.1 µg kg^{-1} and from 0.01 to 0.02 µg kg^{-1}. The air exposure concentrations were for all ENMs and scenarios in very small ng m^{-3} dimensions. However, this model has some limitations as only two scenarios deal with high uncertainties of a broad spectrum of input parameters which cover ENM transfer between several ENM applications and environmental as well as technical compartments. Second, due to missing input data some relevant compartments such as sediments, effluents from STPs, STP sludge and others were not included in this model. It also focused only on two ENM release scenarios and the geographical boundaries of Switzerland.

Subsequent studies on ENM release and exposure[104,105] extended the single scenario analysis by using a new methodological approach and by considering further ENMs and environmental compartments as well as geographical regions. A probabilistic/stochastic methodology was used to deal with the high uncertainty and variability in the model input data. The stochastic material flow analysis (PMFA) model used[116] was fully based on Monte Carlo (MC) computations for the input and output of the whole complex material flow system that covers ENM production and application in products, distribution to product consumption dynamics, environmental release as well as inter compartmental transfer (natural and technical compartments) kinetics. This approach allows one to compute probability distributions for all the parameters throughout the whole material flow system by incorporating a large number of material transfer processes (see Figure 12.6). In addition to the fact that such a model may consider a whole spectrum of the target material flow events for potential contaminants, it also provides insight into the likelihood of those flows. Another advantage of this model is that if sufficient data are available, prior input and output information may be transformed into posterior results by embedding Bayesian updating into the MC computations. Furthermore, a sensitivity analysis method is provided that estimates the influence of an individual input parameter on the model output variables by considering the variability and/or uncertainty of the investigated input parameter. Environmental release and exposure on a regional scale for the USA, Europe and Switzerland were modeled for Ag,

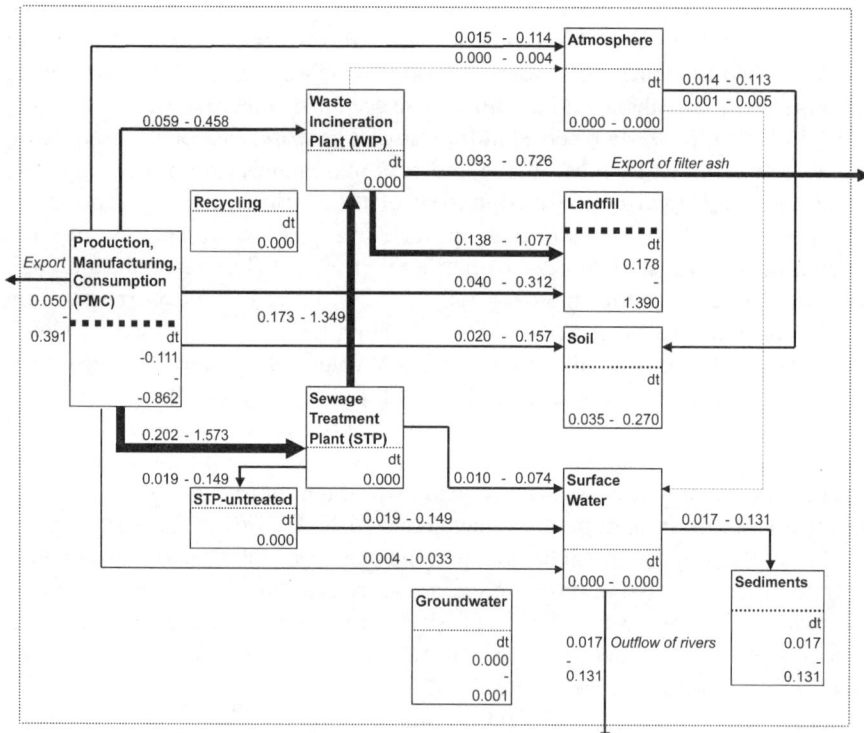

Figure 12.6 Emission and transmission of nano-Ag flows (>0.0005 tonnes *per annum*), elimination and accumulation rates modeled as the range of the lower and upper quantiles: Q(0.15) and Q(0.85). The thickness of the arrows shows the proportions of the ENM flows. The thickness of the horizontal lines in the boxes indicate the proportion of the annual accumulation as well as elimination. Figure reproduced with permission from ref. 105. Copyright (2011) SETAC.

ZnO,TiO$_2$ and ENM, fullerenes and CNTs. For surface water, STP effluents and air this model produced ENM concentrations; for sediments, soils and sludge-treated soils, the annual increase of ENM concentration. The rounded modal surface water values modeled for Europe were 0.02 µg L^{-1} for nano-TiO$_2$, 0.01 µg L^{-1} for nano-ZnO, 0.8 µg L^{-1} for nano-Ag, 0.004 µg L^{-1} for CNT and 0.02 µg L^{-1} for fullerenes. The corresponding results (annual accumulation) for soils were 1.3 µg kg^{-1}, 0.09 µg kg^{-1}, 22.7 µg kg^{-1}, 1.5 ng kg^{-1} and 0.06 ng kg^{-1}, those for sediments 360 µg kg^{-1}, 2.9 µg kg^{-1}, 0.95 µg kg^{-1}, 240 ng kg^{-1} and 17 ng kg^{-1}. The air exposure concentrations were in line with the previous study, again for all ENMs in a very small ng m^{-3} range. In Figure 12.5 regional (Switzerland) results for the STP effluents and STP sludge are also shown.

All these bottom-up studies have some limitations since they are faced with a complete lack of information regarding the ENM emissions from the ENM production process and their incorporation into industrial and consumption products. Another problematical point with all these exposure studies at regional scale is the homogeneous contaminant material distribution (in time and space) within aggregated environmental compartments. In order to overcome such model weaknesses, the use of higher spatial and temporal resolutions is required.[58]

Musee[117] concentrated on ENM release (nano-Ag and nano-TiO$_2$) into aquatic and terrestrial environments from cosmetic products in Metropolitan Johannesburg City. A simple mass-flow analysis covering treated and untreated streams from STPs, landfilling and deposition of sewage sludge on agricultural soils was used. The terrestrial PEC values for nano-Ag and nano-TiO$_2$ in Johannesburg City (considering high and low removal efficiency under minimum, probable, and maximum release scenarios) from the cosmetic products ranged in total from 5.33×10^{-6} to 1.93×10^{-3} µg L^{-1} and from 4.76×10^{-6} to 7.73×10^{-4} µg L^{-1}. The aquatic PEC values (in STP effluents) of nano-Ag ranged at high removal efficiency and considering a dilution factor of 1 between 2.8×10^{-3} and 6.2×10^{-1} µg L^{-1} whereas the nano-TiO$_2$ results ranged from 2.7×10^{-3} to 2.7×10^{-1} µg L^{-1}.

Gottschalk *et al.*[110] linked PMFA and graph theory[118] to extend regional exposure analysis to a local scale modeling that considered spatial and time-dependent differences in ENM exposure. The PECs of TiO$_2$, ZnO and Ag ENM were modeled in 543 Swiss river sections by considering the geographical variation and flow measurements of a 20-year time period to account for temporal variability (Figure 12.7). Due to missing or contradicting input data for such generic ENM materials the transport and fate of ENMs in the river network had to be covered by two scenarios: a scenario with no ENM transformation/deposition and an optimistic scenario including complete ENM removal from the water phase. Besides providing updated environmental exposure results for ENMs, this study conducted a comprehensive systemic analysis that distinguished between model input uncertainty, such as nanomaterial release and transport amounts in the river system, and natural (geographical and temporal) variation. The results showed strongly location-

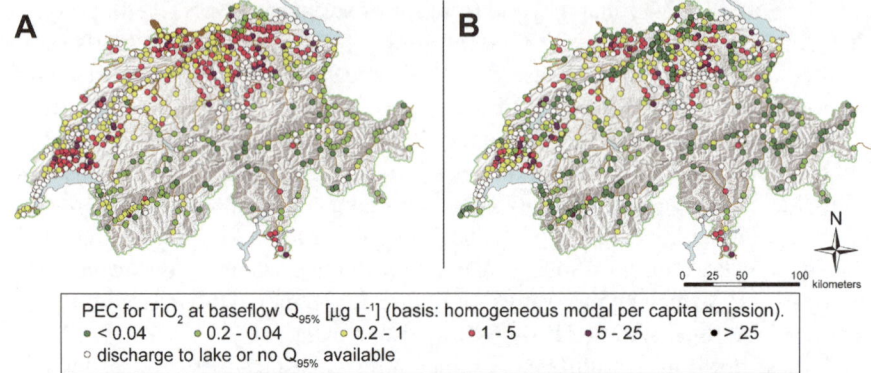

and time-dependent exposure concentrations. Local PEC values varied by a factor of 5 due to the variation of ENM emissions (15%–85% quantiles of ENM release amounts) and up to a factor of 10 when considering time-dependent river flow variability (same quantiles as before) (Figure 12.8). The median PECs for nano-TiO$_2$ varied from 11 ng L^{-1} to 1.623 µg L^{-1} (conservative scenario) and from 2 ng L^{-1} to 1.618 µg L^{-1} (optimistic scenario). The corresponding results of nano-Ag and nano-ZnO were lower by factors of 240 and 14.

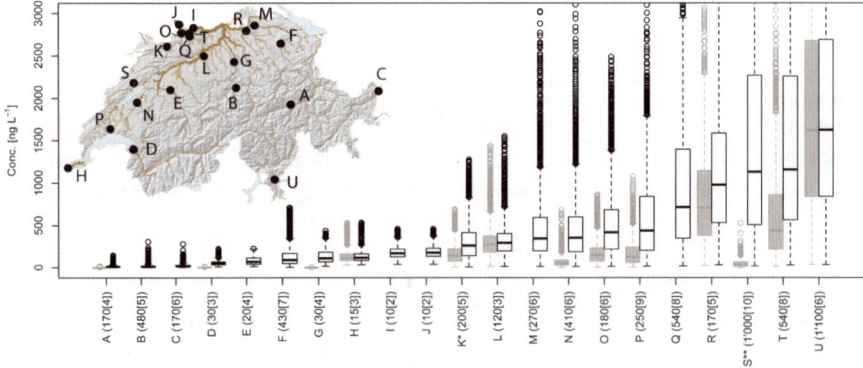

Arvidsson *et al.*[107] suggested a particle flow analysis (PFA) methodology. In PFA the ENM transfer from society to nature is modeled using the particle number instead of mass as the metric; otherwise the model geometry and conceptualization resemble the ones of the earlier ENM multimedia studies. PFA is said to facilitate the consideration of specific nanoparticle characteristics in the material flow analysis, *e.g.* the size of the particles under investigation. However, this study does not cover the broad spectrum of a distinct diffusive ENM release into the environment by only including ENM (nanosilver) release from wound dressings, textiles, and ink in electronic circuitry. However, improvements in the technological diffusion modeling are promised by the authors by means of the use of an explorative scenario method that was propagated by referring to the explorative scenario methods applied by others (the Intergovernmental Panel on Climate Change[119]) in other insecure contexts. No claims were made regarding how likely a particular emission scenario was. However, the main assumptions based on very insecure (at the most valid for very pessimistic scenarios) forecasts: a 100% market share of the ENM within the product groups investigated, a world population increase to 10 billion people by the year 2050 (as predicted by the United Nations[120]), metabolism rates of the considered products (product groups) calculated for high-income regions (USA, EU). Environmental exposure concentrations were not modeled explicitly. However, actual use phase emissions for the three nano-Ag applications were calculated: $<8.5 \times 10^{23}$ (particles per year) from textiles, 4.6×10^{21} (particles per year) from wound dressing, $<6.8 \times 10^{24}$ (particles per year) from electronic circuitry. A significant increase was forecast for ENM emissions from all applications. Those from textiles and electronic circuitry were predicted to be higher than the emission amounts from wound dressings due *e.g.* to the dissipative nature of ENM in textiles and the limited consumption volume for the latter application.

12.3.3 Analytical and Experimental Efforts

Analytical measurements of ENM in natural compartments are very rare or almost not available.[121] However, some results have been published on ENM emission tests that had been performed and quantified nanosized material in environmental samples. These results are also incorporated into Table 12.5.

12.3.3.1 Measurements Under Natural Conditions

Park *et al.*[102] investigated nano-CeO_2 release into air and soils from use as a diesel additive and measured a fourfold increase in cerium concentrations in ambient PM_{10} extracts at a Newcastle monitoring site after the introduction of the CeO_2-based fuel additive: prior results were 10.42 ± 4.61 *versus* posterior ones of 41.83 ± 18.91 ng mg^{-1} PM_{10} ($p < 0.001$) corresponding to 0.145 ± 0.064 *versus* 0.612 ± 0.287 ng m^{-3} ($p < 0.001$). The authors state that these cerium concentrations after the introduction of Envirox PM10 expressed in ng m^{-3} were of the same dimensions as those at the London, Marylebone Road site before such a fuel additive was used.

Kiser *et al.*[122] filtered Ti (< 700 nm) in STP effluent effluents by showing that part of such Ti was at the nanoscale. Hence, evidence was given that emissions of nanosized Ti particles into water bodies *via* STP effluents and sewage treatment sludge may occur. Their results revealed Ti concentrations in STP effluents ranging from <5 to 15 μg L^{-1}. Removed Ti accumulated in settled solids, concentrations were found ranging from 1–6 mg of Ti g^{-1}. The authors imaged solids containing Ti in different samples (sewage, STP effluents, STP biosolids and commercially available nano-TiO$_2$ products). In all samples particles were found ranging from 50 nm to a few hundred nm and were composed of sub-50 nm spheres of Ti and oxygen only (probably TiO$_2$).

Farré *et al.*[123] found in 50% of the samples of effluents from 22 wastewater treatment plants in Catalonia (Spain) C$_{60}$, C$_{70}$ and *N*-methylfulleropyrrolidine C$_{60}$ fullerenes. The concentrations derived for C$_{60}$ varied between 0.5 ng L^{-1} and 19 μg L^{-1}. The corresponding values for C$_{70}$ ranged from 181 ng L^{-1} to 1.7 μg L^{-1}, for functionalized C$_{60}$ from 60 ng L^{-1} to 66 μg L^{-1}. This study is the first work that shows the presence of fullerenes in the environment affected by human activities (in biosolids from sewage treatment plants). The presence of fullerenes in geological samples (*e.g.* layers from meteorite impacts) has been reported earlier.[124–126]

Neal *et al.*[127] reported for the UK on average 2.1 μg L^{-1} nanosized Ti (<0.45 μm filtered fraction) in river water draining different land types (rural/agricultural and urban/industrial land). The averaged results ranged in total from 0.55 to 6.48 μg L^{-1}. However, as the authors state, their contribution is far removed from distinguishing natural colloids and complexes from synthetic and non-synthetic anthropogenic nanoparticles. Also intermediary nanomaterial source scenarios are mentioned where colloids are generated in the water phase itself *e.g.* due to the dilution processes of effluents and chelating agents.

Mitrano *et al.*[128] used inductively coupled plasma–mass spectrometry (ICP-MS) in a single-particle (SP) counting mode to quantify nanosized silver particles. The authors report that between 100 and 200 ng L^{-1} nanosized particulate Ag had been detected in wastewater samples in Boulder, Colorado (USA). However, as the authors emphasize, the SP-ICP-MS analytics they used could not discriminate engineered nanosilver from natural silver-based colloids such as *e.g.* colloid-bound Ag$^+$ ions or precipitates (Ag$_2$S, AgCl and others). The method has its limits at particle sizes of 40 nm and could therefore miss smaller – but, from an ecotoxicological perspective, relevant – material.

12.3.3.2   *Analytical and Experimental Validation of ENM Exposure Modeling*

Analytical validation of the modeled exposure concentrations is crucial to improve future modeling. Unfortunately, these days it is almost impossible to measure nanoscaled fractions of contaminants in environmental compartments at trace concentrations. [121] However, rough comparisons (see Figure 12.5), for example, for modeled nano-TiO$_2$ and nano-Ag PECs to some first filter and spectrometry results have revealed that the exposure models used may produce reasonable results: Kiser *et al.*[122] measured 5–15 µg L^{-1} Ti (<0.7 µm) in effluents from STPs that validate to some extent the modal values of nano-Ti in STP effluents, *e.g.* 3.5 (nano-TiO$_2$) µg L^{-1} for the EU.[104] Neal *et al.*[127] found on average 2.1 µg L^{-1} nano-Ti (<0.45 µm filtered fraction) in rivers (England) draining different land-use types (rural, agricultural, urban and industrial). Modeled median values for nano-Ti (<0.1 µm) in comparable lowland rivers in Switzerland ranged from a few 100 ng L^{-1} to approx. 1.5 µg L^{-1}.[110] Mitrano *et al.*[128] found between 100 and 200 ng L^{-1} nanosized particulate Ag in wastewater samples; modal modeling results for STP effluents were about 43 ng L^{-1}.[104]

12.4   Adequacy of the REACH Release Parameters for ENMs

The Scientific Committee on Emerging and Newly-Identified Health Risks[129] examined how appropriate the exposure (risk) assessment approach of the Technical Guidance Documents of the EU is for nanomaterials.[58] The committee concluded that no completely new methodology is needed and that the risk (exposure) research can build on available knowledge about chemicals. However, REACH has not thus far addressed any specific environmental release parameters for nanosized particles. Nevertheless, the general approach (described in section 12.2) for assessing environmental release may also be applied to ENMs and products containing this material. Release is quantified by means of release rates estimated based on release factors and the use tonnage/amount assigned to a specific chemical source. Hence, such an assessment includes information on the amounts ending up in a particular natural compartment as well as dilution calculations and a standard size of such compartments.

In the Guidance on Information Requirements and Chemical Safety Assessment of the EU,[58] methodological tools and parameter values for quantifying the emissions into the environment (see section R.16 Estimation of Environmental Exposure) are given. R.16 provides ERCs for a first generic estimation of the environmental emissions of chemicals. Emissions to surface water, wastewater, soil and air are described. To what extent such conservative release factors may be adequate for ENMs has been discussed in an EU report[130] in a preliminary way. This report focuses especially on emissions during the production of ENMs and manufacturing (incorporation of ENM into products) processes are discussed. The reason for such a focus on manufacturing and production processes is that, due to missing empirical knowledge, most ENM models[100,101,103,131] do not consider emissions of ENMs to the environment occurring during ENM production and an ENM product's manufacture. However, since worker exposure was shown at factories producing ENM-containing products,[90,91,93,94,132] such releases should no longer be neglected. As confirmed elsewhere,[28] such ENMs released e.g. to indoor air most likely end up in the environment. Unfortunately, worker exposure results normally cannot be used for quantifications of the environmental release since the mass flow per unit time of ENMs reaching outdoor air or the water body are almost never indicated. ERCs[58] for chemicals recommend release factors for emissions during production processes of 6% for emissions to surface waters before reaching an STP, 0.01% for those to soils and 5% for those to the air. Furthermore, ERCs distinguish the release during production of formulations of mixtures embedded and not embedded in a matrix. Factors of 2% for release into water before ending up in STPs and 2.5% for release into air and 0.01% to soils were used for the first case. The respective values for the second scenario are 0.2%, 30% and 0.1%. Initial ENM studies[104,105] used parameters for ENM emissions to the environment during such processes ranging from 0 to 2% of the ENM volume produced.

12.5 Outlook for Future Modeling and Experimental Work

The ambition of future ENM release and exposure research should be to improve current multimedia transfer modeling methodologies to approaches that fully and simultaneously consider (i) time-dependent (intra- and interperiodical variance/uncertainty), (ii) conceptual and (iii) parametric uncertainty and variability at different release levels determined by the nanosized material entity (physicochemical properties and processes), socio-technical drivers (recovery, production, manufacturing, consumption etc.) and political (regulatory) influence factors and that consider (iv) explicitly nano-specific properties.

However, the main practical challenge in such modeling work is that currently none of the ENM multimedia exposure assessment models provides convincing solutions for considering comprehensively the environmental fate of ENMs in the environment.[133] Due to a frustrating lack of data most multimedia models only use mass-transfer simulations between environmental and technical compartments. However, simple mass-transfer functions are very limited when one is accounting for processes of particulate transfer/ degradation such as e.g. dissolution, suspension/resuspension, agglomeration/ aggregation, settling etc. Currently mechanistic parameterization of such processes can hardly be implemented in such models, since almost no empirical data are available that would describe these dynamics.[108] A first effort of environmental fate process modeling was done by Koelmans et al.[109] where the material flow analysis of Mueller and Nowack[103] was linked with a more sophisticated consideration of sedimentation. Others[46] have taken first fate modeling steps by using principles from colloid chemistry and considering particle agglomeration and sedimentation for nano-TiO_2. The results of such modeling reflect the environmental exposure by using the particle number concentration as the metric units. Although we have to consider such results as findings made at an early stage in research, they reveal that a closer look is needed at ENM discharge amounts and collision efficiency in particular as these seem to have a higher influence on exposure than *e.g.* settling and shear flow dynamics. If sufficient data are available, probabilistic material flow analysis[116] may prove very valuable as well for integrating submodels that reflect in detail ENMs' fate and behavior in natural environments.[108]

Acknowledgements

This work was funded by the EU-FP7 project "ModNanoTox – Modelling nanoparticle toxicity: principles, methods, novel approaches" under Grant Agreement 266712.

References

1. A. Nel, T. Xia, L. Madler and N. Li, *Science*, 2006, **311**, 622–627.
2. A. D. Maynard and E. D. Kuempel, *J. Nanopart. Res.*, 2005, **7**, 587–614.
3. K. Donaldson, R. Aitken, L. Tran, V. Stone, R. Duffin, G. Forrest and A. Alexander, *Toxicol. Sci.*, 2006, **92**, 5–22.
4. K. Savolainen, L. Pylkkanen, H. Norppa, G. Falck, H. Lindberg, T. Tuomi, M. Vippola, H. Alenius, K. Hameri, J. Koivisto, D. Brouwer, D. Mark, D. Bard, M. Berges, E. Jankowska, M. Posniak, P. Farmer, R. Singh, F. Krombach, P. Bihari, G. Kasper and M. Seipenbusch, *Safety Sci.*, 2010, **48**, 957–963.
5. A. E. Nel, L. Madler, D. Velegol, T. Xia, E. M. V. Hoek, P. Somasundaran, F. Klaessig, V. Castranova and M. Thompson, *Nat. Mater.*, 2009, **8**, 543–557.
6. G. B. Andreev, V. M. Minashkin, I. A. Nevskii and A. V. Putilov, *Russ. J. Gen. Chem.*, 2009, **79**, 1974–1981.
7. P. J. J. Alvarez, V. Colvin, J. Lead and V. Stone, *ACS Nano*, 2009, **3**, 1616–1619.
8. B. Nowack, *Environ. Pollut.*, 2009, **157**, 1063–1064.
9. C. Som, M. Berges, Q. Chaudhry, M. Dusinska, T. F. Fernandes, S. I. Olsen and B. Nowack, *Toxicology*, 2010, **269**, 160–169.
10. M. R. Wiesner, G. V. Lowry, K. L. Jones, M. F. Hochella, R. T. Di Giulio, E. Casman and E. S. Bernhardt, *Environ. Sci. Technol.*, 2009, **43**, 6458–6462.
11. A. Dhawan and V. Sharma, *Anal. Bioanal. Chem.*, 2010, **398**, 589–605.
12. M. R. Gwinn and L. Tran, *Wiley Interdiscip. Rev.: Nanomed. Nanobiotechnol.*, 2009, **2**, 130–137.
13. M. H. Depledge, L. J. Pleasants and J. H. Lawton, *Environ. Toxicol. Chem.*, 2010, **29**, 1–4.

14. S. J. Klaine, A. A. Koelmans, N. Horne, S. Carley, R. D. Handy, L. Kapustka, B. Nowack and F. von der Kammer, *Environ. Toxicol. Chem.*, 2012, **31**, 3–14.

15. B. Nowack and T. D. Bucheli, *Environ. Pollut.*, 2007, **150**, 5–22.

16. S. J. Klaine, P. J. J. Alvarez, G. E. Batley, T. F. Fernandes, R. D. Handy, D. Y. Lyon, S. Mahendra, M. J. McLaughlin and J. R. Lead, *Environ. Toxicol. Chem.*, 2008, **27**, 1825–1851.

17. M. R. Wiesner, G. V. Lowry, P. Alvarez, D. Dionysiou and P. Biswas, *Environ. Sci. Technol.*, 2006, **40**, 4336–4345.

18. M. Scheringer, *Nat. Nanotechnol.*, 2008, **3**, 322–323.

19. P. J. A. Borm, D. Robbins, S. Haubold, T. Kuhlbusch, H. Fissan, K. Donaldson, R. Schins, V. Stone, W. Kreyling, J. Lademann, J. Krutmann, D. Warheit and E. Oberdorster, *Part. Fibre Toxicol.*, 2006, **3**, 1743–8977.

20. Q. Chaudhry, M. Scotter, J. Blackburn, B. Ross, A. Boxall, L. Castle, R. Aitken and R. Watkins, *Food Addit. Contam.*, 2008, **25**, 241–258.

21. R. D. Handy, F. von der Kammer, J. R. Lead, M. Hassellov, R. Owen and M. Crane, *Ecotoxicology*, 2008, **17**, 287–314.

22. G. Oberdorster, V. Stone and K. Donaldson, *Nanotoxicology*, 2007, **1**, 2–25.

23. A. Helland, H. Kastenholz, A. Thidell, P. Arnfalk and K. Deppert, *J. Nanopart. Res.*, 2006, **8**, 709–719.

24. K. Hund-Rinke and M. Simon, *Environ. Sci. Pollut. Res.*, 2006, **13**, 225–232.

25. Royal Commission Report, *Novel Materials in the Environment: The Case of Nanotechnology*, Royal Commission on Environmental Pollution, TSO, 2008.

26. A. D. Maynard, R. J. Aitken, T. Butz, V. Colvin, K. Donaldson, G. Oberdorster, M. A. Philbert, J. Ryan, A. Seaton, V. Stone, S. S. Tinkle, L. Tran, N. J. Walker and D. B. Warheit, *Nature*, 2006, **444**, 267–269.

27. S. F. Hansen, E. S. Michelson, A. Kamper, P. Borling, F. Stuer-Lauridsen and A. Baun, *Ecotoxicology*, 2008, **17**, 438–447.

28. F. Gottschalk and B. Nowack, *J. Environ. Monit.*, 2011, **13**, 1145–1155.

29. D. R. Boverhof and R. M. David, *Anal. Bioanal. Chem.*, 2010, **396**, 953–961.

30. G. V. Lowry and E. A. Casman, in *Nanomaterials: Risks and Benefits*, ed. I. Linkov and J. Steevens, NATO Science for Peace and Security Series C: Environmental Security, 2009, vol. 2, pp. 125–137.

31. M. Siegrist, M. E. Cousin, H. Kastenholz and A. Wiek, *Appetite*, 2007, **49**, 459–466.

32. M. Siegrist, A. Wiek, A. Helland and H. Kastenholz, *Nat. Nanotechnol.*, 2007, **2**, 67–67.

33. J. Hudson and M. Orviska, *Nano Today*, 2009, **4**, 455–457.

34. M. Fu'hr, A. Hermann, S. Merenyi, K. Moch and M. Möller, *Legal Appraisals of Nanotechnologies*, Umweltforschungsplan des Bundesministeriums für Umwelt, Naturschutz und Reaktorsicherheit, Forschungsbericht 363 01 108 UBA-FB 000996, 2007.

35. G. Hodge, D. Bowman and K. Ludlow, *New Global Frontiers in Regulation: The Age of Nanotechnology*, Edward Elgar Publishing, Cheltenham, 2007. ISBN 978-1847205186.
36. REACH, *REACH and Nanomaterials*, European Commission (Registration, Evaluation, Authorisation and Restriction of Chemicals), http://ec.europa.eu/enterprise/sectors/chemicals/reach/index_en.htm, Brussels, 2010.
37. G. L. Hornyak, H. F. Tibbals and J. Dutta, *Fundamentals of Nanotechnology*, Taylor & Francis Group, Boca Raton, 2008.
38. B. Nowack, in *Nanotechnology:Environmental Aspects*, ed. H. F. Krug, WILEY-VCH Verlag GmbH & Co. KGaA, Weinheim, 2008, vol. 4, pp. 1–15.
39. B. Nowack, *Thema Umwelt*, 2007, **2**, 6–8.
40. A. Koehler, C. Som, A. Helland and F. Gottschalk, *J. Cleaner Product.*, 2008, **16**, 927–937.
41. A. Helland, P. Wick, A. Koehler, K. Schmid and C. Som, *Environ. Health Perspect.*, 2007, **115**, 1125–1131.
42. S. Inui, H. Aoshima, A. Nishiyama and S. Itami, *Nanomed.: Nanotechnol., Biol. Med.*, 2011, **7**, 238–241.
43. R. Loutfy, T. Lowe, A. Moravsky and S. Katagiri, in *Perspectives of Fullerene Nanotechnology*, ed. E. Ōsawa, Springer, Netherlands, 2002, pp. 35–46.
44. EPA, *Nanomaterial Case Studies: Nanoscale Titanium Dioxide in Water Treatment and in Topical Sunscreen*, Federal Register, US Environmental Protection Agency, Report EPA/600/R-09/057F, 2010.
45. E. McCarthy and C. Kelty, *Soc. Stud. Sci.*, 2010, **40**, 405–432.
46. R. Arvidsson, S. Molander, B. A. Sanden and M. Hassellov, *Hum. Ecol. Risk Assess.*, 2011, **17**, 245–262.
47. C. Marambio-Jones and E. Hoek, *J. Nanopart. Res.*, 2010, **12**, 1531–1551.
48. R. Dastjerdi and M. Montazer, *Colloids Surf., B*, 2010, **79**, 5–18.
49. K. Chaloupka, Y. Malam and A. M. Seifalian, *Trends Biotechnol.*, 2010, **28**, 580–588.
50. EPA, *Nanotechnology White Paper*, U.S. Environmental Protection Agency (EPA), Washington, DC, 2007.
51. M. Zhang, T. An, X. Liu, X. Hu, G. Sheng and J. Fu, *Materials Letters*, 2010, **64**, 1883–1886.
52. Z. A. Lewicka, A. F. Benedetto, D. N. Benoit, W. W. Yu, J. D. Fortner and V. L. Colvin, *J. Nanopart. Res.*, 2011, **13**, 3607–3617.
53. S. Kwon, M. Fan, A. T. Cooper and H. Yang, *Critical Reviews in Environmental Science and Technology*, 2008, **38**, 197–226.
54. S. J. Klaine, P. J. J. Alvarez, G. E. Batley, T. F. Fernandes, R. D. Handy, D. Y. Lyon, S. Mahendra, M. J. McLaughlin and J. R. Lead, *Environmental Toxicology & Chemistry*, 2008, **27**, 1825–1851.

55. R. Kaegi, A. Ulrich, B. Sinnet, R. Vonbank, A. Wichser, S. Zuleeg, H. Simmler, S. Brunner, H. Vonmont, M. Burkhardt and M. Boller, *Environ. Pollut.*, 2008, **156**, 233–239.

56. T. M. Benn and P. Westerhoff, *Environ. Sci. Technol.*, 2008, **42**, 4133–4139.

57. L. Geranio, M. Heuberger and B. Nowack, *Environ. Sci. Technol.*, 2009, **43**, 8113–8118.

58. ECHA, *Guidance on information requirements and chemical safety assessment*, chapter R.16: Environmental Exposure Estimation, European Chemicals Agency, 2010.

59. T. G. Vermeire, D. T. Jager, B. Bussian, J. Devillers, K. denHaan, B. Hansen, I. Lundberg, H. Niessen, S. Robertson, H. Tyle and P. T. J. van der Zandt, *Chemosphere*, 1997, **34**, 1823–1836.

60. C. E. Cowan, D. Mackay, T. C. J. Feijtel, D. van de Meent, A. Di Guardo, J. Davies and N. Mackay, *The multi-media fate model: A vital tool for predicting the fate of chemicals*, Proceedings of a workshop organized by the Society of Environ. Toxicol. Chem. (SETAC), Pensacola, FL, SETAC, 1995.

61. D. Mackay, *Multimedia Environmental Models: The Fugacity Approach*, Lewis Publishers, Boca Raton, FL, 2001.

62. T. E. McKone and M. MacLeod, *Annu. Rev. Environ. Resour.*, 2003, **28**, 463–492.

63. D. Mackay, A. DiGuardo, S. Paterson, G. Kicsi and C. E. Cowan, *Environ. Toxicol. Chem.*, 1996, **15**, 1618–1626.

64. J. A. Arnot, *Mass Balance Models for Chemical Fate, Bioaccumulation, Exposure and Risk Assessment*, Nato Science for Peace and Security Series C – Environmental Security, Sofia, Bulgaria, 2009.

65. M. Scheringer, F. Wegmann and K. Hungerbuhler, *Environ. Toxicol. Chem.*, 2004, **23**, 2433–2440.

66. M. MacLeod, D. G. Woodfme, D. Mackay, T. McKone, D. Bennett and R. Maddalena, *Environ. Sci. Pollut. Res.*, 2001, **8**, 156–163.

67. M. Scheringer, F. Wegmann, K. Fenner and K. Hungerbuhler, *Environ. Sci. Technol.*, 2000, **34**, 1842–1850.

68. E. D. Kuempel, C. L. Geraci and P. A. Schulte, in *Nanotechnology: Toxicological Issues and Environmental Safety*, ed. P. P. Simeonova, N. Opopol and M. I. Luster, Springer, Dordrecht, 2007, pp. 119–145.

69. K. A. D. Guzman, M. R. Taylor and J. F. Banfield, *Environ. Sci. Technol.*, 2006, **40**, 1401–1407.

70. Z. D. Ok, J. C. Benneyan and J. A. Isaacs, *2009 IEEE International Symposium on Sustainable Systems and Technology*, 2009, 50–54.

71. Y. Ju-Nam and J. R. Lead, *Sci. Total Environ.*, 2008, **400**, 396–414.

72. P. C. Ray, H. T. Yu and P. P. Fu, *J. Environ. Sci. Health, Part C: Environ. Carcinog. Ecotoxicol. Rev.*, 2009, **27**, 1–35.

73. A. D. Ostrowski, T. Martin, J. Conti, I. Hurt and B. H. Harthorn, *J. Nanopart. Res.*, 2009, **11**, 251–257.

74. K. Aschberger, C. Micheletti, B. Sokull-Kluettgen and F. M. Christensen, *Environ. Int.*, 2011, **37**, 1143–1156.

75. A. Franco, S. F. Hansen, S. I. Olsen and L. Butti, *Regul. Toxicol. Pharmacol.*, 2007, **48**, 171–183.

76. K. D. Grieger, S. F. Hansen and A. Baun, *Nanotoxicology*, 2009, **3**, 222–233.

77. B. Karn, T. Kuiken and M. Otto, *Environ. Health Perspect.*, 2009, **117**, 1823–1831.

78. N. Mueller and B. Nowack, *Elements*, 2010, **6**, 395–400.

79. K. Morgan, *Risk Anal.*, 2005, **25**, 1621–1635.

80. M. Kandlikar, G. Ramachandran, A. Maynard, B. Murdock and W. A. Toscano, *J. Nanopart. Res.*, 2006, **9**, 137–156.

81. D. G. Rickerby and M. Morrison, *Sci. Technol. Adv. Mater.*, 2007, **8**, 19–24.

82. J. M. Davis, *J. Nanosci. Nanotechnol.*, 2007, **7**, 402–409.

83. J. M. Davis and V. Thomas, *Ann. N. Y. Acad. Sci.*, 2006, **1076**, 498–515.

84. C. O. Robichaud, D. Tanzil, U. Weilenmann and M. R. Wiesner, *Environ. Sci. Technol.*, 2005, **39**, 8985–8994.

85. EPA, *Nanomaterial Case Study: Nanoscale Silver in Disinfectant Spray*, US Environmental Protection Agency, EPA report EPA/600/R-10/081, 2010.

86. E. S. Michelson and D. Rejeski, *Falling through the cracks? Public perception, risk, and the oversight of emerging nanotechnologies*, IEEE-ISTAS Conference Proceeding, Queens, NY, 2006.

87. L. K. Limbach, R. Bereiter, E. Müller, R. Krebs, R. Gälli and W. J. Stark, *Environ. Sci. Technol.*, 2008, **42**, 5828–5833.

88. G. Bystrzejewska-Piotrowska, J. Golimowski and P. L. Urban, *Waste Management*, 2009, **29**, 2587–2595.

89. B. Nowack, F. Gottschalk, F. Mueller and C. Som, Life-Cycle Concepts for Sustainable Use of Engineered Nanomaterials in Nanoproducts, in: Boethling, R., Voutchkova, A. (Eds.), Handbook of Green Chemistry, Volume 9: Designing Safer Chemicals. Wiley-VCH., 2012, 227–249.

90. B. Yeganeh, C. M. Kull, M. S. Hull and L. C. Marr, *Environ. Sci. Technol.*, 2008, **42**, 4600–4606.

91. L. F. Mazzuckelli, M. M. Methner, M. E. Birch, D. E. Evans, B. K. Ku, K. Crouch and M. D. Hoover, *J. Occup. Environ. Hyg.*, 2007, **4**, D125–D130.

92. A. D. Maynard, P. A. Baron, M. Foley, A. A. Shvedova, E. R. Kisin and V. Castranova, *Journal of Toxicology and Environmental Health-Part A*, 2004, **67**, 87–107.

93. Y. Fujitani, T. Kobayashi, K. Arashidani, N. Kunugita and K. Suemura, *J. Occup. Environ. Hyg.*, 2008, **5**, 380–389.

94. D. Bello, A. J. Hart, K. Ahn, M. Hallock, N. Yamamoto, E. J. Garcia, M. J. Ellenbecker and B. L. Wardle, *Carbon*, 2008, **46**, 974–977.

95. A. R. Koehler and C. Som, *Hum. Ecol. Risk Assess.*, 2008, **14**, 512–531.

96. R. Hischier and T. Walser, *Sci. Total Environ.*, 2012, **425**, 271–282.

97. C. Som, M. Halbeisen and A. Köhler, *Integration von Nanopartikeln in Textilien – Abschätzungen zur Stabilität entlang des textilen Lebenszyklus,*

Report, Empa und TVS Textilverband Schweiz, St. Gallen (Switzerland), 2009.

98. I. Linkov, F. K. Satterstrom, J. Steevens, E. Ferguson and R. C. Pleus, *J. Nanopart. Res.*, 2007, **9**, 543–554.

99. SCENIHR, *Scientific Committee on Emerging and Newly-Identified Health Risks. Risk Assessment of Products of Nanotechnologies*, Scientific Committee on Emerging and Newly-Identified Health Risks, The SCENIHR adopted this opinion at its 28th plenary on 19 January 2009, 2009.

100. A. B. A. Boxall, Q. Chaudhry, A. Jones, B. Jefferson and C. D. Watts, *Current and Future Predicted Environmental Exposure to Engineered Nanoparticles*, Central Science Laboratory, Sand Hutton, UK, 2008.

101. S. A. Blaser, M. Scheringer, M. MacLeod and K. Hungerbuehler, *Sci. Total Environ.*, 2008, **390**, 396–409.

102. B. Park, K. Donaldson, R. Duffin, L. Tran, F. Kelly, I. Mudway, J. P. Morin, R. Guest, P. Jenkinson, Z. Samaras, M. Giannouli, H. Kouridis and P. Martin, *Inhalation Toxicol.*, 2008, **20**, 547–566.

103. N. C. Mueller and B. Nowack, *Environ. Sci. Technol.*, 2008, **42**, 4447–4453.

104. F. Gottschalk, T. Sonderer, R. W. Scholz and B. Nowack, *Environ. Sci. Technol.*, 2009, **43**, 9216–9222.

105. F. Gottschalk, T. Sonderer, R. W. Scholz and B. Nowack, *Environ. Toxicol. Chem.*, 2010, **29**, 1036–1048.

106. N. O'Brien and E. Cummins, *J. Environ. Sci. Health, Part A: Toxic/Hazard. Subst. Environ. Eng.*, 2010, **45**, 992–1007.

107. R. Arvidsson, S. Molander and B. A. Sanden, *J. Ind. Ecol.*, 2011, **15**, 844–854.

108. J. T. K. Quik, J. A. Vonk, S. F. Hansen, A. Baun and D. Van De Meent, *Environ. Int.*, 2011, **37**, 1068–1077.

109. A. A. Koelmans, B. Nowack and M. R. Wiesner, *Environ. Pollut.*, 2009, **157**, 1110–1116.

110. F. Gottschalk, C. Ort, R. W. Scholz and B. Nowack, *Environ. Pollut.*, 2011, **159**, 3439–3445.

111. EPA, *User's Guide for HIWAY-2: A Highway Air Pollution Model*, EPA-600/8-80-018, Research Triangle Park, NC, 1980.

112. L. Ntziachristos and Z. Samaras, *COPERT III: Computer program to calculate emissions from road transport, methodology and emission factors (Version 2.1)*, Technical report no. 49, http://vergina.end.auth.gr/mech/lat/copert/copert.htm, 2000.

113. M. Giannouli, Z. Samaras, M. Keller, P. deHaan, M. Kallivoda, S. Sorenson and A. Georgakaki, *Sci. Total Environ.*, 2006, **357**, 247–270.

114. N. O'Brien and E. Cummins, in *Nanomaterials: Risks and Benefits*, ed. I. Linkov and J. Steevens, Springer, Netherlands, 2009, pp. 161–178.

115. N. J. O'Brien and E. J. Cummins, *Risk Anal.*, 2010.

116. F. Gottschalk, R. W. Scholz and B. Nowack, *Environ. Modell. Software*, 2010, **25**, 320–332.

117. N. Musee, *Hum. Exp. Toxicol.*, 2010, **30**, 1181–1195.

118. C. Ort, J. Hollender, M. Schaerer and H. Siegrist, *Environ. Sci. Technol.*, 2009, **43**, 3214–3220.
119. L. Borjeson, M. Hojer, K.-H. Dreborg, T. Ekvall and G. Finnveden, *Futures*, 2006, **38**, 723–739.
120. United Nations, *World population prospects: The 2008 Revision, Highlights* Working Paper No. ESA/P/WP.210, New York (USA), 2008.
121. M. Hassellov, J. W. Readman, J. F. Ranville and K. Tiede, *Ecotoxicology*, 2008, **17**, 344–361.
122. M. A. Kiser, P. Westerhof, T. Benn, Y. Wang, J. Pérez-Rivera and K. Hristovski, *Environ. Sci. Technol.*, 2009, **43**, 6757–6763.
123. M. Farré, S. Pèrez, K. Gajda-Schrantz, V. Osorio, L. Kantiani, A. Ginebreda and D. Barcelü, *J. Hydrol.*, 2010, **383**, 44–51.
124. D. Heymann, W. S. Wolbach, L. P. F. Chibante, R. R. Brooks and R. E. Smalley, *Geochim. Cosmochim. Acta*, 1994, **58**, 3531–3534.
125. D. Heymann, L. P. F. Chibante and R. E. Smalley, *J. Chromatogr., A*, 1995, **689**, 157–163.
126. L. Becker, J. L. Bada, R. E. Winans, J. E. Hunt, T. E. Bunch and B. M. French, *Science*, 1994, **265**, 642–645.
127. C. Neal, H. Jarvie, P. Rowland, A. Lawler, D. Sleep and P. Scholefield, *Sci. Total Environ.*, 2011, **409**, 1843–1853.
128. D. M. Mitrano, E. K. Lesher, A. Bednar, J. Monserud, C. P. Higgins and J. F. Ranville, *Environ. Toxicol. Chem.*, 2011, **31**, 115–121.
129. SCENIHR, Scientific Committee on Emerging and Newly-Identified Health Risks, *The appropriateness of the risk assessment methodology in accordance with the Technical Guidance Documents for new and existing substances for assessing the risks of nanomaterials*, 21–22 June 2007.
130. NANEX, *Development of Exposure Scenarios for Manufactured Nanomaterials*, Chapter 5: Report on exposure scenarios and release of nanomaterials to the environment, EU, FP7 Project Number 247794, 2010.
131. B. Park, *Issues Environ. Sci. Technol.*, 2007, **24**, 1–18.
132. J. H. Han, E. J. Lee, J. H. Lee, K. P. So, Y. H. Lee, G. N. Bae, S. B. Lee, J. H. Ji, M. H. Cho and I. J. Yu, *Inhalation Toxicol.*, 2008, **20**, 741–749.
133. EPA, *State-of-the-Science Report on Predictive Models and Modeling Approaches for Characterizing and Evaluating Exposure to Nanomaterials*, EPA/600/R-10/129, National Exposure Research Laboratory Office of Research and Development Washington, DC, 20460, 2010.

CHAPTER 13

Comprehensive Environmental Assessment of Nanotechnologies: a Case Study Using Self-decontaminating Surface Materials

J. A. STEEVENS*[a], A. BEDNAR[a], M. CHAPPELL[a],
K. DONOHUE[a], M. GINSBERG[b], K. GUY[b], D. JOHNSON[a],
A. KENNEDY[a], R. MOSER[a], M. PAGE[b], A. PODA[a] AND
C. WEISS JR.[a]

[a] U.S. Army ERDC, 3909 Halls Ferry Road, Vicksburg, MS 39180, USA;
[b] U.S. Army ERDC, 2902 Newmark Drive, Champaign, IL 60822, USA
*E-mail: Jeffery.A.Steevens@us.army.mil

13.1 Introduction

13.1.1 Life-cycle Approach for Assessing the Risk of Nanotechnologies

A life-cycle approach has been proposed to address concerns regarding the potential human and environmental risks of nano-based technologies. Traditional risk assessments focus on the characterization of contaminants present in the environment and predict potential risks based on exposure levels associated with a toxic response. However, this approach is not valid for developing technologies because the goal is to prevent contamination. A

RSC Nanoscience & Nanotechnology No. 25
Towards Efficient Designing of Safe Nanomaterials: Innovative Merge of Computational Approaches and Experimental Techniques
Edited by Tomasz Puzyn and Jerzy Leszczynski

proactive approach to risk assessment is required to determine the safety of chemical, material, or technology prior to its use. To fully understand the potential adverse effects of a chemical or material, the entire life-cycle of the technology should be considered.[1-3] Life-cycle approaches to assessing risk are different from traditional life-cycle analysis by using the general material flow rather than a mass-based accounting of materials, energy, and other resources. In this approach, the potential release of nanoparticles is considered through the various life-cycle stages including raw materials, production, distribution, use, and end of use. Specific processes in these life-cycle stages can inform the risk assessor and manager about the potential releases and also processes where releases are unlikely to occur so that efforts can be focused on the stages with the highest risk. This approach has been recommended by the National Nanotechnology Initiative Research Strategy.[4]

An approach for providing the necessary environmental and human health safety information for nanomaterial use in Department of Defense (DoD) applications is to conduct a comprehensive environmental assessment (CEA). The CEA approach (Figure 13.1) enables the evaluation of environmental hazards for technologies through the entire life-cycle (*e.g.*, feedstocks, manufacturing, distribution, storage, use, and disposal).[5] Furthermore, the CEA also takes into account the environmental pathways, fate and transport, exposure/dose, and effects of nanomaterials once they enter the environment.

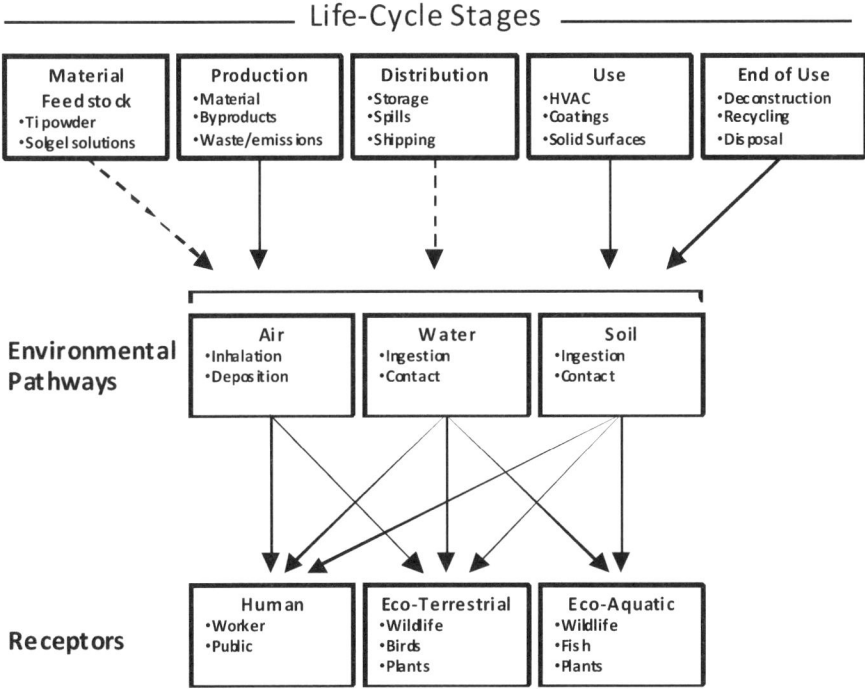

Figure 13.1   General CEA conceptual model framework.

This comprehensive approach allows scientists, risk assessors, and acquisitions personnel to visualize holistically the potential effects of nanomaterials, as well as any data gaps, which will help them make better-informed decisions for developing and acquiring materials containing nanomaterials. The CEA approach has been developed to support several emerging materials and technologies including methyl *tert*-butyl ether (MTBE) fuel additives, nanosized silver, and nanosized titanium dioxide.[6–8]

13.1.2 A Case Study for Comprehensive Environmental Assessment

Self-decontaminating surfaces (SDSs) offer great potential for mitigating biological and chemical agents. Research into these reactive surfaces has included catalysts and enzymatic approaches with the goal of providing a reactive self-decontaminating platform that does not adversely affect the performance of the system. Metals such as titanium and silver have been incorporated into coatings. Recent advances in materials science have enabled these metals to be used at the nanometre scale. The metal-containing surfaces rely on surface-charge and/or photoactivation mechanisms to elicit the desired biocidal activity. Some of the novel approaches include chemicals added to a polymer coating, such as the quaternary ammonium biocides. These materials have demonstrated biocidal activity when incorporated into polyurethane coatings.[9] Specific materials under investigation by Ginsberg and colleagues focused on nanoscale titanium dioxide that, upon photoactivation with ultraviolet light, destroys biological agents. This material has the potential to be incorporated into other matrices such as coatings for non-filter applications.

While there are clear technological benefits of SDSs, the unique attributes of nanoscale titanium dioxide raise questions regarding the environmental safety of the composite. The SDS platforms that incorporate active nanometals have the potential to release particles over time and may be affected by factors such as light, temperature, and abrasion. The release of metal- and carbon-based nanoparticles in composites has been studied for various applications.[10–13]

The goal of this effort was to evaluate the potential risks of nanoscale decontaminating surfaces through a focused examination following the CEA process. Data gaps identified through the CEA process focused primarily on the fate of nanoparticles in SDSs and toxicity in relevant biological models. Discussions with technology developers indicated best management practices and management controls will be implemented for production of the materials to limit potential releases. However, the long-term stability of particles in the surface coating product was identified as a potential concern.

13.1.3 Comprehensive Environmental Assessment Framework

The CEA framework that outlines the life-cycle stages (material flow) for nanosized titanium dioxide for heating, ventilation, and air conditioning

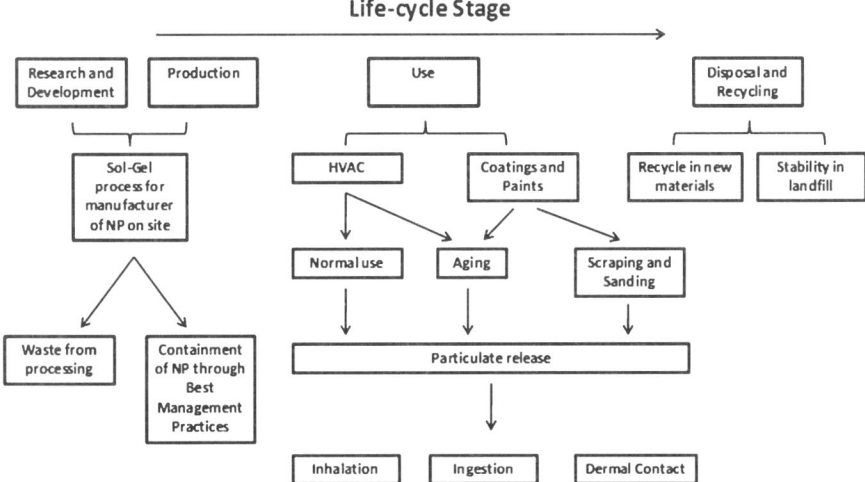

Figure 13.2 CEA framework for self-decontaminating surfaces containing titanium dioxide.

(HVAC) systems and coatings is shown in Figure 13.2. Expert judgment from material developers and a literature review were used to construct a critical relevant flow of the technology and constituent materials through development and use processes. These processes and flows (*i.e.*, use, life-cycle stage, material handling) identify mechanisms where material releases could potentially occur. Further development was used to identify environmental exposure routes associated with particulate releases. In some cases, pathways can be eliminated through an understanding of the processes and handling involved. For example, releases during research and development were identified as unlikely sources because of engineering controls and the mechanisms used to fabricate the surfaces (*i.e.*, the sol–gel process outlined below). Additional details and assumptions used to construct the CEA framework are described below.

13.2 Evaluation of Nanotechnologies

13.2.1 Development and Production

Development and production is unlikely to result in the release of nanoparticles due in part to the processing methods and best management practices. In this study, the SDSs being developed rely on nanosized TiO_2 applied to a stainless steel substrate. It is applied as a composite and is post-treated with a silver salt. The process used is sol–gel *in situ* synthesis directly on the surface. In general, nano-metals created using this wet chemistry approach are synthesized in place and immediately incorporated into the "gel" composite. Although manufacturing the material at a larger scale requires hazardous reagents including hydrochloric acid, titanium tetrachlorate, and phosphoric acid it minimizes the potential release of nanoparticles because the nanoparticles are synthesized and mixed into the composite simultaneously.[1,14]

During the manufacture of nanoparticles and composites containing nanoparticles there is always a potential for unintended release. However, recent research focusing on worker exposures in the laboratory and manufacturing facilities has informed managers of best management practices to minimize release and exposure.[15] These practices include keeping nanoparticles in a liquid matrix to minimize exposure, working in hoods that vent outdoors or through an adequate filtration system, working in containment boxes, specialized personal protection, and the use of high-efficiency particulate arresting (HEPA) vacuum filtration systems for cleaning facilities. The National Institute for Occupational Safety and Health (NIOSH) has published guidance for implementing engineering controls and safety practices for minimizing worker exposure and the release of nanoparticles.[16]

13.2.2 Self-decontaminating Surface Use

A UV-irradiated titanium dioxide purification-based system for air purification was examined for the potential release of TiO_2 nanoparticles.[17] Numerous examples exist where titanium dioxide is being incorporated into coatings for the reduction of pathogens and organic chemicals on surfaces of buildings, air handling systems, and water treatment systems.[18–21] These systems all rely on the stabilization of titanium dioxide on a substrate and enhance the generation of superoxide ions and hydroxyl radicals through exposure to ultraviolet light.[22]

13.2.3 Heating, Ventilation, and Air Conditioning (HVAC) Systems

The use of an SDS in an HVAC system has been proposed by several investigators listed above as a mechanism to reduce pathogen levels in indoor air. Generally, the systems use a reactive coating on the surface of the indoor coils to provide a high level of contact with the air as it moves through the HVAC system. The conditions in the HVAC system are conducive to the aging and/or degrading of the reactive surface.[23] Conditions which can accelerate the surface degradation include the presence of UV light, changes in temperature, high humidity or water, high air flow, and the presence of other chemicals that can react with the surface. In a study examining the release of TiO_2 nanoparticles from coating surfaces, UV light and substrate bonding were the two most critical variables that affected the release from the substrate. These results suggested that after the initial flow of air (<2 min period) the release of particles from the substrate was consistent and ranged from nanosized to micrometre-sized particles.

The most likely exposure route for nanosized titanium dioxide from the HVAC coating is through inhalation (see Figure 13.2). Other routes of exposure include ingestion and dermal exposure. Inhalation is the most likely exposure pathway because of the ability of the particles to move in the air and reach the alveoli of the lungs.[24] Exposure to nanoparticles through dermal contact has been demonstrated to be limited because of the protective ability of the dead layer of skin cells, the epidermis, which has been shown to prevent the uptake of nanoparticles.[25–27] If the epidermis is damaged, such as from an abrasion or a cut, then penetration of nanoparticles through the skin may occur.[28] However, the ability of the nanoparticles to penetrate the skin is dependent on other variables including particle type and surface coating.

13.2.4 Coatings and Paints

Titanium dioxide has been used for pigments in paints and coatings for many years. Over the past 10 to 20 years there has been significant interest in the use of titanium dioxide in paints for the photocatalysis of organic pollutants and control of fungi, bacteria, viruses, and molds. These applications have included paints or coatings on exterior surfaces of buildings and other infrastructure such as roads and bridges.[29] Interior applications have included surfaces in hospitals, bathrooms, and high traffic public areas.

Environmental health concerns regarding the use of TiO_2 in coatings include the generation of toxic degradation products from degradation of other toxic chemicals and through the release of TiO_2 particulates from the coating itself. Photocatalytic TiO_2 coatings are used to degrade many organic contaminants in the air to form reaction products that are relatively harmless. There is the potential for these degradation products to be more toxic. For example, Jacoby *et al.*[30] observed the generation of low levels of dichloroacetyl chloride and phosgene gas from the photocatalytic degradation of trichloroethylene. In general, these reaction products are produced at low levels and are short-lived.

A second potential release mechanism of TiO_2 from a paint coating is through abrasion of the surface through direct contact and by through sanding during refinishing of the surface. Research has focused on the release of TiO_2 and particles from paint coatings. Hsu and Chein[11] observed the emission of 55 nm particles from composite coatings as a result of UV light and scraping with a rubber knife over the surface. The scraping resulted in significant increases in particle release when compared to UV light alone. In a study examining the abrasion of four different nano-based composites, the investigators observed the production of nanoscale and larger particles depending on the method of abrasion.[13] Sanding resulted in larger particles (1 to 80 μm) whereas a Taber abrader produced primarily particles less than 100 nm. Because of this potential for release from the coating, understanding the stability of nanoparticles in the composite matrix and expected abrasion is important for quantifying long-term release from paint.

13.2.5 Disposal and Recycling

The disposal of SDS coatings will contribute two main metals, titanium and silver, to the total metal loading in the landfill. On a total mass basis, it is not anticipated that the disposal of HVAC systems would contribute significantly. However, the renovation of a building and surfaces coated with the SDS could contribute to the overall loading. After placement in a landfill, it is expected that the stability of nano-particulate TiO_2, or its agglomerates, will minimize offsite movement.[31] Although Ag has a relatively high solubility, after proper placement in the landfill it is unlikely to result in releases due to effective management and engineering controls.[32]

The broad range of applications for titanium including paints, aviation composites, electronics, medical devices, and the automotive industry, to name a few, have increased the demand for titanium, resulting in a 2- to 3-fold increase in global consumption and a 2-fold increase in price for raw titanium ore.[33] Because of the increased demand and limited infrastructure for refining the metal, there is increased interest in recycling. The SDSs may be recycled in a manner similar to the high temperature and chemical processing used for the isolation of organic composite components from metals.[34,35] Because nano-based metals and composites are processed in a similar fashion to bulk metals it is expected that practices to isolate the metal (through melting) will eliminate the nanoscale properties and titanium will follow conventional metal properties. Therefore, existing controls during recycling will not result in any additional releases.

13.2.6   Data Gaps and Uncertainty

The CEA framework and discussion focuses the investigation on the most critical processes where exposures and subsequent risk may occur during the life cycle of nanosized TiO_2 HVAC coatings and paints. It also aids in limiting the scope of subsequent investigations, including laboratory-based studies, with the materials developed in this effort. For example, minimal releases are expected to occur during the development, production, and end-of-use life stages. Such minimal releases derive from the use of proper engineering controls and selecting processes (*i.e.*, sol–gel synthesis) that minimize the potential for airborne particles. Several areas were identified in the CEA framework where particles may be released during use of the technology.

Based on the CEA, three main data gaps can be identified:

1. In what form does the nanosized titanium exist after product incorporation? How stable is the nanoparticle in the coating?
2. Is there a potential release of nanosized titanium particles in air handling systems? What is the particle size released? How much material would be released?
3. What are the potential risks from inhalation of the particles? Does the material interact with alveolar fluid in the lungs? What are the potential effects if the particles reach the lungs?

This work investigates a TiO_2 SDS deposited onto stainless steel plates for biocidal applications in HVAC systems to gain additional information and address the data gaps identified in the CEA framework. Stainless steel substrate coupons were coated with titanium dioxide (anatase) having a high surface area and an adsorbed silver ion coating. These coupons were prepared using a sol–gel process to isolate the particles on the stainless steel substrate followed by volatilization and sintering. The novelty of this SDS derives from incorporating a photocatalytic nanosized TiO_2 powder on the stainless steel plates utilized in HVAC systems. As these nanomaterials are to be incorporated into an HVAC system, there comes an immediate concern for the potential release and subsequent respiratory exposure of those persons working in the buildings. Research shows that inhalation of nanoparticles is associated with adverse health effects. Coupons were evaluated using scanning electron microscopy–energy-dispersive X-ray spectroscopy (SEM-EDS) to determine the surface morphology and elemental composition. Additional analysis included leaching studies, air flow release, and cyctotoxicity using alveolar cells in a mixed lung cell culture.

13.3 CEA Conceptual Model to Identify Data Needs

13.3.1 Exposure Scenario Characterization and Analysis

The exposure scenarios identified in the CEA framework focused on the analysis and characterization of the SDS coating. The primary goal of the characterization and analysis was to provide additional information about the potential for the SDS to release particles during use and during activities such as sanding or general abrasion. To determine the potential for the release of nanoparticles from the surface and potential exposure, three types of samples were analyzed: (1) the coating directly on the surface of the stainless steel plate, (2) particles ejected from the coating when tested in a flow cell which simulated the conditions in a typical HVAC system, and (3) loose powders of the coating milled off of the surface of the stainless steel plate. Relevant characterization for the SDS includes the surface morphology on stainless steel after powder incorporation, sizing information on released particulates, the concentration of particulates released (release rate), the particle morphology, and the chemical composition of the SDS and evolved particulates.

Characterization efforts consisted of scanning electron microscopy (SEM), energy-dispersive X-ray spectroscopy (EDS), and inductively-coupled plasma–mass spectroscopy (ICP-MS). SEM imaging was performed using an FEI Nova NanoSEM 630 capable of high-resolution imaging on non-conductive materials. The imaging was performed in low-vacuum mode at pressures of 0.1 to 0.3 mbar and an accelerating voltage of 5 to 15 kV. All images were acquired using a backscattered electron detector to improve the phase contrast. In conjunction with SEM imaging, chemical analysis was performed using a Bruker Quantax EDS microanalysis system and a Perkin Elmer ICP-MS.

13.3.2 Surface Analysis of Intact SDS

The "as-produced" SDS was analyzed to determine the stability of the solid phase. This analysis focused on characterizing the surface morphology and composition as well as the potential for release from the stainless steel substrate. The analysis included SEM and EDS microanalysis of the surface as well as a tape adhesion testing to evaluate the adhesive strength of the coating. The potential for water to affect the presence of nanoparticulates and dissolvable metals was evaluated during a rinsing procedure.

A representative SEM image and EDS analysis of the SDS test coupons is shown in Figure 13.3 and indicates a non-uniform film with an exposed titanium oxide, silver, silicon dioxide adhesion layer and exposed stainless steel substrate material. The initial focus was to determine the TiO_2 coating characteristics to ascertain the extent of the TiO_2 material that remained in the nanometre regime after surface processing. During examination of the surfaces, it was determined that the TiO_2 no longer maintained the characteristics associated with nanopowder, but formed a composite with the nanostructured materials dispersed in a continuous matrix. SEM analysis revealed that the coating appeared to be delaminating in regions, leading to concerns regarding surface adhesion and surface integrity. Figure 13.3b presents a typical elemental map of Si, Fe, Ag, and Ti obtained using EDS. The results of the mapping confirmed the microstructure observed in backscattered SEM micrographs with significant heterogeneity and partitioning between the various elements present in the coating. The bright regions in the SEM micrographs were found to correspond to sub-micrometre-sized agglomerates of Ag particles.

The Scotch tape adhesion testing method (ICP-TM-650) was performed on the steel coupons. This test method uses pressure-sensitive tape to determine the adhesion quality of plated films, marking inks or paints, and other materials used in conjunction with printed boards. A roll of pressure-sensitive

(a) Backscattered SEM micrograph of SDS

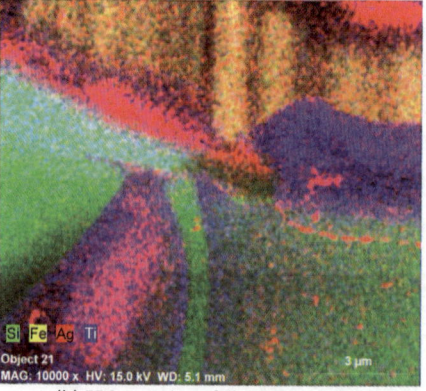

(b) EDS mapping of Si, Fe, Ag, and Ti

Figure 13.3   (a) SEM image of TiO_2 SDS with (b) corresponding EDS elemental map.

tape 3M Brand 600 12.7 mm (0.5 in) wide or a tape as described in (CID AA-113) was used. The procedure involves pressing a strip of pressure-sensitive tape, 50 mm (2.0 in) minimum in length, firmly across the surface of the test area removing all air entrapment. The time between application and removal of tape was less than one minute. The tape is removed by a rapid pull force applied approximately perpendicularly (at a right angle) to the test area. The tape and test area were visually examined for evidence of any portion of the material tested having been removed from the specimen. The tape adhesion testing of the surfaces shown in Figure 13.4 revealed that portions of the SDS coating could be removed easily. This method proved to be a non-quantitative way of measuring accurate particle size and/or for determining particle release rates and was used for an initial screening of the SDS coatings.

An additional qualitative measure of SDS adhesion was performed on the test coupons, as observed by elemental analysis before and after a gentle water rinse. Three SDS coupons and a blank were immersed in 30 ml of deionized (DI) water with gentle agitation for 24 h. After 24 h, the coupons were removed and analyzed by EDS to determine changes in surface composition. Preliminary screening with bulk EDS analysis revealed a limitation in sensitivity, as no differences were observed in the pre-washed surfaces when compared with those gently washed in 30 ml of DI water for 24 h.

To improve sensitivity, the analysis of dissolved metals recovered in the deionized water rinse was performed using ICP-MS. Metal analysis required that the rinsate be filtered through a 0.45 μm quartz filter and acidified to pH < 1.5 with ultrapure HNO_3 prior to analysis by ICP-MS following a modified US Environmental Protection Agency (EPA) test method 6020. Scandium, yttrium and rhodium were added on-line prior to the introduction of the sample into the ICP-MS equipment as internal standards to correct for instrumental drift. Two isotopes were used to provide confirmation of analyte detection. The instrument, a Perkin Elmer Elan DRC II, was calibrated with National Institute of Standards and Technology (NIST)-traceable standards and the calibration was verified with a second-source NIST-traceable standard. The analytical results from the analysis are shown in Table 13.1. This technique revealed low-level detection of Ti found in the rinsate from the surfaces. Table 13.1 shows the elemental analysis results after water rinsing on three separate coupons as compared to a blank stainless steel plate without the SDS coating. Iron was not part of the analytical method and therefore was not analyzed in the samples.

Figure 13.4   Image of a TiO_2 SDS which was tape adhesion tested on the bottom half.

Table 13.1   ICP-MS analysis of 30 ml DI water rinsate after 24 h of a gentle rinse.

	Blank		*Water I*		*Water II*		*Water III*	
	mg L^{-1}	*Total μg*	*mg L^{-1}*	*Total μg*	*mg L^{-1}*	*Total μg*	*mg L^{-1}*	*Total μg*
Ag	0.001	0.03	4.170	125.1	0.366	10.98	2.647	79.41
Mn	0.019	0.57	0.053	1.59	0.091	2.73	0.070	2.10
Ni	0.001	0.03	0.001	0.03	0.234	7.02	0.006	0.18
S	0.174	5.22	0.208	6.24	0.153	4.59	0.176	5.28
Si	4.941	148.2	4.856	145.7	4.736	142.1	4.216	126.5
Ti	0.002	0.06	0.017	0.51	0.0	0.0	0.015	0.45

It should be noted that for the results presented the silver values were extremely elevated due to a processing error during the manufacturing of the plates and are not valid. This processing error was associated with a missing final rinse of the coupon after silver nitrate incorporation.

13.3.3   Analysis of Particles Released During Laminar Flow

In order to obtain accurate release rates and sizing information comparable to HVAC system parameters, the coupons were subject to laminar flow conditions and particles were collected and analyzed. To accomplish this, a hexagonal laminar flow reactor was constructed of aluminum with inside rectangular dimensions of 76 mm width, 24 mm height, and 246 mm length. This system was modeled after air flows consistent in a scaled up HVAC system. Laminar flow conditions were maintained and the flow conditions were 0.014 m^3 min^{-1} (0.5 ft^3 min^{-1}) corresponding to a Reynolds number of 320. Figure 13.5 shows the interior of the laminar flow reactor with a sample coupon recessed inside the lower wall. For these experiments, the reactor was place inside a nitrogen glove box to minimize the collection of contaminant particles. In addition, for the calculations and for the experiments dry air was used as the test gas. The experiments were performed at 25 °C. Particles were collected as they exited the reactor on a collector surface (25 mm × 75 mm [1 in. × 3 in.] glass slide lined with SEM mountable carbon tape). This surface was placed at a distance of 2.5 cm away from the outlet of the reactor so as not to interfere with air flows exiting the reactor. The carbon tape was examined

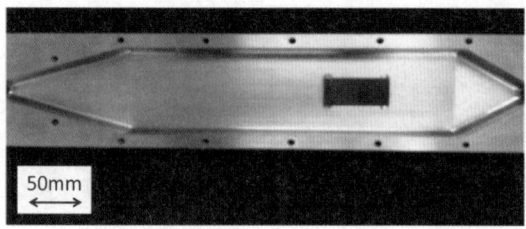

Figure 13.5   Laminar flow reactor modeled after HVAC air flows.

by SEM analysis to quantify the particle size and particle shape. Three test coupons were weighed before and after a 24 h exposure to air flows. The weight change recorded after testing revealed that under laminar flow conditions an average of 5.33 μg of weight loss (s.d. of 4.618 μg) on the coupons over a 24-hour period. This large variation in weight loss made quantitative evaluations difficult. The level of uncertainty with the weighing technique in the μg range is large. Although a more sensitive balance was available, the mass of the stainless steel coupon exceeds its weight limits.

Figure 13.6a and b show typical groupings of particles ejected from the coating which were found adhered to the carbon tape during SEM imaging. Small particles (typically less than 10 μm) were found to be sparsely distributed across the surface of the carbon tape. Chemical analysis by EDS has confirmed that the particles contained Fe and Si (consistent with the major constituents of the plate and coating) with only trace concentrations of Ag and Ti.

The image analysis was performed using the National Institutes of Health (NIH) software ImageJ to determine particle size distributions. SEM micrographs were processed in order to threshold out phases of interest and form a binary (black and white) image. Once in the binary form, the ImageJ particle analysis algorithm was used to calculate the particle size distributions. This algorithm determines the mean diameter of each particle in the binary image by calculating the area of irregularly shaped particles and equating it to the diameter of a circle with the same area. Figure 13.7 illustrates the typical image analysis process used. This image analysis procedure was found to work well for smaller disperse particles but less so for commonly clustered larger particles which overlapped together in the analysis.

Image analysis performed to determine particle size distributions on the particle collected from the laminar flow reactor found a mean particle size of

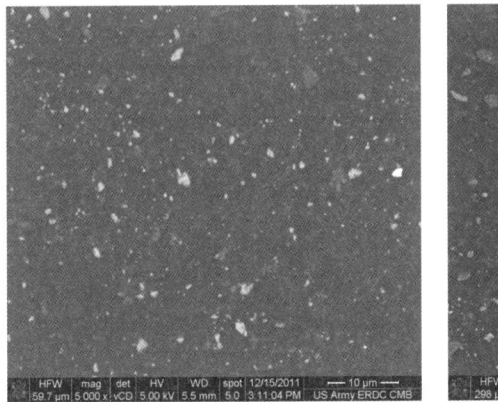
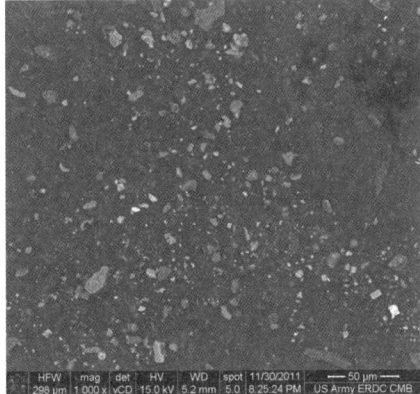

(a) Typical particles at high magnification   (b) Typical particles at low magnification

Figure 13.6   SEM micrographs of particles ejected from coating and adhered to carbon tape.

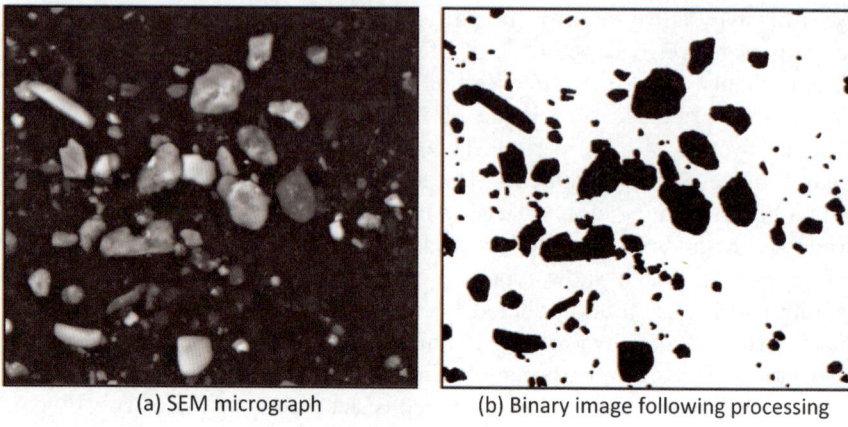

(a) SEM micrograph   (b) Binary image following processing

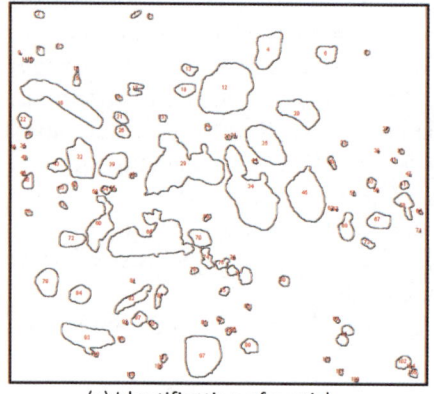

(c) Identification of particles

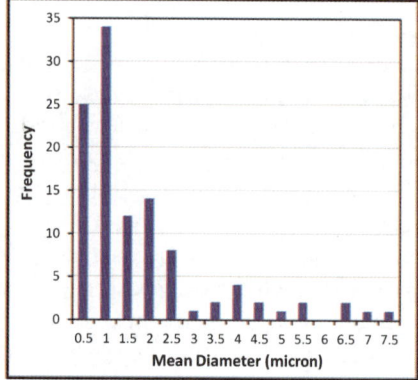

(d) Resulting particle size distribution

Figure 13.7   Image analysis results for determining particle size distribution in a replicate. Samples were analyzed through image capture (a) followed by monochromatic processing to identify primary particles (b). These particles were assigned individual identification numbers (c) and each measured individually to develop frequency histogram (d).

1.9 µm. The overall particle size distribution obtained from 5 images and a total of approximately 2400 particles is shown in Figure 13.8.

13.3.4 Analysis of Particles Expected During Abrasion or Sanding of the Self-decontaminating Surface

To evaluate a worst-case scenario for release of particles that would be representative of sanding, scraping, or general abrasion, SDS was removed from four 200 mm × 200 mm (8 in. × 8 in.) stainless steel panels and milled to a small size. The surface was scraped using a razorblade and combined. The

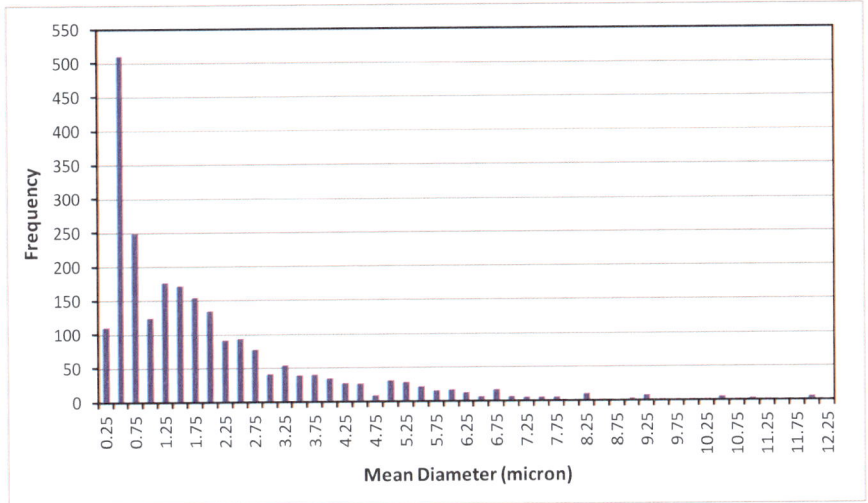

Figure 13.8 Particle size distribution of particles ejected from an SDS coating in flow cell.

SDS removed from multiple coupons was dry milled by vigorous agitation in a 50 ml polypropylene cylinder containing 5 mm glass beads (4 h, 300 rpm, horizontal tube orientation). The particles were recovered by dissolution in ultrapure water, separation of the beads, and evaporation of the water at 100 °C. Only 0.4 g of material was collected during this process such that only alveolar dissolution studies and alveolar cell exposures were performed with this material.

Figure 13.9 shows representative SEM images of the material collected from the milling process. The milled material exhibited a large distribution in particle size from larger "chunks" of matrix material in the 5 to 20 μm range, to agglomerates of TiO_2 and Ag particles in the 0.5 to 5 μm range, and single nanosized particles. The mean particle size was 1.73 μm and less than 0.57%

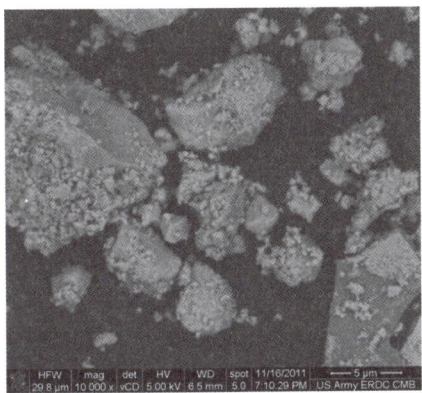 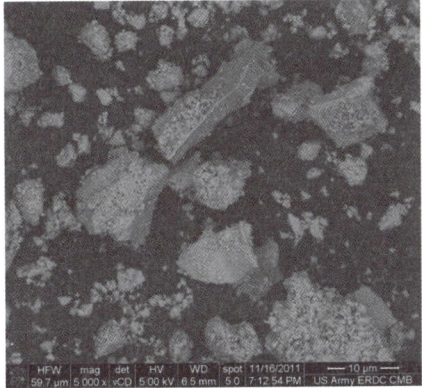

Figure 13.9 SEM images of particles collected from milling process.

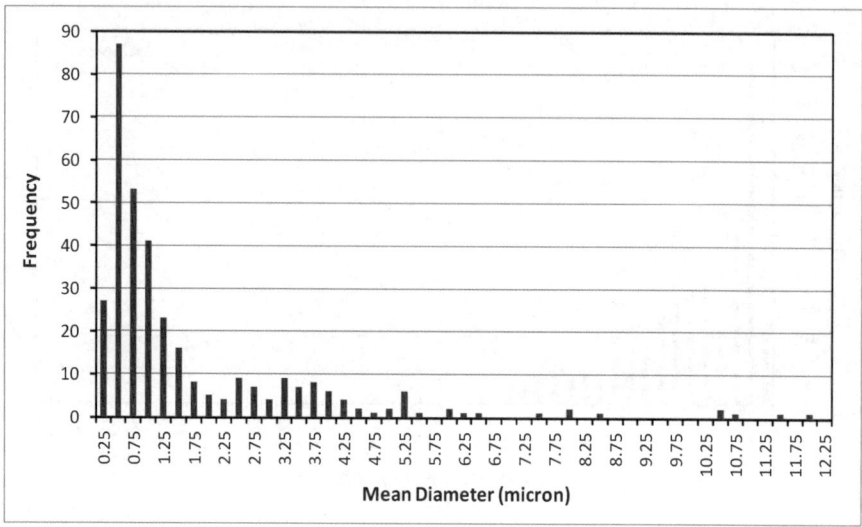

Figure 13.10   Particle size distribution of particles scraped from SDS and milled.

were 100 nm or smaller. Larger particles of matrix material were typically covered with adsorbed particulates of TiO$_2$ and Ag. A particle size distribution obtained by image analysis of the SEM micrographs is shown in Figure 13.10. The beginning of a bimodal distribution was observed with the mean particle size of the larger "chunks" centered around 4 µm and the smaller agglomerates of nanoparticles centered around 750 nm.

13.4   Toxicity of Milled Surface Materials

13.4.1   Background Information Supporting Toxicity Data Gaps in Conceptual Model

Humans are likely to be exposed to nanoparticles (NPs) and nanocomposites through the air and thus the respiratory tract is a potential organ system of exposure and effects. Air and particles travel through the respiratory tract into the bronchial tubes and into the lung lobes which are subdivided into bronchioles and ultimately end in alveoli. Alveoli are multi-lobed air sacs with large surface-area-to-volume ratios where oxygen and waste gases are exchanged with capillaries. Alveoli need to stay moist in order to function properly. Alveoli are composed of type I pneumocytes, type II pneumocytes, and alveolar macrophages. Type I pneumocytes are simple squamous cells responsible for the gas exchange within the alveolus. Type II pneumocytes are cuboidal epithelial cells responsible for secreting a surfactant that reduces the surface tension of water so the alveolus does not collapse upon itself. Alveolar macrophages are resident immune cells that engulf foreign entities such as particles and microorganisms.

Particle size is critical for determining the potential for deposition along the respiratory tract. Larger particles (>2.5 µm) tend to deposit within the tracheobronchial tract where they are cleared out by the mucociliary escalator. As particles get smaller (<2.5 µm), they tend to deposit deeper in the lungs (*i.e.*, in an alveolus) where gas exchange occurs. However, once particles get even smaller (<100 nm), diffusion processes tend to cause these particles to deposit in the upper respiratory tract similarly to larger particles. Therefore, the generally accepted nanoparticle definition (<100 nm) is a particle size most likely to deposit in the alveoli and potentially translocate into the blood circulation.

The properties of inhaled nanoparticles and nanocomposites may change in the extracellular environment of the lungs. Specifically, lung fluid components (*i.e.*, salts, surfactants, and proteins) may alter the agglomeration and suspension state of the particles, causing them to deposit more quickly along the respiratory tract. Nanoparticle deposition may trigger an inflammatory reaction within the alveoli. Alveolar macrophage may try to engulf these nanoparticles, yet their function may be affected by the nanoparticles, resulting in the reduced removal of foreign bodies.[36,37] Furthermore, the presence of nanoparticles within the alveoli may cause alveolar macrophages to release inflammatory mediators to recruit additional immune cells to fight off the foreign particles. Such an immune response within the lungs could result in pulmonary toxicity, pulmonary inflammation and scarring, and reduced lung function. It could also result in a disruption of the alveolus barrier, resulting in increased numbers of particles crossing into the blood.

In these sets of experiments, the properties of the SDS were determined in an artificial alveolar fluid. Milled nanocomposite particle dispersions in lung fluids were analyzed using dynamic light scattering (DLS) analysis (Malvern Zetasizer Nanoseries Nano-ZS) and by field flow fractionation–inductively coupled plasma–mass spectroscopy (FFF-ICP-MS) analysis (Post Nova Analytics and Perkin Elmer ELan DRC2). These studies provide information for predicting the behavior and properties in inhalation exposures. Secondly, the SDS particles were used to develop dose–response relationships in a respiratory cell co-culture containing type II pneumocytes (A549; cells that produce lung surfactant) and alveolar macrophages (U937; resident phagocytic cells in lung). Endpoints focused on relevant lung toxicity including cell survival, phagocytosis activity (*i.e.*, engulfing foreign particles), and inflammation responses. These data will be used to compare the current TiO$_2$-based composite to other particulates and metal data from the literature.

13.4.2 Nanocomposite Particulate Size and Dispersion in Alveolar Fluid

The SDS in this report consists of titanium dioxide (TiO$_2$) and silver (Ag) particles; hence, we examined the dispersion of TiO$_2$ and Ag nanoparticle and SDS particles in an artificial alveolar fluid and interstitial fluid to represent the two fluids present near the alveolus–capillary interface. Because it was initially uncertain what size the milled SDS particles would be, both 20 and 80 nm Ag NPs were used as model NPs and as a worst-case scenario. Raw Ag NPs are extensively agglomerated so citrate- and polyvinylpyrrolidone (PVP)-coated Ag NPs were used to improve particle dispersion prior to addition to the artificial biofluids.

Particle sizes were first examined over 20 min following their addition to the artificial alveolar fluid. At 10 mg l^{-1} (parts per million, ppm), the Ag-citrate NP size increased almost instantly. As soon as the 20 nm Ag-citrate NP (measured 33 nm in-house by DLS) was added to the alveolar fluid, the particles increased $\sim$2-fold in size to 60 nm (Figure 13.11, left). Larger particles were also detected, as indicated by the tailing effect on the FFF-ICP-MS chromatogram. It is likely that the interaction of the high salt content of the alveolar fluid and high nanoparticle concentration caused the rapid particle agglomeration. Additional incubation (30 min) in the alveolar fluid produced a similar chromatogram, showing that the Ag-citrate NPs were relatively stable after the initial incubation in the biofluid (Figure 13.11, right).

The particle dispersion was next analyzed over a 24 h period in alveolar fluid to see how dispersion changes over time. Ag-citrate (20 nm) was initially highly agglomerated ($\sim$200 nm) compared to the stock solution (33 nm), then decreased to $\sim$100 nm from 1–14 h (Figure 13.12, left). After 14 h, particle sizes began to increase up to $\sim$650 nm at 20 h then decreased to $\sim$580 nm at 24 h. The polydispersivity index (PDI) showed an inverse response to the mean particle size spectra (Figure 13.12, right). When combined, these data suggest that though the particle sizes were relatively small from 1–14 h, there was significant particle size heterogeneity. Conversely, though particle sizes increased after 14 h, the particle size became more consistent within the larger size range. This appears to be consistent with the raw data showing a "red shift" over time, indicating that particle diffusion in the alveolar fluid was slowing down due to larger average particle sizes being formed (data not shown).

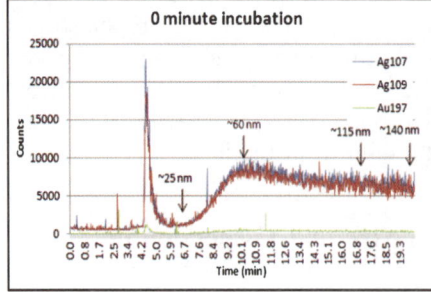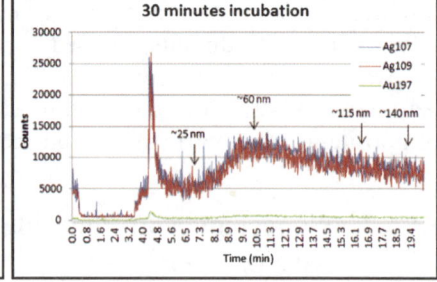

Figure 13.11   Particle sizes of 20 nm citrate-coated silver nanoparticles (Ag-citrate; 10 mg l^{-1}) in artificial alveolar fluid at 0 min (left) and 30 min (right) incubations.

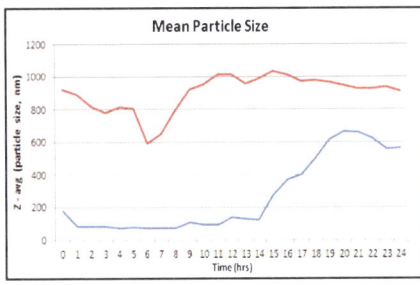
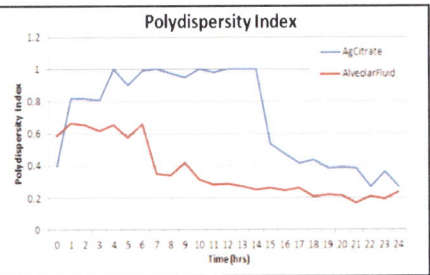

Figure 13.12   DLS analysis of Ag-citrate NPs in artificial alveolar fluid over 24 h: (left) mean particle size; (right) polydispersivity index.

One complication with using the alveolar fluid was the surfactant properties of the phospholipid phosphatidylcholine (PC), a constituent in the alveolar fluid. The presence of PC (Figure 13.12, red line) increased particle size throughout the experiment and yet, similarly to Ag-citrate particles, the alveolar fluid PDI decreased with time. It is known that PC can create micelles (a lipid monolayer) at lower concentrations and liposomes (an encapsulating lipid bilayer) at higher concentrations. It is possible that liposome formation and size settled out over time to a more consistent size by 24 h. The implications of alveolar fluid liposomes will be discussed below.

Larger Ag-citrate particles (10 mg l^{-1}) were also examined in alveolar fluid at time 0. Though the stock solution of 80 nm Ag-citrate was consistently measured at approximately 80 nm, addition to the alveolar fluid resulted in high agglomeration above 140 nm particles (data not shown). The particle size was not analyzed beyond 30 min because the particles are heavily agglomerated and not suited for fine-tuned analyses with FFF-ICP-MS. For that reason, we did not analyze 80 nm Ag-citrate for 30 min in alveolar fluid.

The dispersion of PVP-coated Ag nanoparticles was also examined. PVP is an organic compound that is more stable than citrate, resulting in a more stable coated NP in various solutions. Unlike Ag-citrate, the 20 nm Ag-PVP particle size was stable at time zero in alveolar fluid (Figure 13.13, top). This stability remained over a 30 min incubation in alveolar fluid (Figure 13.13, middle). In contrast, 80 nm Ag-PVP particles appeared to agglomerate quickly in alveolar fluid, resulting in very few primary particles and most agglomerated particles well above the 140 nm threshold (Figure 13.13, bottom). These data demonstrate that the particle coating and size are critical for dispersion stability in biofluid suspensions.

Size measurements for both TiO$_2$ NPs and the SDS particles were not conducted with the DLS or FFF-ICP-MS because of size limitations. TiO$_2$ NPs were highly agglomerated, even after sonication, resulting in particle sizes > 1 μm. The milled SDS particle range was primarily above the 140 nm particle size threshold for the FFF-ICP-MS. Similarly, DLS may have had difficulty accurately analyzing the particles sizes because of the masking of smaller particles by larger particles. An alternative approach used was adding

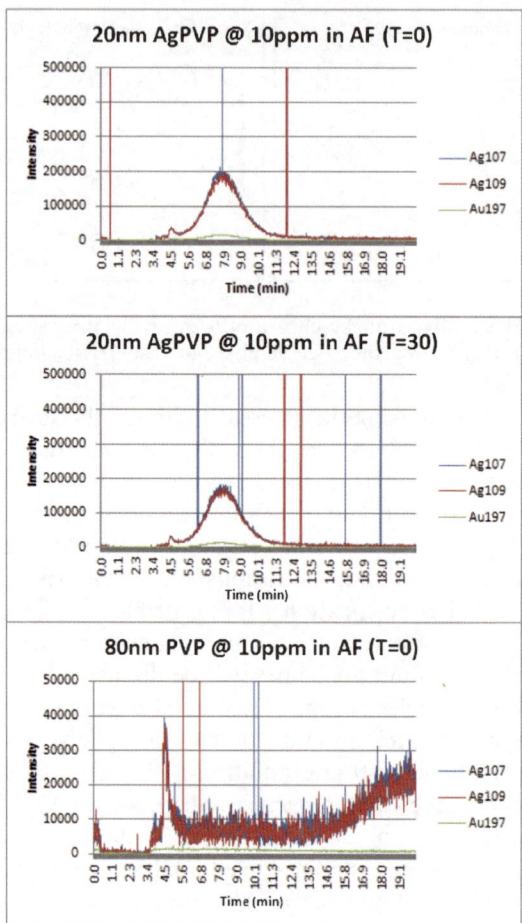

Figure 13.13   Silver particle sizes of in artificial alveolar fluid (AF). (top) 20 nm PVP-coated silver nanoparticles (Ag-PVP; 10 mg l^{-1}) at 0 min; (middle) 20 nm Ag-PVP (10 mg l^{-1}) at 30 min; (bottom) 80 nm Ag-PVP (10 mg l^{-1}) at 0 min.

the NPs to alveolar fluid and imaging the particles by SEM. While this may result in some methodological artifacts (*e.g.*, salt crystallization of culture media and artificial biofluids), this approach gave a better indication of size and dispersion than was detected using other techniques. Figure 13.14 shows the size and dispersion of TiO$_2$ NPs and the SDS particles when in cell culture media for 6 h. TiO$_2$ NPs appeared agglomerated as expected (Figure 13.14, left). In addition, cell culture salts could be seen forming dendritic salt crystals in the background (diffuse cloudy structures). The SDS particles also agglomerated in the cell culture media at 6 h, forming agglomerates of many sizes (*e.g.*, linear, spherical).

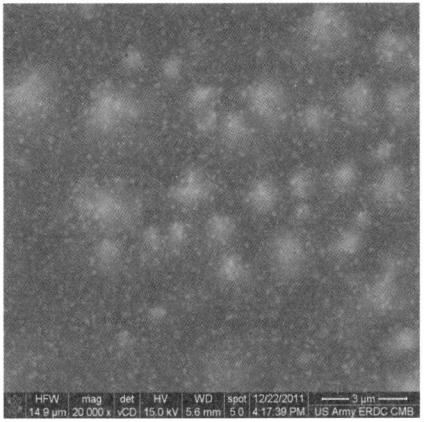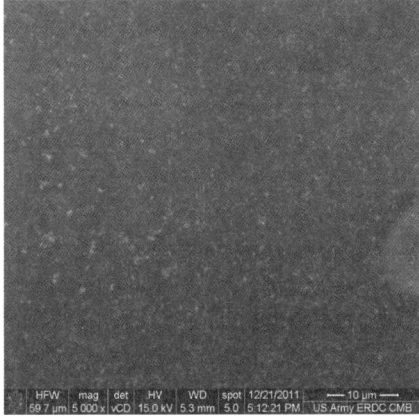

Figure 13.14 Nanoparticle dispersion in cell culture media. (left) TiO$_2$ nanoparticles (0.015 μg ml^{-1} [1 day TiO$_2$ permissible exposure limit (PEL) level]). (right) SDS particles (0.097 ng ml^{-1} [1 day Ag PEL level]). Size bars located in bottom right corners of pictures.

13.4.3 Nanoparticle Settling in Alveolar Fluid

Next, the settling of NPs in alveolar fluid was examined. These settling studies were used to determine the resultant effects of NP dispersion in artificial biofluids. Since Ag-citrate NPs rapidly agglomerated in alveolar fluid, these NPs were expected to settle rapidly in alveolar fluid. In fact, both 20 and 80 nm Ag-citrate remained suspended in the alveolar fluid with only a 5–6% decrease in absorbance during the 30 min study (Figure 13.15 (top), 20 nm: red line, 80 nm: green line). For Ag-PVP at 20 nm (purple line) and 80 nm (blue–green line) absorbance only decreased 2% and 7%, respectively. The absorbance for both TiO$_2$ NPs (orange line) and bulk TiO$_2$ (lavender line) did not change from the control. Interestingly, Ag-citrate showed the most reduced absorbance in this study, possibly due to AgCl precipitation caused by the high salt content of the alveolar fluid. Longer incubations in alveolar fluid resulted in reduced absorbance, but at a slower rate than might be expected. Ag-citrate (20 nm) showed a 6%, 9%, and 17% decrease in absorbance at 30 min, 8 h, and 24 h (Figure 13.15, bottom). These results suggest that the presence of surfactant (PC) may be playing a role in reduced loss of absorbance (*i.e.*, reduced settling). It is possible that the NP agglomerates are being surrounded by PC liposome vesicles. These liposomes not only appear to keep the NP agglomerates stable (Figure 13.12) but also suspended in alveolar fluid despite the high salt content (Figure 13.15). If this occurs *in vivo*, the liposome coating may also help protect cell membranes from NP-induced damage.

As mentioned above, there is the potential for NPs encountering interstitial fluid when moving from the alveolus to pulmonary capillaries. Therefore, the same size and dispersions were conducted using artificial interstitial fluid. The only difference between alveolar fluid and interstitial fluid is the presence of

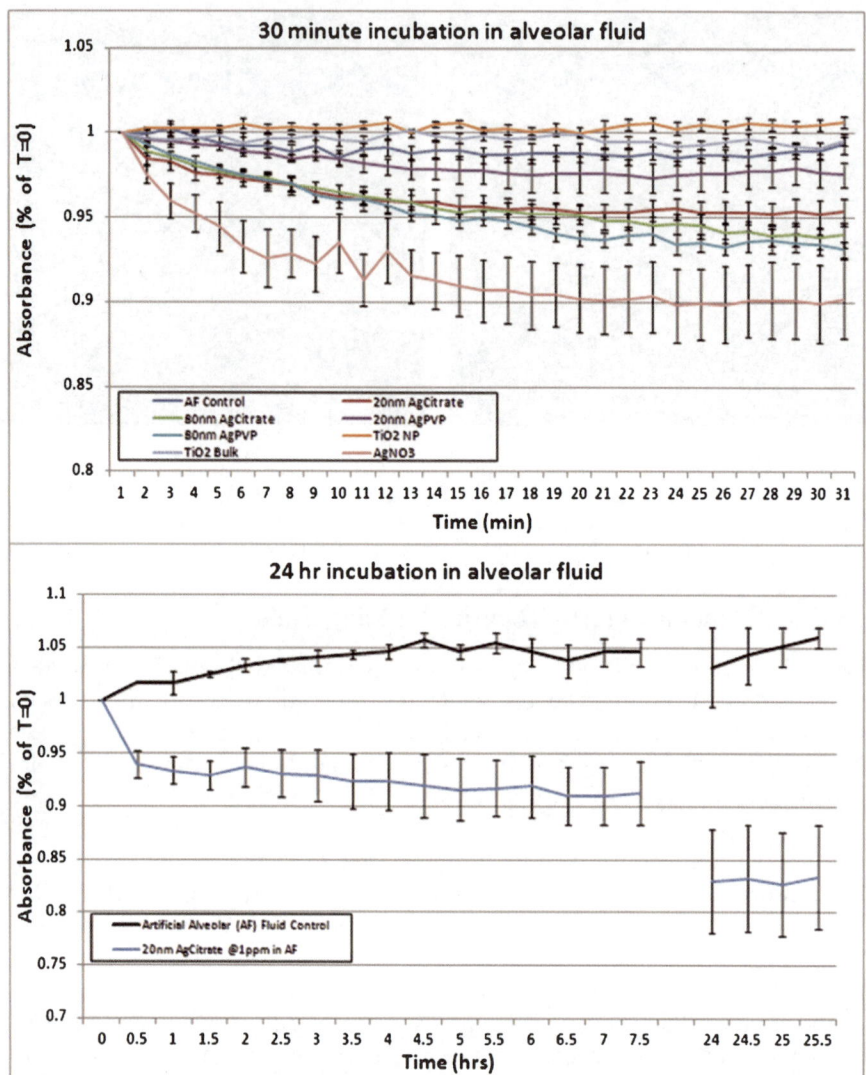

Figure 13.15   Absorbance change due to NP settling in artificial alveolar biofluid. (top)
Change in NP (1 mg l^{-1} suspension) absorbance during a 30 min study;
(bottom) change in NP (1 mg l^{-1} suspension) absorbance over 24 h.

PC in the former. Particle size changes were analyzed for 60 min using the
DLS. All particles examined showed fairly stable particle size at 0 min, with the
exception of TiO$_2$ NPs which were 350 nm (primary particle size from the
manufacturer was 25 nm but the hydrodynamic diameter measured in-house
was consistently 150–250 nm) (Figure 13.16). Particles got progressively bigger
over 60 min, with increases ranging from 20% (TiO$_2$) to 800% (20 nm Ag-
citrate). Particle size confirmation with FFF-ICP-MS at time zero has not been
conducted due to instrument availability.

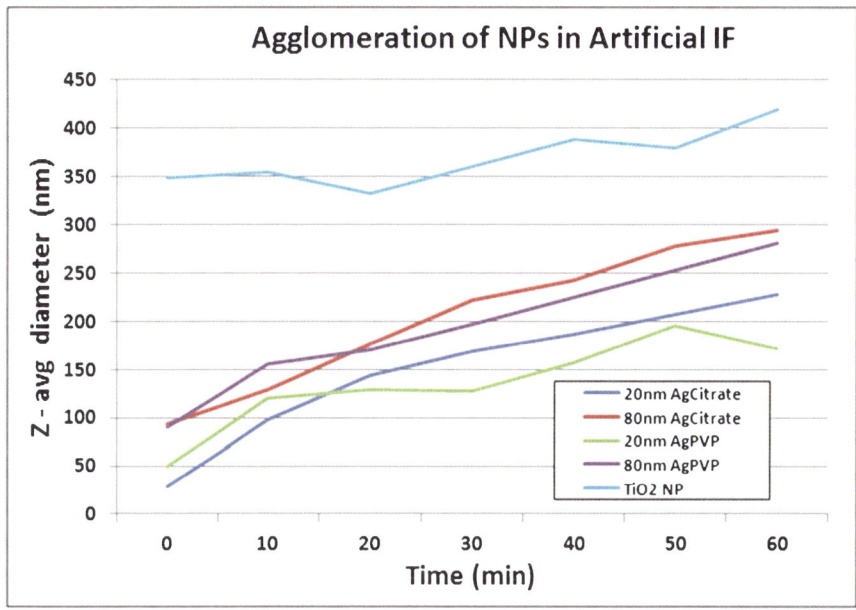

Figure 13.16 Change in particle size when incubated in artificial interstitial fluid (IF) for 60 min.

Next, the NP settling in interstitial fluid over 30 min was examined. Similar to the anticipated effects of particles in alveolar fluid, the particles were expected to rapidly settle because of the high agglomeration in interstitial fluid (Figure 13.17). Ag-citrate particles produced more settling in interstitial fluid than alveolar fluid (16% reduced absorbance for 20 nm Ag-citrate (red line) and 23% reduced absorbance for 80 nm Ag-citrate (green line)). Ag-PVP showed mixed results, with only 2% reduced absorbance for 20 nm Ag-PVP (purple line) and 25% reduced absorbance for 80 nm Ag-PVP (blue–green line). Both bulk and nanosized TiO_2 showed no difference from the control. In contrast to alveolar fluid, silver nitrate showed increased absorbance in interstitial fluid over 30 min. These data suggest size-related settling with Ag particles as well as a potential coating-related settling at smaller particle sizes. These data also suggest that the presence of PC in alveolar fluid kept 20 nm Ag-PVP, 80 nm Ag-citrate, and 80 nm Ag-PVP suspended.

The settling of the nanocomposite was also examined over 30 min. Figure 13.18 shows that the SDS particles remained stable over 30 min, with less than 5% loss of absorbance. This stability was similar to what was seen with 20 nm Ag-PVP and TiO_2. While these data provide insight into particle behavior in artificial biofluids, additional studies should be considered. Anecdotal laboratory observations indicated that several of the particles which stayed suspended in artificial biofluids for 30 min actually began settling on the bottom of the stock solution containers. Since it is possible for particles to reside in the lung for longer periods of time prior

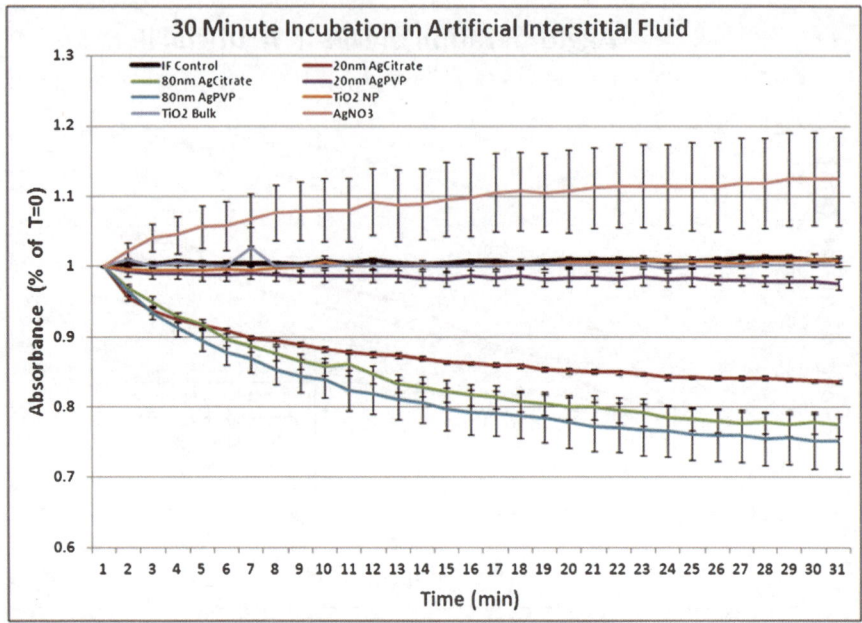

Figure 13.17   Change in particle settling in artificial interstitial fluid. NP (1 mg l^{-1} suspension) absorbance during a 30 min study.

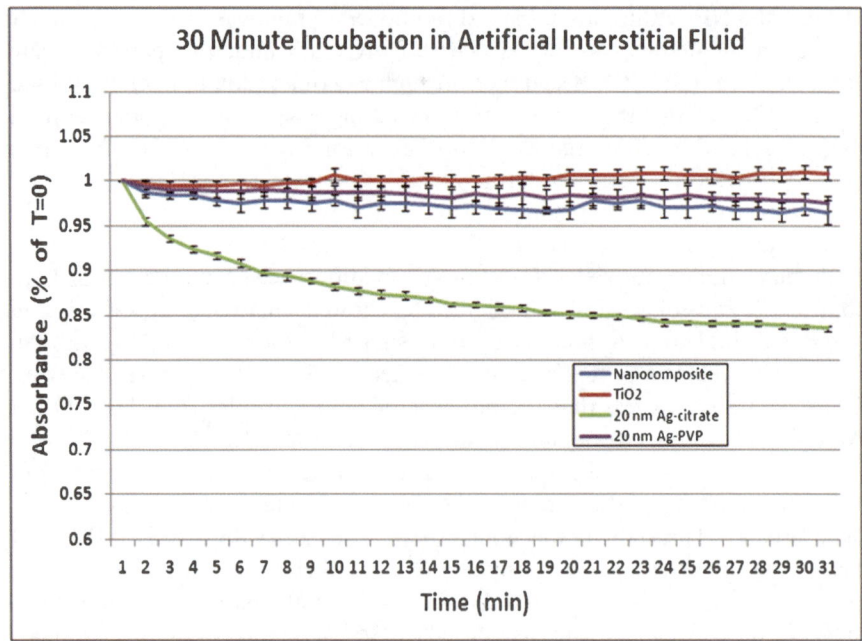

Figure 13.18   Change in milled nanocomposite particle settling in artificial interstitial fluid. NP (1 mg l^{-1} suspension) absorbance during a 30 min study.

to macrophage engulfment or transmembrane movement into the blood circulation, longer settling studies (*e.g.*, 24 h) need to be performed to corroborate these observations. Furthermore, particle deposition on the bottom of the test system can be analyzed for particle size and agglomeration.

13.4.4 SDS Dissolution in Alveolar Fluid

Ag NPs are known to dissolve in environmental media. As a result, it is possible that the Ag from the SDS material may also dissolve and ionize in lung fluid. TiO_2, in contrast, is virtually insoluble but there is a possibility that it too could dissolve in the lung fluids. Therefore, SDS dissolution was studied in both DI water and artificial alveolar fluid. This study was designed to measure the bioavailability of the milled TiO_2 material by assessing its dissolution potential in simulated lung fluid. Due to the limited amount of material the rate of dissolution was not obtained and only duplicate repetitions were performed. The methodology was based on EPA SW 846 Method 1312 (Synthetic Acid Rain Leaching Procedure) and in accordance with this method dissolution time was set to 24 h. The dissolution was evaluated at a concentration of 500 mg l^{-1} using known masses (in the range 2.42 to 2.52 mg) and determined gravimetrically using a calibrated balance capable of reading to 10 μg (Orion Cahn C-35, Thermo Electron Corporation). The weighed powder was mixed into 5 ml of artificial lung fluid in a polypropylene vial and gently agitated at room temperature for 24 h. Following extraction, the liquid extract is separated from the solid phase by filtration through a 0.45 μm glass fiber filter prior to acidification and analysis. The analysis of the dissolved metals was performed using inductively coupled plasma–mass spectroscopy (ICP-MS) with sample digestion. The analytical method EPA SW 846 Method 6010 was used to measure 31 metals in the dissolved fraction and resulted in reporting limits for all of the metals ranging from 20 to 50 μg l^{-1}.

The analytical limit of detection (LOD) for titanium was 0.01 mg l^{-1} and the limit of quantification (LOQ) was 0.02 mg l^{-1}. After 24 h exposure at a concentration of 500 mg l^{-1} of powder, 67.0 μg l^{-1} of Ti metal and 103 μg l^{-1} of Ag was found to dissolve in the synthetic alveolar fluid. Additional elements found included phosphorus (2.5 mg l^{-1}), chromium, zinc and boron (all at less than 100 μg l^{-1}). Blank alveolar fluid was not examined with ICP-MS. The dissolved TiO_2 (or particles less than 0.45 μm) was measured at 133 μg l^{-1}. These values are 3700–4800-fold lower than the starting material. These data demonstrate that very little Ag or TiO_2 dissolved from the SDS in artificial lung fluid.

13.4.5 Toxicity of SDS Particles

SDS particle toxicity to lung cells was examined using a co-culture of type II pneumocytes (A549, cells that produce lung surfactant) and alveolar macrophages (U937, resident lung macrophages). Cells were exposed to the following dosing regimen (concentrations listed below): control media, TiO_2

(bulk), AgNO$_3$ (ionic Ag), coated Ag NPs, TiO$_2$ NP, and the milled SDS particles at concentrations presented in Table 13.2. SDS particles were dosed at the same level as TiO$_2$ to correspond with TiO$_2$ 1 day permissible exposure limits (PEL) values and Ag 0.01 day, 1 day, and 1 week PEL values. Coincidentally, these concentrations also have been shown to be non-toxic to A549 cells and U937 cells.[38,39]

Cells were cultured in a 3:1 ratio of A549:U937 cells for 24 h at 37 °C. Cell co-cultures were cultured in both standard media and alveolar fluid in order to examine any biochemical responses to lung alveolar fluid. Alveolar macrophages (U937 cells) were activated with PMA for 48 h prior to particle exposures. Cells were then exposed to bulk and ionic salts, NPs, and the SDS particles for 24 h. After exposure, the cells were trypsinized from the plates and analyzed by flow cytometer (Gallios system, Beckman Coulter, Miami, FL) for cellular marker identification (distinguishes between A549 and U937 cells) and cytokine marker expression. Media was also analyzed by enzyme-linked immunosorbant assay (ELISA) for cytokine release. Phagocytosis was analyzed by Vybrant Green Phagocytosis Assay.

As anticipated, the chemical concentrations used in this study did not cause significant cytotoxicity to either cell type in the co-culture (data not shown). Flow cytometry data show that there was differential cytokine expression between alveolar macrophages and type II pneumocytes, as well as between cells cultured in media *versus* in alveolar fluid (Figure 13.19). Macrophages showed comparable increases in IL-1β expressing cells exposed to both nanosized TiO$_2$ and AgNO$_3$ at all concentrations regardless of culture media (Figure 13.19a and b). The SDS particles also increased IL-1β expressing macrophages, though at lower expression levels than nanosized TiO$_2$ and AgNO$_3$. No dose responses were seen with the dose ranges used in this study. In contrast, the percentage of macrophages expressing TNF-α, another pro-inflammatory cytokine, and TGF-β, an anti-inflammatory cytokine, were both

Table 13.2   Dosing regimen for nanoparticles and SDS particles.

Dosing agent	Dose	Permissible exposure limit (PEL)[a] equivalent
Ag	0.97 pg ml^{-1}	1/100 day
	0.097 ng ml^{-1}	1 day
	0.63 ng ml^{-1}	1 week
TiO$_2$	1.5 ng ml^{-1}	1/100 day
	0.15 μg ml^{-1}	1 day
	1.02 μg ml^{-1}	1 week
Nanocomposite	0.97 pg ml^{-1}	1/100 day (Ag)
	0.097 ng ml^{-1}	1 day (Ag)
	0.63 ng ml^{-1}	1 week (Ag)
	0.15 μg ml^{-1}	1 day (TiO$_2$)

[a]PELs are Occupational Safety and Health Administration (OSHA)-enforcable regulatory limits on the amount or concentration of a substance in the air. OSHA PELs are based on an 8 h time-weighted average (TWA) exposure. (Source: http://www.osha.gov/SLTC/pel/)

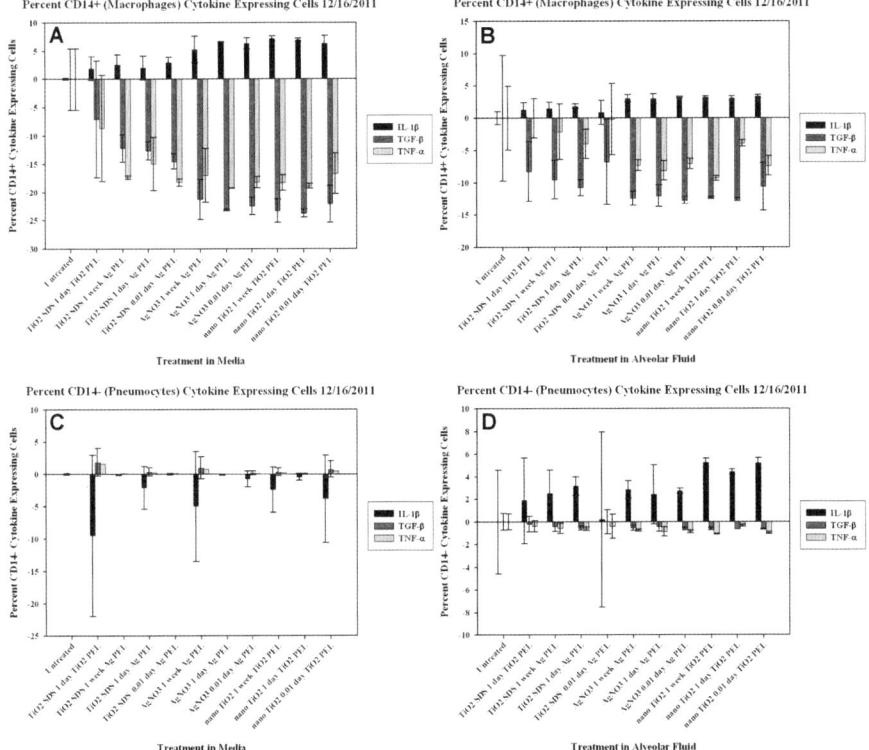

Figure 13.19 Cytokine expression in lung cell co-cultures. Cells exposed for 24 h to TiO₂ SDS particles, silver nitrate (silver ion positive control), and nanosized titanium dioxide particles at concentrations (see Table 13.2) corresponding to OSHA permissible exposure limit (PEL) levels for silver and titanium dioxide. (A) Cytokine expression in U937 cells (alveolar macrophages) in cell culture media. (B) Cytokine expression in U937 cells (alveolar macrophages) in artificial alveolar fluid. (C) Cytokine expression in A549 cells (type II pneumocytes) in cell culture media. (D) Cytokine expression in A549 cells (type II pneumocytes) in artificial alveolar fluid.

decreased by nanosized TiO₂ and AgNO₃ at all concentrations. Type II pneumocytes showed a different profile from alveolar macrophages. The percentage of pneumocytes expressing IL-1β was decreased at several concentrations tested in media (Figure 13.19c). In contrast to media, the percentage of cells expressing IL-1β was increased at all doses and chemical concentrations in alveolar fluid (Figure 13.19d). This suggests that there is a chemical component in the alveolar fluid (*e.g.*, phosphatidylcholine surfactant) that may be activating scavenging receptors which activate IL-1β production.[40] The SDS particles, nanosized TiO₂, and AgNO₃ did not affect the percentage of pneumocytes expressing TGF-β and TNF-α regardless of culture media.

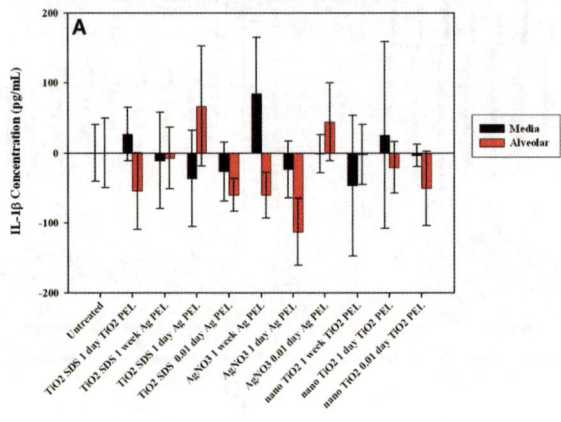

IL-1β Concentrations in Media and Alveolar Fluid from 12/16/2011 as Measured by ELISA (Untreated Set to 0)

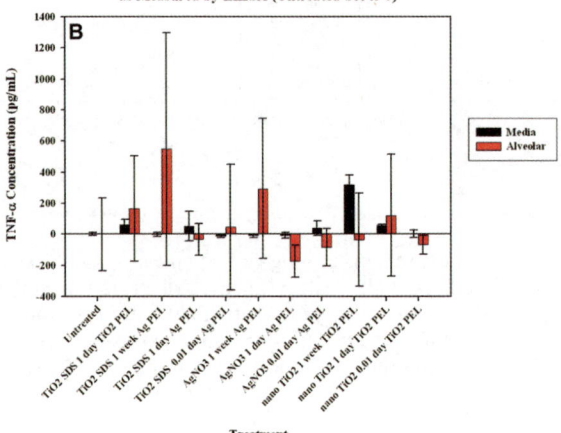

TNF-α Concentrations in Media and Alveolar Fluid from 12/16/2011 as Measured by ELISA (Untreated Set to 0)

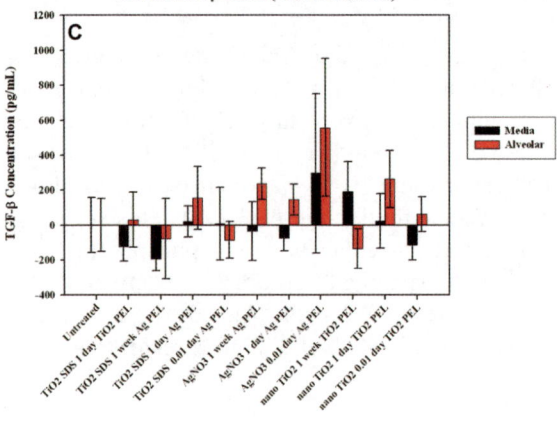

TGF-β Concentrations in Media and Alveolar Fluid from 12/16/2011 as Measured by ELISA (Untreated Set to 0)

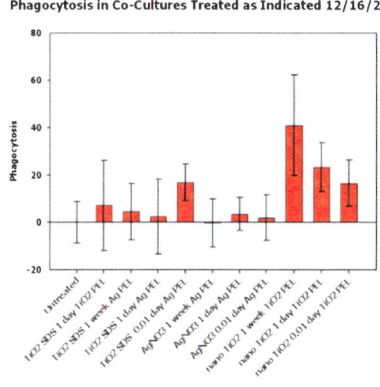

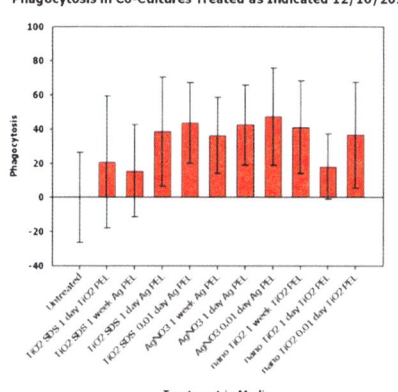

Figure 13.21 Phagocytic activity of U937 alveolar macrophage cells from lung co-culture. Cells exposed for 24 h to SDS particles, silver nitrate (silver ion positive control), and nanosized titanium dioxide particles at concentrations (see Table 13.2) corresponding to OSHA PEL levels for silver and titanium dioxide. Phagocytosis indicative of the net amount of fluorescent beads engulfed by alveolar macrophages. (left) Cells cultured in cell culture media; (right) cells cultured in artificial alveolar fluid.

Cytokine release into culture media was also examined (Figure 13.20). Despite the percentage of IL-1β expressing macrophages being increased (Figure 13.20a), there was no clear trend in IL-1β concentrations regardless of dose, chemical treatment, or culture media. The TNF-α showed a slight dose-dependent increase with nanosized TiO$_2$ treatment in media, but not with other chemical treatments in culture media (Figure 13.20b). In contrast, TNF-α concentrations appeared to trend toward a dose-dependent increase with both AgNO$_3$ and SDS particles in alveolar fluid (Figure 13.20b). TGF-β expression showed a trend toward a dose-dependent increase with nanosized TiO$_2$ treatment and dose-dependent decreases with AgNO$_3$ and SDS particle treatments in media (Figure 13.20c). Overall, the amounts of cytokines released by chemical treatments were very low (pg ml^{-1}). These data, in combination with Figure 13.18, demonstrate that neither the SDS particles nor the Ag and TiO$_2$ positive controls elicit much of an inflammatory

Figure 13.20 (A) Interleukin 1-beta (IL-1β) release into media (black bars) or alveolar fluid (red bars) from lung cell co-cultures. (B) Tumor necrosis factor alpha (TNFα) release into media (black bars) or alveolar fluid (red bars) from lung cell co-cultures. (C) Transformation growth factor beta (TGF-β) release into media (black bars) or alveolar fluid (red bars) from lung cell co-cultures. All cells exposed for 24 h to SDS particles, silver nitrate (silver ion positive control), and nanosized titanium dioxide particles at concentrations (see Table 13.2) corresponding to OSHA permissible exposure limit (PEL) levels for silver and titanium dioxide. Cytokine expression was detected by ELISA immunoassays.

response regardless of concentration or culture media used. This may be due to the low concentrations used in this study compared to other reports where much higher concentrations were used. However, these *in vitro* lung cell co-culture results are more in line with limited pulmonary inflammatory effects seen *in vivo* than other *in vitro* cell culture studies.[41–46]

Co-culture phagocytosis activity in culture media was increased to comparable levels for nanosized TiO_2, $AgNO_3$, and SDS particles (Figure 13.21). There was no dose-response for any of the chemical treatments, though there appeared to be a dose-dependent decrease with SDS particles treatment. Phagocytic activity in alveolar fluid showed a dose-dependent increase trend with nanosized TiO_2 treatment, no effect with $AgNO_3$ treatment, and increased activity with the SDS particle treatment. These data demonstrate that, in general, phagocytosis is increased by chemical particle treatment, regardless of particle size and dissolution potential. Phagocytosis may decrease as particle concentrations increase, as seen both *in vitro* and *in vivo*.[36,37,47]

13.5   Conclusions

13.5.1   General Conclusions

The results of our study show the potential for release of particulate titanium coatings from the SDS coating. Long-term use of these coatings may result in the release of particles over time.[10,14] This release may be mitigated through the improvement of composites and coating technology. Current studies that will be described include the results of the leaching, air flow release of particles, and cytotoxicity and inflammation responses. Coatings include adsorbed metals which, when released, have the potential to cause toxicity in *in vitro* systems; however, some evidence suggests particulate and ionic silver may cause inflammation and cell toxicity *in vivo*.[24] These results integrate information described above on the potential for release with material-specific toxicity information.

13.5.2   SDS Airborne Exposure: a Conservative Worst-case Scenario

The NIOSH has develop recommended exposure limits (REL) for ultrafine (<100 nm) and larger TiO_2. NIOSH recommends airborne exposure limits of 2.4 mg m^{-3} for fine TiO_2 (>100 nm and collectable with respirable particle sampling) and 0.3 mg m^{-3} for ultrafine TiO_2 (<100 nm including engineered nanoscale), as TWA concentrations for up to 10 h per day during a 40 h work week. These recommendations represent levels that over a working lifetime are estimated to reduce risks of lung cancer to below 1 in 1000. The recommendations are based on using chronic inhalation studies in rats to predict lung tumor risks in humans. Using data generated in this study and several assumptions (*i.e.*, TiO_2 to treat 1 m^3 of air, fraction TiO_2 released per hour, and particle size) the potential exposure received through use of the SDS can be estimated. It should be recognized that some of the assumptions are conservative and the actual exposure will be less than the levels predicted here. Based on these calculations, it is estimated the highest levels in the air are expected to be up to 0.14 mg m^{-3} over a 10 hour period. These values are below the NIOSH standards previously mentioned from of 2.4 mg m^{-3} for fine TiO$_2$ (> 100 nm) and 0.3 mg m^{-3} for ultrafine TiO$_2$ (< 100 nm), as TWA concentrations for up to 10 h per day during a 40 h work week. In addition, the particle size analysis shows that a very small fraction of particles 0.42% are considered ultrafine (<100 nm).

13.5.3 Summary of Findings

The results of the laminar flow study show a wide size range of particles released from the SDS. However, a very small fraction of these particles (less than 0.1%) were "nano" sized (<100 nm). Therefore, the exposure to particles from the surfaces is likely to be from micrometre-sized particles that have been previously assessed. Removal of the SDS coating to mimic a worst-case scenario (scraping and milling) resulted primarily in particles around 750 nm to 4 μm. These particles represent what would be expected if the surface were sanded or scraped during a restoration or reconstruction of a painted building surface. SDS particles are unlikely to persist as individual particles in lung fluid. In this study SDS nanocomposite particles were agglomerated in artificial biofluids and culture media. Furthermore, very little Ti and Ag dissolves from the SDS in artificial lung fluid (dissolved Ti and Ag fractions were 3700–4800-fold lower than the mass of the starting material). The milled SDS particles did not cause lung cell toxicity at concentrations or levels corresponding to the relevant Ti and Ag PEL values. These milled nanocomposite particles did not elicit an inflammatory response but it did cause a dose-dependent decrease in phagocytotic activity within the lung cell co-culture. This response is typical and a compensatory response to particulate or contaminant exposure.

The conceptual analysis followed up by laboratory-based studies of the SDS coating suggests four key points. Stability of the SDS coating controls the release of particles and unintended exposure to particles from the surface. Improvements to the surface stability and durability will make the surface last longer and minimize risk associated with the use of the coating. However, these improvements must be made to insure the efficacy of the coating is retained. When particles are released from the surface, they are primarily as agglomerates or fused micrometre-sized particles. These particles are less likely to penetrate the lung and elicit effects associated with nanosized particles. Subsequent modifications to the SDS should also consider the effect the modification may have on the size of the particles released from the surface. Any modifications to the surface charge of the particles may have significant effects on the particles released during laminar flow or abrasion.

While the current study shows little to no toxicity in response to the milled SDS coating, any significant modifications to the particles in the SDS should be followed up with additional *in vitro* bioassays to screen for changes in particle agglomeration and inflammatory activation that may be signs of potential health concerns. Predicted exposure concentrations, based on the current data, from the use of the SDS coating in a HVAC system are below current NIOSH guidance for fine and ultrafine particles in air. Based on these estimates the use of the SDS is not expected to pose any signficant additional risk to workers in a building.

13.5.4   Conclusions on the Use of CEA

This case study has shown the use of a CEA approach to better understand the potential risks of a nanotechnology. By using the CEA process the investigation was able to focus research on the most important information requirements; release of particles from the surface during laminar flow. Investigations focused on the stability of the surface, properties of particles released (size, morphology, composition), and toxicity of the particles. The use of this approach can be integrated with materials developers to improve the safety of the technology. This approach can also be iterative. When potential releases are identified and result in an unacceptable level of exposure and/or risk, the technology can be modified and the life cycle re-evaluated.

Traditionally, assessments of an assessment are often made in an undocumented *ad hoc* approach. A benefit of the CEA life-cycle-based approach is that it considers the potential releases across the value chain. As a result it is likely that *ad hoc* assumptions will be avoided and unacceptable releases and/or risks can be avoided.

Acknowledgements

This work was funded, in part, through the Defense Threat Reduction Agency (DTRA), U.S. Army Engineer Research and Development Center, Center Directed Research (CDR, Dr Jeffery Holland, Director), and U.S. Army Environmental Quality Research Program (Dr Elizabeth Ferguson, Technical Director).

References

1. C. O. Robichaud, D. Tanzil, U. Weilenmann and M. R. Wiesner, *Environ. Sci. Technol.*, 2005, **39**(22), 8985–8994.
2. J. C. Bare, *Hum. Ecol. Risk Assess.*, 2006, **12**(3), 493–509.
3. C. Som, M. Berges, Q. Chaudhry, M. Dusinska, T. F. Fernandes, S. I. Olsen and B. Nowack, *Toxicology*, 2010, **269**(2–3), 160–169.
4. *National Nanotechnology Initiative*, 2011, http://nano.gov/sites/default/files/pub_resource/nni_2011_ehs_research_strategy.pdf

5. J. M. Davis, *J. Nanosci. Nanotechnol.*, 2007, **7**(2), 402–409.
6. J. M. Davis and V. M. Thomas, *Ann. N. Y. Acad. Sci.*, 2006, **1076**, 498–515.
7. U.S. Environmental Protection Agency, (Report No. EPA/600/R-10/081). Washington, DC: U.S., 2010.
8. U.S. Environmental Protection Agency, (Report No. EPA/600/R-10/042). Research Triangle Park, NC, 2010.
9. M. B. Harney, R.R. Pant, P.A. Fulmer and J.H. Wynne, *Appl. Mater. Interfaces*, 2009, **1**(1), 39–41.
10. L. Reijinders, *Polym. Degrad. Stab.*, 2009, **94**, 873–876.
11. L. Hsu and H. Chein, *J. Nanopart. Res.*, 2006, **9**(1), 157–163.
12. D. Bello, B. L. Wardle, N. Yamamoto, R. G. deVilloria, E. J. Garcia, A. J. Hart, K. Ahn, M. J. Ellenbecker and M. Hallock, *J. Nanopart. Res.*, 2009, **11**(1), 231–249.
13. W. Wohlleben, S. Brill, M. W. Meier, M. Mertler, G. Cox, S. Hirth, B. Vacano, V. Strauss, S. Treumann, K. Wiench, L. Ma-Hock and R. Landsiedel, *Small*, 2011, **7**(16), 2384–2395.
14. L. Reijinders, *J. Ind. Ecol.*, 2008, **12**(3), 297–306.
15. D. Johnson, M. Methner, A. Kennedy, and J. Steevens, *Environ. Health Perspect.*, 2010, **118**, 49–54.
16. Department of Health and Human Services Centers for Disease Control and Prevention National Institute for Occupational Safety and Health, Current Intelligence Bulletin 63 2011, http://www.cdc.gov/niosh/docs/2011-160/pdfs/2011-160.pdf.
17. J. Zhao and X. Yang, *Build. Environ.*, 2008, **38**(5), 645–654.
18. J. C. Yu, W. Ho, J. Lin, H. Yip and P. K. Wong, *Environ. Sci. Technol.*, 2003, **37**(10), 2296–2301.
19. E. J. Wolfrum, J. Huang, D. M. Blake, P. Maness, Z. Huang and J. Fiest, *Environ. Sci. Technol.*, 2002, **36**(15), 3412–3419.
20. J. Chen and C. Poon, *Build. Environ.*, 2009, **44**(9), 1899–1906.
21. R. E. Baier, C. M. Izzo and P. J. Nicotera, ADA409494 Proceedings of the 2001 ECBC Scientific Conference on Chemical and Biological Defense Research, 6-8 March, Marriott's Hunt Valley Inn, Hunt Valley, MD. University of Buffalo, 2002, 1–7.
22. A. L. Linsebigler, G. Lu and J. T. Yates, Jr, *Chem. Rev.*, 1995, **95**(3), 735–758.
23. S. Hager and R. Bauer, *Chemosphere*, 1999, **38**(7), 1549–1559.
24. M. E. Quadros and L. C. Marr, *Environ. Sci. Technol.*, 2011, **45**(24), 10713–10719.
25. J. G. Rouse, J. Yang, J. P. Ryman-Rasmussen, A. R. Barron and N. A. Monteiro-Riviere, *Nano Lett.*, 2007, **7**(1), 155–160.
26. J. P. Ryman-Rasmussen, J. E. Riviere and N. A. Monteiro-Riviere, *Toxicol. Sci.*, 2006, **91**(1), 159–165.
27. L. Zhang, W. W. Yu, V. L. Colvin and N. A. Monteiro-Riviere, *Toxicol. Appl. Pharmacol.*, 2008, **228**(2), 200–211.
28. L. W. Zhang and N. A. Monteiro-Riviere, *Skin Pharmacol. Physiol.*, 2008, **21**(3), 166–180.

29. L. Frazer, *Environ. Health Perspect.*, 2001, **109**, 174–177.
30. W. Jacoby, M. Nimlos, D. Blake, R. Noble and C. Koval, *Environ. Sci. Technol.*, 1994, **28**(9), 1661–1668.
31. A. A. Keller, H. Wang, D. Zhou, H. S. Lenihan, G. Cherr, B. J. Cardinale, R. Miller and Z. Ji, *Environ. Sci. Technol.*, 2010, **44**(6), 1962–1967.
32. M. J. Eckelman and T. E. Graedel, *Environ. Sci. Technol.*, 2007, **41**(17), 6283–6289.
33. S. Burns, *Titanium Demand Exceeding Supply*, 2011, http://agmetalminer. com/2011/10/25/titanium-demand-exceeding-supply/
34. R. E. Allred and R. M. Salas, Adherent Technologies, Inc., report (AT-93-5133-FR-001) Albuquerque, NM 87111, 1995.
35. Y. Yang, *et al*, *Chem. Eng. Process.*, 2012, **51**, 53–68.
36. A. J. Wagner, C. A. Bleckmann, R. C. Murdock, A. M. Schrand, J. J. Schlager and S. M. Hussain, *J. Phys. Chem. B*, 2007, **111**(25), 7353–7359.
37. L. K. Braydich-Stolle, J. L. Speschock, A. Castle, M. Smith, R. C. Murdock and S.M. Hussain, *ACS Nano*, 2010, **4**(7), 3661–3670.
38. R. Foldbjerg, D.A. Dang and H. Autrup, *Arch. Toxicol.*, 2011, **85**, 743–750.
39. C. I. Vamanu, M. R. Cimpan, P. J. Hol, S. Sornes, S. A. Lie and N. R. Gjerdet, *Toxicol. In Vitro*, 2008, **22**(7), 1689–1696.
40. T. Palkama, *Immunology*, 1991, **74**(3), 432–438.
41. D. B. Warheit, T. R. Webb, C. M. Sayes, V. L. Colvin and K. L. Reed, *Toxicol. Sci.*, 2006, **91**(1), 227–236.
42. V. H. Grassian, P. T. O-Shaughnessy, A. Adamcakova-Dodd, J. M. Pettibone, P. S. Thorne, *Environ. Health Perspect.*, 2007, **115**, 397–402.
43. D. B. Warheit, T. R. Webb, K. L. Reed, S. Frerichs and C. M. Sayes, *Toxicology*, 2007, **230**(1), 90–104.
44. L.V. Stebounova, A. Adamcakova-Dodd, J. S. Kim, H. Park, P. T. O'Shaughnessy, V. H. Grassian and P. S. Thorne, *Part. Fibre Toxicol.*, 2011, **8**, 5.
45. C. Monteiller, L. Tran, W. MacNee, S. Faux, A. Jones, B. Miller and K. Donaldson, *Occup. Environ. Med.*, 2007, **64**(9), 609–615.
46. I. Iavicoli, V. Leso, L. Fontana and A. Bergamaschi, *Eur. Rev. Med. Pharmacol. Sci.*, 2011, **15**(5), 481–508.
47. L. C. Renwick, K. Donaldson and A. Clouter, *Toxicol. Appl. Pharmcol.*, 2001, **172**(2), 119–127.

Subject Index

References to figures are given in *italic* type. References to tables are given in **bold** type.